FINANCIAL INTELLIGENCE IN HUMAN RESOURCES MANAGEMENT

New Directions and Applications for Industry 4.0

FINANCIAL INTELLIGENCE IN HUMAN RESOURCES MANAGEMENT

New Directions and Applications for Industry 4.0

Edited by

Gurinder Singh, PhD

Hardeep Singh Dhanny

Vikas Garg, PhD

Silky Sharma

First edition published 2022

Apple Academic Press Inc.
1265 Goldenrod Circle, NE,
Palm Bay, FL 32905 USA

4164 Lakeshore Road, Burlington,
ON, L7L 1A4 Canada

CRC Press
6000 Broken Sound Parkway NW,
Suite 300, Boca Raton, FL 33487-2742 USA

2 Park Square, Milton Park,
Abingdon, Oxon, OX14 4RN UK

Apple Academic Press exclusively co-publishes with CRC Press, an imprint of Taylor & Francis Group, LLC

Library and Archives Canada Cataloguing in Publication

Title: Financial intelligence in human resources management : new directions and applications for industry 4.0 / edited by Gurinder Singh, PhD, Hardeep Singh Dhanny, Vikas Garg, PhD, Silky Sharma.

Names: Singh, Gurinder, editor. | Dhanny, Hardeep Singh, editor. | Garg, Vikas (Assistant director), editor. | Sharma, Silky, editor.

Description: First edition. | Includes bibliographical references and index.

Identifiers: Canadiana (print) 20200386778 | Canadiana (ebook) 20200386964 | ISBN 9781771889346 (hardcover) | ISBN 9781003083870 (ebook)

Subjects: LCSH: Personnel management—Technological innovations. | LCSH: Personnel management—Economic aspects.

Classification: LCC HF5549.A27 F56 2021 | DDC 658.300285—dc23

Library of Congress Cataloging-in-Publication Data

Names: Singh, Gurinder, editor. | Dhanny, Hardeep Singh, editor. | Garg, Vikas, editor. | Sharma, Silky, editor.

Title: Financial intelligence in human resources management : new directions and applications for industry 4.0 / edited by Gurinder Singh, PhD, Hardeep Singh Dhanny, Vikas Garg, PhD, Silky Sharma.

Description: First edition. | Palm Bay, FL : Apple Academic Press, 2021. | Includes bibliographical references and index. | Summary: "Financial Intelligence in Human Resources Management: New Directions and Applications for Industry 4.0 familiarizes readers with the very relevant concepts of human resources and finance in Industry 4.0. The book looks at the adoption of current fast-moving computers and automation in the workplace and its impact on the financial aspects of human resources and how HR can be enhanced with smart and autonomous systems fueled by data and machine learning. The chapters offer case studies that provide firsthand knowledge of real-life problems, solutions, and situations faced by the industry and highlight the thought process in resolution of the complex problems. Topics include HR management approaches, global HR challenges, behavioral finance for financial acumen, corporate social responsibility, women empowerment in the HR industry, emotional intelligence in the era of Industry 4.0, and more. The book will be very informative academicians, students, research associates, and entrepreneurs, as well as for industry professionals and those in the corporate sector"-- Provided by publisher.

Identifiers: LCCN 2020050415 (print) | LCCN 2020050416 (ebook) | ISBN 9781771889346 (hardcover) | ISBN 9781003083870 (ebook)

Subjects: LCSH: Personnel management--Technological innovations. | Personnel management--Economic aspects.

Classification: LCC HF5549.5.T33 F56 2021 (print) | LCC HF5549.5.T33 (ebook) | DDC 658.3--dc23

LC record available at https://lccn.loc.gov/2020050415

LC ebook record available at https://lccn.loc.gov/2020050416

ISBN: 978-1-77188-934-6 (hbk)
ISBN: 978-1-77463-815-6 (pbk)
ISBN: 978-1-00308-387-0 (ebk)

About the Editors

Gurinder Singh, PhD
Group Vice Chancellor, Amity Universities, Uttar Pradesh, India

Gurinder Singh, PhD, is the Director Executive Programs Management Domain at Amity University Uttar Pradesh, Greater Noida Campus, India. He has extensive experience of more than 27 years in institutional building, teaching, consultancy, research, and industry. He holds the distinction of being the youngest Founder Pro Vice-Chancellor of Amity University for two terms as well as the Founder Director General of the International Business School and the Founder CEO of the Association of International Business School, London, UK. He has conducted many international conferences, including the International Business Horizons Conference, a landmark conference, for the last 20 years. He has also conducted IEEE conferences in India and London. He has established international campuses in many countries and has also made numerous contributions to the academic field, including writing Scopus-indexed books and research papers. He is the recipient of more than 25 international and national awards and has been invited on many TV talk shows. A renowned scholar and academician in international business, Dr. Singh earned his PhD at the Indian Institute of Foreign Trade.

Brig. Hardeep Singh Dhanny, PhD

Hardeep Singh Dhanny, PhD, is the Dean of Management and Allied Programs at Amity University Uttar Pradesh, Greater Noida Campus, India. As an active researcher, he has written a number of Scopus-indexed books and articles. He has conducted many prestigious national and international conferences with illustrious internationally acclaimed academicians and thought leaders. As a part of the Board of Industry at the university, he has taken strong initiatives to bridge the gap between academia and industry, thus helping to shape the future of budding professionals. An active mentor and leader, he is working on path-breaking initiatives to enhance the quality of education and to standardize business schools in India.

Vikas Garg, PhD

Vikas Garg, PhD, is the Director Executive Programs Management Domain at Amity University Uttar Pradesh, Greater Noida Campus, India. He is UGC NET qualified with 15 years of academic experience. His areas of specialization are accounting and finance, and his major interests are in financial markets, financial reporting, and analysis. He is a lifetime member of the Indian Commerce Association, Indian Accounting Association, and Indian Management Association. He is certified in Customer Relationship Management from the Indian Institute of Management Bangalore, India. He has published numerous research papers in various Scopus- and ABDC-indexed international and national journals. He holds many copyrights and patents and has organized many national and international conferences. He is associated with several universities as an external guide for research scholars and has conducted many workshops at institutions.

Ms. Silky Sharma

Ms. Silky Sharma is the Program Director at Amity Business School, Amity University Uttar Pradesh, Greater Noida Campus, India. She has considerable experience in corporate HR and academic industry. Her area of specialization includes managing talent for superior organizational performance. She is UGC-NET qualified and is currently writing articles and research papers for Scopus-indexed Springer books and journals in the area of finance, HR, and social sciences. She has organized many national and international conferences and is a member of several professional bodies, including the All India Management Association. Ms. Sharma was awarded a Young HR Manager Award by the Professional Network Group of India.

Contents

Contributors

Divya Agarwal
Jesus and Mary College, University of Delhi, New Delhi, Delhi 110021, India

Shalini Aggarwal
Chandigarh University, Chandigarh 160036, India

Mehdi Benfadel
Iveco France, 14 Avenue du 24 Août, 69960 Corbas, France

Vartika Chaturvedi
Jaipuria School of Business, Ghaziabad, Uttar Pradesh 201012, India

Ankita Dhamija
Lingaya's Lalita Devi Institute of Management & Sciences, Delhi 110047, India

H. S. Dhanny
Amity University, Noida, Uttar Pradesh 201308, India

Ahmed. A. Elngar
Beni-Suef University, Beni Suef City 62511, Egypt

Vikas Garg
Amity University, Noida, Uttar Pradesh 201308, India

Richa Goel
Amity International Business School, Amity University, Uttar Pradesh 201313, India

Anil K. Gupta
Division of Environment and Disaster Risk Management, National Institute of Disaster Management, New Delhi, India 110001, India

Anubhuti Gupta
Amity University, Greater Noida, Uttar Pradesh 201308, India

Divyansh Gupta
Birla Institute of Technology and Science, Pilani, Rajasthan 333031, India

K. P. Kanchana
Jaipuria School of Business, Ghaziabad, Uttar Pradesh 201012, India

Shikha Kapoor
Amity International Business School, Amity University, Uttar Pradesh, 201313, India

Nitin Kulshrestha
Chandigarh University, Chandigarh 160036, India

Guneet Kaur Mann
Jaipuria Institute of Management, Ghaziabad 201014, Uttar Pradesh, India

Stephen Mckenna
Curtin University, School of Management, Bentley, WA 6102, Australia

Saurabh Mittal
GL Bajaj Institute of Management and Research, Greater Noida 201306, Uttar Pradesh, India

Seth Shyirakera Munyanziza
University of Kigali, KG 541 St, Kigali, Rwanda

Gaurav Nagpal
Birla Institute of Technology and Science, Pilani, Rajasthan 333031, India

Singh Satyendra Narayan
University of Kigali, KG 541 St, Kigali, Rwanda

Renu Paisal
Gautam Buddha University, Greater Noida, Uttar Pradesh 201308, India

Neha Puri
Amity University, Noida, Uttar Pradesh 201313, India

Sonam Rani
GL Bajaj Institute of Management, Greater Noida 201306, Uttar Pradesh, India

Shiv Ranjan
Gautam Buddha University, Greater Noida, 201308, Uttar Pradesh, India

Seema Sahai
Amity International Business School, Amity University, Uttar Pradesh 201313, India

Teena Saharan
Department of HR & OB, Doon Business School, Dehradun 248001, Uttrakhand, India

Anshika Sharma
IFTM University, Moradabad 244001 Uttar Pradesh, India

Dinesh Kumar Sharma
Gautam Buddha University, Greater Noida, 201308, Uttar Pradesh, India

Lalit Kumar Sharma
Jaipuria Institute of Management, Ghaziabad 201014, Uttar Pradesh, India

Silky Sharma
Amity University, Noida, Uttar Pradesh 201308, India

Gurinder Singh
Amity University, Noida, Uttar Pradesh 201308, India

Harjit Singh
Amity University, Noida, Uttar Pradesh 201313, India

Shalini Srivastav
Amity University, Greater Noida Campus, Uttar Pradesh 201308, India

Pooja Tiwari
ABES Engineering College Ghaziabad, Uttar Pradesh 201009, India

Preeti Tewari
Lloyd Law College, Greater Noida 201306, Uttar Pradesh, India

Anita Venaik
Amity Business School, Amity University, Uttar Pradesh 201313, India

Abbreviations

AI	artificial intelligence
API	application programming interface
AWS	Amazon Web Server
BPOs	business process outsourcings
CEO	Chief Executive Officer
CFP	Corporate Financial Position
CPS	cyber-physical systems
CSR	corporate social responsibility
DSS	decision support systems
EDA	Executive Development Associates
EI	emotional intelligence
EQ	emotional quotient
ERP	enterprise resource planning
FTE	full-time-equivalent
GAAP	Generally Accepted Accounting Principles
GBM	gradient boosting method
GLM	generalized linear model
HPWS	high-performance work system
HR	human resource
HRIS	human resource information system
HRM	human resource management
IAS	International Accounting Standards
IASB	International Accounting Standards Board
IASC	International Accounting Standards Committee
ICAI	Institute of Chartered Accountants of India
IFRS	International Financial Reporting Standard
IoT	Internet of Things
IT	information technology
KPI	key performance indicator
MCA	Ministry of Corporate Affairs
ML	machine learning
MMORPG	Massively Multiplayer Online Role Playing Games
MNEs	multinational enterprises
NBFC	Non-Banking Financial Company

NFP	nonfinancial planning
NGO	nongovernmental organization
OD	organization development
P2P	peer-to-peer
PAT	profit after tax
PE	private equity
PLC	Public Limited Company
PMS	performance management system
RBI	Reserve Bank of India
REST	representational state transfer
ROI	return on investment
RPA	robotic process automation
SDG	sustainable development goal
SEBI	Securities and Exchange Board of India
SHRM	strategic human resource management
SMEs	small- and medium-enterprises
SP	succession planning
UN	United Nations
UNCCD	United Nations Convention to Combat Desertification
VCF	venture capital funds
VR	virtual reality
VUCA	volatility, uncertainty, complexity, ambiguity
WBCSD	World Business Council for Sustainable Development
WOW	Wealth Out of Waste

Preface

This world has seen three industrial revolutions till now. The first revolution was due to utilization of steam as power, the second came about by electricity, and the third happened due to the introduction of IT and electronics. We are now into the Fourth Industrial Revolution termed as Industry 4.0, which has been possible due to the introduction of newer technologies like artificial intelligence and the Internet of Things. The Fourth Industrial Revolution started in a European country, Germany, which involved the coordinated working of the Internet, artificial intelligence, and various sensors with better output in the Industry.

In the domain of Industry 4.0 context, we are now at the brink of Finance 4.0 and Smart HR, and it should be understood what it really signifies.

What makes Finance 4.0 different from the earlier version is the arrival of blockchain or distributed ledger technology. This has led to an innovation called cyber currencies, which made possible peer-to-peer (P2P) transactions across a distributed ledger system (commonly known as blockchain. Cyber currencies or tokens were invented because of the lack of trust in the official system.

The cyber currencies emerged because financial regulations have increased the number of rules on which the official sector can regulate, which is actually the banking system. It is but natural that money is moving to outside the banking system into the asset management industry.

Various sectors have dramatically changed the method of storing data and then retrieve it in a required format to achieve the best results.

While we analyze the effects of Industry 4.0 on human resources management, two factors appear to have the major impact on the number of jobs available for people and also the availability of skilled manpower required to perform the emerging jobs:

Firstly, with the introduction of artificial intelligence and robots in our daily life, it is estimated that more than 60% of primary school-going children will be doing jobs that do not exist today.

Secondly, the mobile Internet and cloud-based technology, Big Data, and 3D printing will have a major impact on manufacturing. It is estimated that this will affect the jobs of millions of persons.

Both the above-mentioned factors will affect jobs of millions of low-skilled persons and create many jobs for highly skilled persons, who will be very hard to find and retain; thus this will be a great challenge for human resource departments.

Because of extensive automation and tasks of humans being taken over by robots and other artificial devices, the management of personnel and their performance will be a major aspect that will require attention. Regular upskilling of personnel and ensuring that they continue to perform to their optimum capacity will require innovative approaches to the whole issue of manning of organizations. Companies that cater to human resource software will have to adopt to the ever-changing requirements in Industry 4.0.

Human resource will find it very difficult to find the required talent who can quickly adapt to Industry 4.0 requirements. HR will have to adopt talent-driven approach and will have to look for qualified persons from all over the world. Such persons will have to be suitably rewarded to ensure their continued contribution to the organization. These persons will have to be given flexibility of working with flexible hours and flexible locations. Their competencies may have to be shared with other organizations in terms of hours and locations.

With various HR software available to monitor and manage the performance of personnel in a company, the HR analytics will play a very important role in optimizing the performance of workforce of a company. The data and vital information of employees will be important for data analytics to arrive at the best performance of all employees. This vital data will attain even bigger value for the employees and for companies. Hence, protection of this information will be of utmost importance for all companies.

Social media will acquire even greater importance, and it will become a central tool for employees as well, as for the organizations. Being a democratic platform, offering transparency to both employees and management, social media allows workers to participate at all levels, including complaints, pleasures, concerns, their rights, and interactions with each other offering.

The future role of HR and finance is thus going to change dramatically and be highly innovative. Talent-driven approaches that are very flexible will have to be adopted for companies to survive and reap the benefits of Industry 4.0.

Introduction

As per the current trends and developments going on in the industry, it is being considered that the fourth industrial revolution has already begun, leading to the emergence of Industry 4.0. Emerging technologies such as Internet of Things, Big Data, and artificial intelligence will automate most of the HR processes, resulting in efficient and leaner HR teams. Also, the digital technologies will drive the next wave of disruptive transformation in finance. The role, size, and shape will change—finance will be smaller, more efficient, and cost less, but with a higher proportion of highly skilled people.

Both organizational structure and leadership style changes will be required for efficient HR 4.0 implementation that would allow HR departments to play a more strategic role in the overall organization growth.

Financial acumen involves gathering of information about the financial affairs of entities of interest, basically to understand their nature and capabilities and predict their intentions. Generally the term applies in the context of law enforcement and related activities. Financial acumen is a type of business intelligence constituted of the knowledge and skills gained from understanding finance and accounting principles in the HR business world. For this purpose, human resources 4.0 and financial acumen will act as a catalyst in the disruption process in the human resource domain as well as in the entire business world globally.

CHAPTER 1

HR Acumen in Industry 4.0: Managing Talent and Achieving Balance in Life

GURINDER SINGH,[1*] H. S. DHANNY,[2] VIKAS GARG,[3] and SILKY SHARMA[4]

[1]*Amity University, Noida, Uttar Pradesh 201308, India*

[2,3,4]*Amity University, Noida, Uttar Pradesh 201308, India*

[*]*Corresponding author: E-mail: ssharma4@gn.amity.edu*

ABSTRACT

As per current trends and developments in the industry, it is considered that fourth industrial revolution has already begun, leading to the emergence of Industry 4.0. Emerging technologies such as Internet-of-Things, big data, and artificial intelligence are leading to automate most of the HR processes, resulting in efficient and leaner HR teams. Also, the digital technologies will drive the next wave of disruptive transformation in finance. The role, size, and shape of HR and finance function will change to a smaller, more efficient, and costless function, but with a higher proportion of highly skilled people. The challenges will be posed in front of HR in terms of managing talent and balance in life of their employees while sustaining financial outcome and impactful performance for the business. The acumen of HR will be assessed critically in managing complexities of identifying, developing, and managing talent side by side ensuring a high performance culture and balance in life of its workforce. Managing human, machine, and finance by HR experts in order to become revenue-oriented function for the organization would be a daunting challenge for them in the organization. Together, smart HR and financial acumen will act as a catalyst in the disruption process in the entire business world globally.

1.1 INTRODUCTION

A challenge is dependably equalized by chance. The present far reaching market instability, combined with cutbacks, ability deficiencies, and quick moves in innovation, focuses to increased difficulties for human resource professionals. Whereas fittingly utilizing human resource function has been key points of dialog for as long as decade, establishments still brawl with a way to build this vital move as a responsible unit.

What will assist human resource managers most in administering the human aspects of business and occupations? Throughout the following decade, the HR professionals will confront a wide scope of people issues that can be merged into a rundown of chief obstacles. While the obstacles talked about here probably will not be comprehensive, tending to these prevalent issues gives driving force to transform difficulties into new prospects. At the very beginning, a standout among the most imperative difficulties for an establishment is to position itself as a preferred company by the people. It includes having a favorable working culture where individuals wish to report happily at workplace and contribute till the time they are available and prefer to stay with the organization by their remarkable contributions. The indicators of HR's performance and achievements are reflected in transcending its difficulties are proven by return on investment in the form of monetary outcomes, but also in terms of practicing the benchmark practices, overall organizational efficiency and effectiveness. It is certain that technology and innovations assume an essential job in combating any HR challenge by encouraging arrangements that result in sure shot success.

Most of the profound organizations search and implement for the most ideal approaches to execute HR-related tasks. Considering HR in isolation limits the wide scope and capacity for HR function that may lead to saturation. The manner, in which, budget and finances are connected and concerned with each and every aspect of a business—typically as a financial plan—similarly, human capital substantially influences the companies in their totality. Hence, companies that aim to become "Great place to work" concentrate on people processes that have an effect at enterprise level. People processes and systems should be established in such a way that it should be helping employees in achieving a balanced and stress-free life so as to become a great place to work in true sense.

Because of changing requirements in Industry 4.0, a number of factors are putting pressure on employees from multiple sides creating a strong need for a support system from their organization that can help them in achieving a balanced life. A 6F model of balance in life has its special significance and applicability in reference to Industry 4.0. Industry 4.0 will present ambiguous and complex situations to all the stakeholders associated; be it managers or employees.

In Industry 4.0, utilization of machines and knowledgeable human resources will play a vital role in manufacturing and service sector. Since multiple devices will be interacting simultaneously through web, a number of unique issues will occur for the first time. Hence, it will be a great need for all humans involved in Industry 4.0 to handle these complex situations appropriately to ensure that there is no negative impact on these six factors which maintain balance in their life. The aim of "6F Model of Balance in Life" is the ***Mantra*** for everyone to achieve stability and balance in their professional and personal life in Industry 4.0.

A balanced life in Industry 4.0 can be achieved by taking care of a number of important factors necessary for a good to great life. To live a quality balanced life, six key areas (6Fs) have been identified, which are as follows (Figure 1.1):

1. Finance *(The Rock Bed of Stability in Life).*
2. Family *(The Immediate Environment That Nurtures Your Growth).*
3. Friends *(The Support System That Will Stand by You).*
4. Figureheads *(The Leading Lights in Your Life).*
5. Fitness *(The Body Mind and Soul Connection).*
6. Fun *(The Destressing and Rejuvenating Element).*

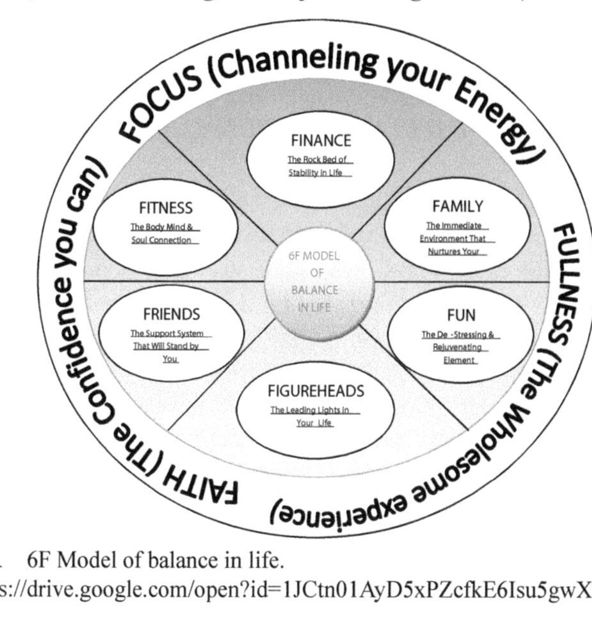

FIGURE 1.1 6F Model of balance in life.
Source: https://drive.google.com/open?id=1JCtn01AyD5xPZcfkE6Isu5gwXwt5dMJL

These 6Fs need to be practiced as an essential part of one's life in balanced proportion by individuals to have a good personal and professional life. These are represented in a circular wheel as shown above encircled by a binding thread of three more Fs (focus, faith, and fullness). All these 6Fs are interconnected, interrelated, and interdependent. Any variation in one factor will have an impact on other factors playing role more prominently.

1.1.1 FINANCE (THE ROCK BED OF STABILITY IN LIFE)

Financial strength means a sustained source of income to meet short-term expenses while taking care of long-term savings. Financial strength helps you to focus on higher order needs, keep you away from daily worries, makes you feel secure and confident, and keeps focused toward their career goals and aspirations. Financial robustness helps people to have freedom in making their career of choice. Financial independence especially empowers women candidates to reach to their full potential both in professional and personal life.

Financially stable people are more confident to take bigger risks in life and accept challenges where more growth opportunities may lie and further good prospects to compound their financial wealth. People who are assured of their financial wellness demonstrate higher levels of confidence, high self-esteem leading to higher levels of emotional wellness. They are not trapped in the web of over thinking.

1.1.2 FAMILY (THE IMMEDIATE ENVIRONMENT THAT NURTURES YOUR GROWTH)

Family is the first place for us to develop a strong foundation in the form of value system in our early years. We make strongest bond with our family members and it is this strong bond that holds us in all good and bad times. Our family members act as an immediate support system for growth and nourishment in our life and give us a sense of security.

All reputed industrial houses try to create the spirit of family in their organizational culture. Such organizations build an open culture providing their employees with listening mechanisms, recognition, appreciation acknowledgement, and reward for the good work done. Companies with strong value system care for their employees even in bad times like business downturns and recession. In such organizations, employees work with a sense of belongingness toward their organization. Thus, this association makes them feel with sense of security, stability, and experience balance in their life.

1.1.3 FRIENDS (THE SUPPORT SYSTEM THAT WILL STAND BY YOU)

After the family, good friends provide the best support system to each other, thus giving confidence and belief to perform well even in adverse circumstances. True friends take out their time and pay attention to our problems. In the company of best friends, we feel free to share our innermost thoughts about dreams, desires, and aspirations. With them, we experience joy and freedom.

At work places as well, colleagues cum good friends may help you in maintaining good work-life balance. Happy colleagues make work interesting by adding fun element in routine and mundane tasks. Team formed of people with similar mind-set or intellectual abilities performs better in the organization. As the saying goes, "You are the average of five people around you with whom you interact the most." Hence, the importance of friends is even more than it appears to be.

1.1.4 FIGUREHEADS (THE LEADING LIGHTS IN YOUR LIFE)

Figureheads are the key people who have the potential to guide you and provide wisdom for you to lead a better personal and professional life. They are the mentors with unmatchable knowledge, experience, and power that can help you in developing right professional skills. They help in developing a better sense of how to navigate challenges and successes in workplace.

They provide the feedback in a constructive manner and radiate goodness. They instill "CAN DO" attitude in the minds of their mentee. They can sow the seeds of strong potent ideas in one's mind and can convert rough uncut piece of carbon into the bright sparkling diamond. They virtually light up the path one should tread to reach the best destination without deviations. Thus, they can act as leading lights in our life and assist you in discovering new opportunities. HR initiatives for executive development and succession planning: The concepts of coaching or mentoring help employees in choosing their role model for faster learning and growth.

1.1.5 FITNESS (THE BODY MIND AND SOUL CONNECTION)

In today's stressful environment, "Fitness" has become one of the most sought after goals of people. Intellectual people know the benefits of having total fitness, a state of physical, mental, emotional, and spiritual well-being. All the people in our social circle like family members, friends, figureheads, colleagues, neighbors, and other people have a huge impact on our social and emotional well-being. When we ignore ourselves and our needs, our wellness gets affected. Thus, a deliberate focused effort in social interactions may lead to positive outcomes.

The fitness has to be seen in totality and we have to ensure that we create conditions for our physical, mental, emotional, and spiritual fitness. Each of these dimensions will require certain activities. Taking care of one dimension, say physical fitness, also helps achieving fitness in other dimensions like mental and emotional as well. Practicing meditation keeps our mind cool, calm, and peaceful to analyze the situations well. And, spirituality directs and motivates us to look for bigger purpose and meaning in our life. The fitness, thus, remains one of the very important factors to get the best out of your life.

1.1.6 FUN (THE DESTRESSING AND REJUVENATING ELEMENT)

Often ignored, fun is the most important element in our life to have balance. This is because of a number of benefits it offers in personal and professional life. Doing activities of our interest keeps stress away from our life as it leads to generation of positive hormones in our body. People who share their feelings and laugh more generally experience less negative emotions or mood swings. Hence, it is very important to take out time for fun and play by identifying the activity that you like and enjoy the most. It is advised to keep some minutes on daily basis to practice the same. *Fun acts as fuel for our body, mind, and soul balance.*

In corporate world, Fun @ Work has its own importance. Companies have started realizing the importance of fun at work. HR department keeps on organizing team building exercises to promote healthy mindsets of their employees. The strong bonds developed with colleagues' lead to happy culture at workplaces that further helps organizations to achieve higher productivity. Also, it helps to attract new people and retain their existing talent in the organization.

1.2 THE BINDING THREAD OF FOCUS, FAITH, AND FULLNESS

1.2.1 FOCUS (CHALLENGING YOUR ENERGY)

Focus on all the 6Fs is essential for desired balance in our life. However, due to demanding work schedules and a number of distractions available in the environment, we are unable to prioritize and may lose focus easily. Without focus in life, ability to think suffers a great deal. Working without focus and concentration generally leads to inefficiency and ineffectiveness. Hence, a focused approach is essential for quality output.

It is very important from time to time to develop focus and concentration through meditation, yoga, or exercise. Prepare "To do list" so as to not to lose focus and keep track of tasks finished or pending. This can be done by maintaining the right balance of work, fun, and spirituality in life. It can help channelize your energy in right proportions in right direction.

1.2.2 FAITH (THE CONFIDENCE YOU CAN)

Growth in personal and professional life requires making deliberate well-planned efforts at crucial moments in life. Any new ideas will require new initiatives to be taken, building a team, and giving them the roadmap to achieve the desired objective. The only thing that holds you back in taking these initiatives is the absence of "Faith in Yourself." The *Quality of Trusting Yourself* that you have the ability to achieve any insurmountable task is very important for faster growth.

Besides the trust in yourself, you also need to have the trust in the abilities of your team which will be implementing your ideas to achieve the designated objective. In case, you find any team member less than the job, we need to reinforce his abilities thus making him worthy of trust.

1.2.3 FULLNESS (THE WHOLESOME EXPERIENCE)

The sense of achievement is one of the most motivating factors in life. This is possible after we work toward an objective and are able to achieve all the required parameters of the outcome. This contributes to the enhancement of other factors also like faith and focus.

The sense of achievement also leads to social and professional respect besides destressing a person and leading to more motivated performance for future projects. Required focus in all of the parameters of "6 Fs" will also add to achieving the sense of fullness in life.

6F model of balance in life in totality brings a number of positive emotions in our life in the form of satisfaction, joy, accomplishments, motivation, energy, and enthusiasm. It further helps in developing a positive outlook toward life with peace and calmness inside. It helps in making individual having inner strength developed to deal with any external turbulence felt in Industry 4.0. It helps in developing an aura of positive energy.

Industry 4.0 has turned out to be common pattern in business market as far back as it was presented (Heng, 2014). Indeed, nowadays it is being considered as the topmost priority in the business market at the global level (Rüßmann et al., 2015). It is the reason for both productive as well as deconstructive deliberations, yet at the end it can be concluded that some transformation is required (Zhou et al., 2015). With the advancements and improvements in the kind of hardware, software, and human wisdom in practice and on the top of that decreasing prices of technology has made it evident to bring changes by fourth industrial revolution to happen (Brettel et al., 2014). Hence, it is imperative to study and explore the preparedness of the organizations for Industry 4.0 and to understand their considerations.

In addition to be ready with assets and technology, the human resource in the organization has likewise be set up for the transition in order to take care of human jobs and expected conduct in the Industry 4.0, yet particularly in the changing period needs to be contemplated. During principal stage, futuristic HR in Industry 4.0 needs to take care of the expectations and ventures of things to inquire about in coming future.

It is always the human who are initiators to think and bring about any change in any commerce surrounding. Humans play the dominant role in inception of progress (Hermann et al., 2016). Because of this reason, the viability of real progress and ultimate success depends just on them. Obviously, the first to request change must be the top level executives, yet so as to make the change progress and ensure that the framework is supportable; each and every employee of the organization must be made a party to the process with a faith in its beneficial outcomes (Bonekamp and Sure, 2015). It is quite natural to resist change everywhere and is generally considered as the fundamental deterrent in some way or another, so it must be kept away from (Kagermann, 2014). HR managers must adjust to new difficulties, similar to change of job as we all are aware of these days. Human role inside the organization changes and progress must be done effectively with appropriate knowledge and instruction (Schuh et al., 2015). Besides educating and training the contemporary workforce in the organizations, a new educational framework must be crafted at schools, colleges, and university level to identify, develop, and nurture talent.

Individual factors are unique to every individual that let the individual adopt his own ways to reach to his performance and potential levels in life. The degree of each factor varies in individuals letting them adopt and create ways to interact with external factors and learn from various sources of learning in his social, religious, and spiritual circle.

Individual factors solely or while interacting with factors in external environment leads to generation of emotions in individuals toward specific goals and abilities to be acquired. Also, if interest and need are very high in individuals, they create the opportunities to perform to become successful. Similarly, key life moments and challenges help in identifying the opportunities and importance of those.

Availability of various sources of learning affects individuals' factors by reshaping their personality, altering belief system. Also, these learning sources help individuals develop their aspirations and expectations levels and helps in creating unique experiences and interpretation of those experiences at different stages of life. However, the initial years of age, education, learning, and experience are crucial and quite fruitful period for an individual.

There are external factors that impact the individual and his learning from the environment. Often, the stimulus in the form of a situation or a person comes across to individual that helps him in identifying and developing the talent. Financial support is crucial for investment and furthering the effort to develop, refine, and shape the talent in an individual at expert and professional level required to compete with others and perform at superior levels of performance. Financial aid is mostly received by family member; however, sometimes is received from any other volunteer. An individual comes across various opportunities and challenges during various stages of life and many a times also created in life by mentor or lucky time frame. It is significant for an individual to identify those opportunities and challenges and take a constructive approach making best use for greater benefit. Here, individual factors like personality, belief system, grit, commitment, aspirations, and expectations play a significant role helping individual for decision-making.

Having understood the importance of opportunities and making use of learning's, the individual gets propelled and positive emotional state is achieved wherein interest in aroused toward that particular activity. The resourceful individuals consciously create the need or scarcity and work for their strengths and abilities. Another set of individuals are propelled significantly out of the strong need felt to remove some of the deficiencies in their life. It thereby motivates them to generate emotions of interest as they are left with no choice. The need reaches to hunger level for such individuals.

Rewards and punishments given during education time period or at workplace at different life stages give individuals motivation and drive to perform and excel. In today's time period of so many options and distractions, individuals opting to implement the means and learning's from religion and spiritual area perform with focus and concentration that are essential for extraordinary performance be it sports, academics, or any other recreational activity. Many a times talent is not utilized properly and one of the main reason is the lack of opportunity or no opportunity received throughout the life of an individual. Sometimes, opportunity comes once or twice and then not encountered again leading to diminishing of interest levels in individuals.

Thus, regular number of opportunities and of progressive level are needed to nurture and make use of talent to the fullest at organization and societal level in Industry 4.0. Performing at some opportunity and getting success generates interest and need to taste the success again and again. Emotions of interest, need, and success create a continuous loop for individuals to enhance performance and succeed leading to the full realization of talent, which is so invaluable and rare.

Thus, talent (T) can be defined as a function of emotion (interest and need) denoted by E and opportunity to perform (O) wherein both are multiplicative and if anyone these are missing, the talent cannot be harnessed

$$T = f(E \times O),$$

where T stands for talent, E stands for emotion (interest and need), and O stands for opportunity to perform (Figure 1.2).

1.3 DECENTRALIZATION AND STANDARDIZATION

There is no single spot to check the current production systems. The majority of the organizations have generally speaking data about system and its parts in entire establishment, yet they lack incomprehensive data as of now. This handicaps the likelihood to have total neglect at the methods that are being implemented inside the organization (Pfeiffer, 2016). Regardless of it is about strategic exercises in assembling area or regulatory segment or neglect to procedures and innovation occurring during assembling. With the implementation of technology and digitization which is a mandate as per Industry

4.0, thorough review of each procedure and subprocess, each equipment and programming with its utilization and exercises must be controlled and must be accessible at a particular place (Varghese and Tandur, 2015). This is just not a single spot, but instead decentralized spots are created to empower various stakeholders to get data about all aspects of the procedure, establishment, and every event happening in the organization (Almada-Lobo, 2015).

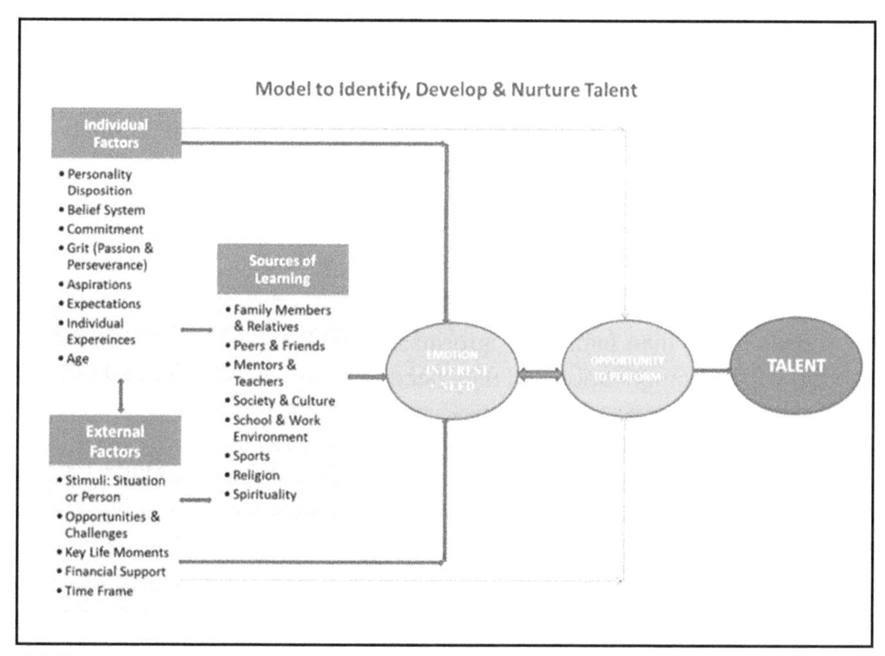

FIGURE 1.2 IDN model to identify, develop, and nurture the talent at individual level.
Source: https://drive.google.com/open?id=11ITnj7tCarvVlMfU2Ip6cKk-r5TPAPfi

Inputs of the customers should be collected at a central spot, so that this information can then be exchanged to other. Further, the collected information can be transmitted to other decentralized areas where everybody ought to possess knowledge in occasions at different pieces of framework and utilization of data assembled and exchanged.

Generally speaking, regulation of framework makes employee's job increasingly mind boggling, in light of the fact that human must process progressively critical data in the continuous (Becker and Stern, 2016). Additionally, human needs to settle on right choices from it, which implies that customary way to deal with work assignments will not be required soon. Manual work and straight forward exercises will in general be mechanized and streamlined requiring no human intervention any longer in given piece of operating circle. In any case, human stays are very much needed for basic leadership in emergent situations and for innovation and creativity division (Tseng and Lee, 2009). Hence, employees must remove a stage from the physical procedure, to happen above particular pieces of general framework. Fresh aptitudes are required, distinctive mindset and increasingly complex undertakings must be unraveled (Paelke, 2014). Long-haul strategies are not drift, transient tactics should have been characterized all the more frequently, yet additionally new thoughts must be made each day (Turner et al., 2016).

It can become a reason for obstruction from present workforce that needs to adjust to different workplace and to leave present working inclinations. The way to deal with training is imperative for this situation; the center ought to be to enhance the characteristic motivation and inspiration of the worker (Gorecky et al., 2016). While experiencing significant change developments, human should feel secure, without dread of his or her job loss, and yet human must make certain that new learning will have more advantage for sooner rather than later. The information accumulated must be utilized to build worker's motivation and they need to experience self-assurance and satisfaction with the present circumstance, yet additionally propelled for the future enhancements.

1.4 DECISION-MAKING

At conventional workplaces, decisions are taken at central point, and are typically taken by the top management (Power and Sharda, 2009). Since businesses have to take care of enormous amount of real-time data and numerous requests that must be settled in as shorter time as could be expected under the circumstances, thereby decision process needs to be changed radically (Lee et al., 2015).

Initial modification is decentralization. Information and data collection from different sections of organization, central decision-making process turns out to be increasingly complex, ultimately of no value. Decentralized decision-making helps in quick and accurate resolution and generating quick response to system needs (Lee et al., 2017). Regardless of this is about client requests, fabricating innovation or plan requests, choice must be made as fast and as apt as could be expected under the present circumstances. The information gathered from the pieces of the framework is unpredictable and criteria for each littler subsystem must be characterized, so as to settle on best possible outcome (Lee et al., 2014). Careful consideration ought to be put on the littler choices that have been made with the measure of the data which were handled to evade probability of subideal outcome. Humans are responsible to pursue the story of escalation of difficulty of multifaceted nature of human activities. Once the norms are decided, Industry 4.0 mandates to find out the precise meaning of information (Weyer et al., 2015). That is the reason which makes it imperative to use decision support systems (DSS) on regular premise (Schuh et al., 2014). They must be actualized in the framework to make choice rapidly and precisely. DSS should to be convenient to understand and make use of by the user for the simplification of the complex tasks assigned to human. Sort of DSS must be characterized by the part of the framework they are expected.

1.5 LEARNING WORKPLACES AS THE FUTURE OF INSTRUCTION

Employees are required to have supplementary knowledge and new abilities so as to become capable of taking decisions at decentralized (Prinz et al., 2016). Ordinary instructive framework generally is not able to meet prerequisites of the information and abilities level required for the manufacturing plants of things to come. It just gives more of hypothetical information (Hold, 2017). This is the reason the majority of the graduates, when they leave their schools and colleges require time to get adjusted in their workplaces. And, few good organizations, for the same reason give a half to a yearlong orientation cum training instructive period for the apprentices. Since these graduates need to possess extra skills and aptitude, this time period is to understand and become acquainted with norms to work as per that particular framework. This poses a question on the credibility of formal instruction system when so much of time and money is devoted in getting theoretical knowledge which is of not much significance at workplace or insignificant until and unless ground work is done.

Major level of changes is required in formal instructive framework. The progressions must be done in each part of instructive framework, most fundamentally even in the first phase of human training. Stress must be given to acquire practical knowledge and creating innovative and creative mindset. Space for imagination and changed outlook is to be built in mind and is quintessential for Industry 4.0. As a matter of significance, employees are required to constantly adopt new ways of doing things, adopting new systems and methods of carrying and must prepare themselves for consistent changes of the strategies, of the workplace with the need of steady improvement (Blöchl et al., 2017). With steady changes, new thoughts, inclines available, in the field of innovation, equipment, and programming; human must be prepared to adjust to new circumstances and assemble new abilities in all aspects of their profession (Reuter et al., 2017).

Emphasis on practical knowledge and hands on experience is already being given and practiced at learning plants (Abele et al., 2015). Thereby, labs have been created to carry out practical sessions integrated with the course curriculum at colleges and university instructive frameworks.

With recreation (both in programming and physical way) of the assembling and hierarchical framework, individual is been set in the consistent, regular circumstances that are possible in the organizations. They figure out how to respond and settle on appropriate choices as per necessities of Industry 4.0 philosophy. It becomes quite essential to acquire skills related to "lean/kaizen." This helps in inculcating in one's normal approach in solving problem and adopting a persistent improvement mindset on regular basis. Thinking and implementing continuous improvement and consistent change becomes a routine and comes natural to their mind (Kolberg and Zühlke, 2015).

There have been instances of existence of learning workplaces in Europe and other countries, where these factories have placed their systems and methods with cutting edge equipment and programming, yet in addition with learning that speaks to the genuine framework worked by the requisites of Industry 4.0 (Jäger et al., 2014). There are different sorts of learning processing plants in which the articulation was put on machining innovations, its exploration, advancement, and instruction about its functionalities. The principle exercises are understudy instruction, organizational trainings, and research works in the area of mechanical designing. With equipment and programming measured quality, it is conceivable to reproduce different circumstances from the truth and become familiar with the clients how to respond in specific circumstances. The enhancement and profitability of clients is being observed by means of key performance indicators. Their principle objective is to create progress from the lean learning manufacturing plant to Industry 4.0 learning production line—entertainment of cyber–physical systems, as this has been perceived as the primary pattern available (Karre et al., 2017).

The point is to make learning industrial facilities adaptable, exceptionally versatile to modifications so that each pattern and requirement from outside insistences can be satisfied. Other than the equipment and programming, fresh callings are being made yet in addition that these can be utilized as exploration and trial grounds for genuine organizations whose R&D division are not capable of performing similar tests inside the organization.

1.6 INDIVIDUAL–MACHINE, MACHINE–MACHINE, AND INDIVIDUAL–INDIVIDUAL INTERACTION

Numerous specialists are worried that machines and AI will assume control above their work areas which could lead to loss of their employments (Weiss et al., 2016). Actually, human work which is manual in nature can be done through machines, will be supplanted soon in near future, yet individuals will not become pointless. Such individuals are the ones who need to realize how to communicate with machines, command them, and enhance their job yet in addition to find the new areas and circumstances for them to utilize so as to make most ideal outcomes. The connection among human and robot to some may be unsavory at first. The arrangement is an appropriate instruction. Becoming acquainted with how the robot functions, their potential outcomes, undertakings, prerequisites, confinements, preferences, and position in the framework is critical, particularly inside the more seasoned specialists that are not as acquainted with Hitech equipment or machinery as more youthful workers are (Gorecky et al., 2014).

Workers must almost certainly speak with robots by means of extraordinary cell phones, to design and to anticipate the required conduct. That is empowered by uncommon programming and learning databases, uniquely made for a solitary organization (Wittenberg, 2016).

The correspondence with different representatives is done by means of exceptional informal organizations which are beneficially utilitarian, yet additionally give another, social viewpoint to the organization. Bigger number of individuals are associated, better societal connections may be framed, that can lead to beneficial outcome on characteristic inspiration of the worker. Once content with present working environment, persuaded and collectively fulfilled laborer will not indulge in negative feelings of being useless and with supplementary help of technology in profoundly mechanized condition (Bedolla et al., 2017).

As per the Industry 4.0 idea, "machine–machine" connection turns out to be vital. By means of general programming framework, robots ought to have the capacity to examine current circumstance in the assembling procedure, settle on choices autonomously, yet additionally to design the assets, or if nothing else caution the human workers or managers with the proposal of the required prerequisites. The machine–machine correspondence will lead to automation of procedure arranging inside the organization, yet additionally the automation of several strategic and arranging exercises (Wang et al., 2016). By automatized process arrangement, fabricating innovations for the item would be made with utilization of man-made brainpower, yet the job of human in the process stays basic—they some way or the other need to manage the framework that ensures that the automation of jobs does not lead to more disadvantages than advantages.

1.7 SUSTAINABILITY OF THE EDUCATIONAL CONCEPT

Much the same as in the equipment and programming, organizations will at the earliest reference point of execution of the latest Industry 4.0 idea inside the organizations need to contribute higher measure of cash so that many people may discover excessively testing and can maintain a strategic distance from some streamlined arrangements given (Sommer, 2015). The best possible and consistent training of the employees is most critical advance in actualizing Industry 4.0. Likewise, transitions in the general instructive framework will bring about creating the fresh callings, youthful, and spurred individuals that will almost certainly adapt to requesting working condition once they enter the activity advertise. Subsequently, any more additional instruction for the single organization will not be required, yet in addition new attitude will empower to make expanding results later on.

The acknowledgment of deep rooted learning is fundamental and must be executed inside each organization. Additionally, this sort of framework must be practical and furthermore part of the ceaseless enhancement process. Workers need to be instructed about fresh patterns and developments consistently; yet, additionally the educators must be the ones to accumulate progressed instructive strategies, so that the new information can be gained and exchanged as effectively and proficiently as conceivable (Hecklau et al., 2016).

The supportability of the general framework lies in the collaboration of each individual engaged with it. The fourth industrial revolution may appear as current pattern that requests major changes, however up until this point, the forecasts for progressively removed future are going similarly as current pattern, yet just with its redesigned varieties. That is the reason new learning and new developments ought to be gained and actualized daily, so the extreme modifications will not be a piece of the organization's prospect any longer. The progressions will be simpler to actualize yet additionally will need lower ventures. As the learning plays the key role in achieving the leading place, joint effort among instructors and working staff is vital, just as their steady progress in their regarded areas (Bauer et al., 2015).

1.8 CONCLUSION AND FUTURE WORK

Human resource is a vital resource of each organization. By means of change by the tenets of fourth industrial revolution, individuals could easily compare to in last-three industrial revolutions. A large portion of the open has an assessment that total automatization of assembling process and other hierarchical undertakings inside the organization may bring about losing the positions; however, the fact of the matter is direct inverse. Employments will be there however they are going to experience several changes that need to be acknowledged. Resisting change is the greatest deterrent inside the organization and, other than money-related speculations; it may moderate or ruin the change procedure. Fresh roles will request progressively difficult assignments; however, labor-intensive jobs work would be totally automated and streamlined. Employees assume the responsibility for the procedure and improve it with accessible learning. Expertise will be the main criterion of progress in the business. Basic leadership will be decentralized with vertical and level combination inside the authoritative framework. Employees must figure out how to function, yet in addition how to convey in latest workplace. The individual–machine, machine–machine, and individual–individual connections will be ordinary regular movement, yet during automation and intelligent machine robots implementation of procedure, employees should feel sheltered and inspired. It will generate new and inventive thoughts; probability of persistent enhancements makes more prominent esteem. With incredible significance of instructive framework inside the organization as well as in the general instructive framework, costs for additional training will be decreased and new callings are prepared to confront each test in market.

1.9 WORKING ATMOSPEHERE

In upcoming investigations and related initiatives endeavors, it is essential to perceive how the efficiency of the laborer could be estimated. The profitability is essential for the some kind of control, yet in addition to streamline the procedure and increment the dimension of inspiration that will have constructive outcome in the close, yet in addition in far off eventual fate of the organization. Likewise, the advancements inside the organization floor, as the principle well spring of improvement, yet in addition of the specialists' inspiration, their significance and the board are another intriguing and imperative area for future research.

KEYWORDS

- **Industry 4.0**
- **smart HR**
- **financial acumen**
- **talent**
- **balance in life**

REFERENCES

Abele, E. et al.: Learning factories for research, education, and training, *Procedia CIRP*, (32) 2015, 1–6.

Almada-Lobo, F.: The Industry 4.0 revolution and the future of manufacturing execution systems, *Journal of Innovation Management*, (3) 2015, 16–21.

Bauer, W. et al.: Transforming to a hyper-connected society and economy—Towards an "Industry 4.0," *Procedia Manufacturing*, (3) 2015, 417–424.

Becker, T.; Stern, H.: Future trends in human work area design for cyber-physical production systemss, *Procedia CIRP*, (57) 2016, 404–409.

Bedolla, J.S.; D'Antonio, G.; Chiabert, P.: A novel approach for teaching IT tools within learning factories, *Procedia Manufacturing*, (9) 2017, 175–181.

Blöchl, S.J.; Michalicki, M.; Schneider, M.: Simulation game for lean leadership —Shop floor management combined with accounting for lean, *Procedia Manufacturing*, (9) 2017, 97–105.

Bonekamp, L.; Sure, M.: Consequences of Industry 4.0 on human labour and work organisation, *Journal of Business and Media Psychology*, (6) 2015, 33–40.

Brettel, M. et al.: How virtualization, decentralization and network building change the manufacturing landspape: And Industry 4.0 perspective, *International Journal of Information and Communication Engineering*, (8) 2014.

Gorecky, D. et al.: Human–machine-interaction in the industry 4.0 era, *12th IEEE International Conference on Industrial Informatics*, Porto Alegre, Brazil, September 07, 2014.

Gorecky, D.; Khamis, M.; Mura, K.: Introduction and establishment of virtual training in the factory of the future, *International Journal of Computer Integrated Manufacturing*, (30) 2016, 182–190.

Hecklau, F. et al.: Holistic approach for human resource management in Industry 4.0, *Procedia CIRP*, (54) 2016, 1–6.

Heng, S.: Industry 4.0: Upgrading of Germany's industrial capabilities on the horizon, Deutsche Bank Research, 2014.

Hermann, M.; Pentek, T.; Otto, B.: Design Principles for Industrie 4.0 Scenarios, *49th International Conference on System Sciences*, Koloa, HI, USA, January 5–6, 2016.

Hold, P.: Planning and evaluation of digital assistance systems, *Procedia Manufacturing*, (9) 2017, 143–150.

Jäger, A, Ranz F., Sihn W., Hummel, V.: (2014) Implications for Learning Factories from Industry 4.0—Challenges for the Human Factor in Future Production Scenarios. *4th Conference on Learning Factories*, Stockholm, Sweden, 1–35, 2014.

Kagermann, H.: *Change Through Digitization—Value Creation in the Age of Industry 4.0. Management of Permanent Change*; Springer: Heidelberg, Germany, 2014, pp. 23–45.

Karre, H. et al.: Transition towards an Industry 4.0 state of the LeanLab at Graz University of Technology, *7th Conference on Learning Factories*, Darmstadt, 2017.

Kolberg, D.; Zühlke, D.: Lean automation enabled by Industry 4.0 technologies, *IFAC-PapersOnLine*, (48) 2015, 1870–1875.

Lee, J.; Bagheri, B.; Kao, H.A.: A cyber-physical systems architecture for Industry 4.0-based manufacturing system, *Manufacturing Letters*, (3) 2015, 18–23.

Lee, J.; Kao, H.A.; Yang, S.: Service innovation and smart analytics for Industry 4.0 and big data environment, *Procedia CIRP*, (16) 2014, 3–8.

Lee C.K.M.; Zhang, S.Z.; Ng, K.K.H.: Development of an industrial Internet of things suite for smart factory towards re-industrialization, *Advances in Manufacturing*, (5) 2017, 335–343.

Paelke, V.: Augmented reality in the smart factory: Supporting workers in an industry 4.0. environment, *IEEE Emerging Technology and Factory Automation*, Barcelona, Spain, September 16–19, 2014.

Pfeiffer, S.: Robots, Industry 4.0 and humans, or why Assembly work is more than routine work, *Societies*, (6) 2016, 1–26.

Power, D.; Sharda, R.: In *Decision Support Systems*; Shimon, Y.N., Ed.; Springer Handbook of Automation; Springer: Berlin, 2009, pp. 1539–1548.

Prinz, C. et al.: Learning factory modules for smart factories in Industrie 4.0, *Procedia CIRP*, (54) 2016, 113–118.

Reuter, M. et al.: Learning factories' trainings as an enabler of proactive workers' participation regarding Industrie 4.0, *Procedia Manufacturing*, (9) 2017, 354–360.

Rüßmann, M. et al.: Industry 4.0: The future of productivity and growth in manufacturing industries, Boston Consulting Group, 2015.

Schuh, G. et al.: Collaboration mechanisms to increase productivity in the context of Industrie 4.0, *Procedia CIRP*, (19) 2014, 51–56.

Schuh, G. et al.: Promoting work-based learning through Industry 4.0, *Procedia CIRP*, (32) 2015, 82–87.

Sommer, L.: Industrial revolution–industry 4.0: Are German manufacturing SMEs the first victims of this revolution? *Journal of Industrial Engineering and Management*, (8) 2015, 1512–1532.

Tseng, Y-F.; Lee, T-Z.: Comparing appropriate decision support of human resource practices on organizational performance with DEA/AHP model, *Expert Systems with Applications*, (36) 2009, 6548–6558.

Turner, C. J. et al.: Discrete event simulation and virtual reality use in industry: New opportunities and future trends, *IEEE Transactions on Human–Machine Systems*, (46) 2016, 882–894. DOI: 10.1109/THMS.2016.2596099

Varghese, A.; Tandur, D.: Wireless requirements and challenges in Industry 4.0, *International Conference on Contemporary Computing and Informatics*, Mysore, India, November 27–29, 2015.

Wang, S. et al.: Towards smart factory for industry 4.0: A self-organized multi-agent system with big data based feedback and coordination, *Computer Networks*, (101) 2016, 158–168.

Weiss, A. et al.: First application of robot teaching in an existing Industry 4.0 environment: Does it really work? *Societies*, (6) 2016, 1–21.

Weyer, S. et al.: Towards Industry 4.0—Standardization as the crucial challenge for highly modular, multi-vendor production systems, (48) 2015, 579–584.

Wittenberg, C.: Human–CPS interaction—requirements and human-machine interaction methods for the Industry 4.0, *International Federation of Automatic Control*, Kyoto, Japan, August 30–September 02, 2016.

Zhou, K.; Liu, T.; Zhou, L.: Industry 4.0: Towards future industrial opportunities and challenges, *12th International Conference on Fuzzy Systems and Knowledge Discovery*, Zhangjiajie, China, August 15–17, 2015.

CHAPTER 2

HR Practices for Financial Acumen

TEENA SAHARAN

Department of HR & OB, Doon Business School, Dehradun 248001, Uttrakhand, India

**Corresponding author. E-mail: dr.teenasaharan@gmail.com*

ABSTRACT

Human resource (HR) plays a very important role in the success of an organization. These are the people who provide right manpower to the business and develop and motivate them according to company requirements. However, majority of the HR professionals are reactive than being transactional. They are incapable of becoming active participants in the strategic planning of the businesses and lack of financial acumen is the main reason. Financial acumen is the ability of an individual to understand the financial situation of business and capability of making responsible and fiscally sound decisions. It requires comprehensive knowledge of the factors driving profitability and cash flow in the organization, business operations, and entire understanding of organizational processes. The profile of HR business partner is a buzz word in recent years and organizations require HR specialists with financial astuteness that can actually bring organizational value. Looking at the new skills requirement, 41% of the CHROs desired financial acumen as the most important ability. Financial acumen is there in requirement of business from years and people have a stronghold on that, but for HR, it is a challenge. HR has to understand and introduce the HR practices that can help its people develop economic acuteness. In a competitive and economically challenging environment, financial literacy is important. Though most of the HR professionals do not carry any formal training in financial management and merely a basic understanding of financial terminology like return of assets, return on investment, cash flow, etc. is not sufficient. For this, one has to truly understand the meaning of these terms not to deceive someone but to actually improve business outputs by minimizing threats and susceptibility. The HR practitioner has to have solid business acumen to understand the financial side of the business so that appropriate actions can be taken for utilizing the limited resources of the company to produce worthy results. For developing this business acumen, professional has to focus on imbibing the HR practices that can develop good financial sharpness among them. In today's era, the knowledge and hands-on experience of HR metrics and analytic, HR accounting, workforce balanced scorecard, and HR benchmarking can help HR professionals in developing financial intelligence. The focus of this chapter is to understand the importance and application of these HR practices to help people management department in determining the effectiveness of important HR activities and initiatives.

2.1 INTRODUCTION

Every organization requires managers who are able to analyze business situations and capable of making profitable decisions for the company. The managers with strong business acumen are the necessity at all levels. The human resource (HR) department was the one which was viewed as a department without business acumen, as a loss-making department where the basic function was hiring and firing of employees. However, the department sensitized the change and is now evolving as a business partner, contributing a lot to organizational growth and building the future of work. Instead of a cost making department, it is evolving as a profit enabling center by maximizing the gains on every penny invested on employees.

Today, HR department has evolved as a multifaceted center that is majorly focusing on employee engagement and empowerment. It has introduced latest technological interferences for maximizing the effectiveness of HR activities. The financial acumen is improved and department has started taking decisions based on available data on the human resource information system (HRIS) with the help of HR metrics and analytics and HR accounting. Instead of taking decisions based on intuition and emotions, HR is now relying on data for better and profitable decisions. HR is utilizing every bit and piece of information available on information system to draw trends and predicting future for drafting best possible strategies in the interest of employees as well as investors.

At present, the HR department understands the depth of its business operations, latest trends within the industry, and is capable of scanning the competitive environment. By the amalgamation of collected information and effective business acumen, HR is able to make effective business decisions. They are better able to present their strategic importance as a core business function. HR is leveraging the available information and technology to solve complex people problems.

The major activities that HR performs are related to strategic HR planning; recruitment, selection and onboarding; training and development; payroll and benefit management; employee wellbeing and relationship management; legal compliances; diversity and risk management; and engaging people and managing change. All these jobs require financial acuity for profitable decisions. The financial astuteness in HR practices provides holistic insight of the business and eventually helps HR professionals to bring proficiency and effectiveness in their jobs.

In the era of data, analytics, and technology, HR can no more function in a traditional set-up. To earn the trust and faith of the organization in itself, good financial acumen is a must in HR. The director of the Wharton School's center for HR, Professor Peter Cappelli stated in an article that "very often, corporate leaders perceive HR as being out of stride with the rest of the business." To break this conception, it is very important for HR to develop business acumen to ponder and proceed strategically in alignment with organizational objectives. Not only from a business perspective but also for better career prospects and successful business leader, it is legitimate for the HR professionals to be proficient in those HR practices that can help them in developing good financial acumen.

2.2 CHALLENGES

The reality is much different than expected as most of the HR professionals lack the expertise of handling HR practices required to develop financial acumen. Studies indicated that although the need to understand business dynamics is required at all levels; however, many employees are not prepared for transforming organization during challenging business scenarios. In one of the recent report of Business Training Systems, it was clearly mentioned that companies are lacking the business intelligence required to execute strategies. Even critical people have limited business acumen and do not know how companies are making money. Due to this reason only, the companies have started bringing a transnational shift of many CFOs to get involved in functions of HR, diversity, and inclusion. Companies are identifying in the real "assets" who are capable of bringing a shift in HR function to make it a profitable option to invest. In the service-driven economy and tough economic conditions, the major concern of the higher level executives shifted toward an increase in productivity and minimizing the staffing cost. People with good business acumen such as CFO have more chances of getting the coveted seat among the board members by identifying and improving the return on investment (ROI) on HR and related activities. However, it can become a risk for HR professionals as well as for the organizations. It is risky because the people with a financial mindset will not be able to understand the depth and subtleties of HR functions which can become expensive for the organization in the long run. It is an opportunity for the HR professionals to enter in the game and get acquainted with the finances of the company to become effective counselors for strategic business decisions. Though most of the HR professionals do not carry any formal training in financial management and merely a basic understanding of financial terminology like return of assets (ROA), ROI, cash flow, etc. is not sufficient. For this, one has to truly understand the meaning of these terms not to deceive someone but to actually improve business outputs by minimizing threats and susceptibility.

The HR practitioner has to have solid business acumen to understand the financial side of the business so that appropriate actions can be taken for utilizing the limited resources of the company to produce worthy results. For developing this business acumen, professional has to focus on imbibing the HR practices that can develop good financial sharpness among them. It is noticed that organizations expect HR to have digital edge, innovativeness and creativity in program designing, and change management, etc. over all other HR practices. Though, in today's era, the knowledge and hands on experience of HR metrics and analytic, HR accounting and HR benchmarking can help HR professionals in developing financial intelligence.

2.3 HR PRACTICES FOR FINANCIAL ACUMEN

2.3.1 HR ANALYTICS AND METRICS FOR FINANCIAL GAINS

HR Metrics: To determine the effectiveness of HR functions, HR metrics are used as measurement tools to calculate the values related to HR initiative such as training effectiveness, cost per hire, absenteeism rate, employee turnover rate, and return on human investment. The metrics help HR professionals to make better decisions for driving maximum output from its employees. It leverages the available data to measure the performance from different aspects that helps them track and predict future outcomes.

Metrics are not just about numbers, it is to go beyond that for demonstrating its impact on bottom line. Ross Sparkman at Facebook, head strategic work for planning said that, "metrics essentially give us a way of qualifying the health of our organization, and, to that end, HR metrics are now different." HR professional has to meld HR metrics with the available business data to measure the effectiveness of HR functions. Metrics are a very important tool in the identification and effective management of human capital. Human capital is the people who actually put their dedication and efforts for achieving the organizational objectives. These are the competitive advantage of the business and build the core competencies of any organization. That is why it is important for any business to effectively manage their human capital to have an upper hand in the market.

It is very crucial for HR to review their practices for measuring the effectiveness of the HR practices on human capital in an organization. The HR metrics not only assess the efficiency and effectiveness of human capital but also measures other indicators of organizational health with respect to organizational culture and strategic business objectives. Unfortunately, most of the organizations are unaware of the number of people working with them, let alone the human capital. HR has never been held accountable for the programs that they promote and run across the company and no one in the organization including top management are worried about measuring the impact of these programs. Due to this ignorance, HR managers do not try to calculate the investment and outcomes of these programs by applying suitable metrics. In today's tough economic times and challenging environment, it is important to calculate the value of HR activity and their outcomes. To have a seat at the table and not to be undermined and sidelined by the organization, HR professionals have to improve the effectiveness of their decisions by capturing and analyzing the available data. To be a better strategic partner, it is important for HR practitioners to understand the benefits of HR metrics and analytics in their day to day operations. It is required for them to develop their expertise in the HR metrics and analytics that can become instrumental to them for solving critical people and business-related problems.

Before the introduction of HR metrics, it was very difficult to quantify any HR activity. This made it always hard for HR professionals to understand and calculate the real cost associated with each employee. For example "if somebody wanted to know the cost per hire, it was a two-week project." With the introduction of HR metrics, the HR managers are capable of calculating the financial impact of various determinants such as employee turnover rate, absenteeism rate, or cost of external recruitment. These calculated figures indicate the cost per employee and cost per activity to help companies in creating effective plans and preventing talent loss. Metrics also enable HR managers and executives to make available efficient and effective delivery of their services to the people and organization at large.

Alexis Fint defined HR metrics as “the operational measures, addressing how efficient, effective, and impactful an organization’s HR practices are.” It is very important to understand the definition of HR metrics. Klipfolio very well explained that any HR metric helps business organizations in quantifying, tracking, and assessing the status of specific Human Resource Management (HRM) processes. More than 40 HR metrics are being utilized by different organizations at present depending upon the nature of demand. Academy to Innovate HR has listed a number of HR metrics for calculating the following:

1. Cost per hire
2. Time to hire
3. Revenue per employee
4. Potential and promotion
5. Employee turnover rate
6. Ratio of HR business partners per employee
7. Ratio of HR professionals to employees
8. Timesheet and matching schedule
9. Time since last promotion
10. Cost of HR per employee

HR metrics are not limited to the above-mentioned list. It also provides a window to look at the soft skills present in HR. Although soft skills such as comfort of gelling with others, social skills, diligence, etc. are difficult to quantify but it is possible to convert them into meaningful numbers by measuring the effect of these skills on the performance of bottom line of a company. A metric not only provides the current output of any variable it presents many other things that are impossible to measure directly. For example, absenteeism rate not only provides the number of working days missed by a person but also represents the happiness status quo of an employee or the issues related to health and home. Absenteeism is calculated by summing up the employees missed working days and dividing it with the total scheduled working days. For example, an employee has missed four working days out of 90, the absence rate will be four divided by 90 (4/90 = 0.21), that is, 21%. Although by looking at the attendance sheet it looks that the employee has just missed working days; however, the metrics are incredibly high. The output clearly presents that employee has missed one out of five working days and it may be due to some professional or personal reasons. The metrics can help HR to have a closer look at the absenteeism rate of the employee and can take some effective measures if the absence persists overtime. FitSmallBusiness provided a list of other people metrics to measure soft aspects of the job such as job satisfaction rate, innovation, health care cost per employee, training cost per employee, and return on training investment.

HR metrics help organizations in measuring the value of time and money spend on different HR activities and enable it to track changes and trends in the critical variables. These metrics are not just able to quantify the associated cost and output, it is a vital way to measure the impact of various HR programs and processes on employees. Metrics also help HR to improve their function by screwing the programs wasting organizational money, time, and efforts and motivating HR to introduce an initiative that helps to improve the organizational culture and its functioning. Some metrics are presented below that measure the efficiency and effectiveness of various HR practices and their functions:

1. Time to fill the open position
2. HR expense factor
3. Training ROI
4. Absent rate
5. Cost per hire
6. Revenue factor
7. Defect rate
8. Percentage of performance goal met and exceeded
9. Percentage of top performers
10. Turnover rate of top performing employees
11. Turnover rate of low performing employees
12. Turnover rate of employees resigned for compensation and related issues
13. Involuntary turnover rate at important position.

It is very important to understand the story behind the data points to take corrective measures for better implementation of business strategies. HR metrics provide a special tool to measure human capital. It enables HR to measure variables such as employee engagement, talent management, and high performers. This tool is popularly known as a Key Performance Indicator (KPI). KPIs are melded with the strategic goals to promote partnership at all organization levels for achieving the desired financial and nonfinancial objectives. KPI is an alignment of individual and departmental objectives with the corporate objectives. The most common HR metrics used by HR analyst to influence the KPI's are:

1. Training efficiency of teams
2. Time to fill replacement
3. Absenteeism rate
4. Involuntary turnover rate
5. Voluntary turnover rate
6. Time to hire an employee
7. Offer acceptance rate
8. Renege rate
9. Training expenses per employee
10. Revenue per employee
11. Time to hire an employee
12. Human capital risk.

Organization expects HR to provide palpable evidences to assist business executives in taking important decisions that affect the bottom line and overall business. With the help of facts and figures, HR needs to convince the leaders to not promote activities where the business will waste resources and lose its money. By deploying the organization-specific metrics business can save a wholesome amount on their important HR functions related to hiring and firing. It is important to identify and select the right metrics as per the overall business requirement.

HR Analytics: The analysis of other important organizational functions like marketing, finance, operations, supply chain, and IT provides an insight into their forward-looking approach and thrust for the development of continuous improvement measures. These functions have adopted some of the latest techniques and technologies such as going digital to stay connected, calculating and presenting the ROI on every dollar spent, and measuring the impact of analytics-based strategic business decisions. To improve the performance and outcomes of HR activities, it is critical for HR to understand and develop analytical intelligence. HR analytics is the area that applies statistical tools on data related to human capital to have a better understanding of their perspectives.

HR analytics is applied in the organization for improving employee performance and talent retention to better the ROIs. According to Techopedia, HR analytics measures the correlation in people's data with respect to business data and tries to establish a relationship or connection between the two. The output of HR analytics provides an insight into the current situation and measures the impact of HR practices on the overall performance of employees and organization. HR analytics establishes a relationship between the HR practices and its business outcomes to formulate effective business strategies for the coming future.

HR analytics helps business to dig deep into the existing functional problems of HR mostly related to talent acquisition, compensation, talent development, and optimization. Analytical workflows provide hands-on information to guide HR managers to gain insight of the agenda and help them in making informed decisions and designing suitable actions. For the sustainable growth of any business, it is imperative to attract, manage, groom, and retain the right talent. However, it is not possible for HR department to understand the available data without the application of HR analytics. To utilize the resources to the optimum and for the long-term success of the organization, it is essential to understand trends and predict the future for designing effective and profitable strategies.

For example: The proposal for conducting employee training may be rejected by the CEO. Contrary to it, if the data available is analyzed and HR is able to present the profit company will earn with the training, the probability of getting a green signal will be 100%. If the HR presents the current turnover rate of the company which is 14% above the competitors and it has increased by 4% in the previous quarter. Exit interview analysis highlighted that the managers lack the skill of dealing with conflicts effectively. If the leadership training is conducted for managers at all levels, the turnover will drop to 6% and that way, the profit will go up by 3%. By presenting the number with adequate logic may help HR in getting quick acceptance of their practices because now it is working in the same direction as of the organization.

In this data driven world, HR analytics helps organizations of all types and sized in guiding people, talent management, and hiring of suitable talent. According to Startup Focus, it is profitable for companies to analyze captured data to have a better and more accurate understanding of targeted customers. It will help companies in defining the competencies and skill-set required in talent, which will eventually decrease the training cost and improve organizational profits. HR analytics can be applied in varied areas of the business to make profitable decisions such as employee turnover analysis. The turnover analysis enables HR to predict the peril of employee turnover at different levels, locations, and functions. The analytics professionals can model the scenarios of employee turnover in advance to minimize risk and loss with the help of predictive analytics.

People analytics is applicable to the functions and processes of HR department. It helps business in taking smarter decision by predicting employee turnover on the basis of function, position, and department; deploying retention strategies accordingly; can forecast high performers at the time of hiring and preparing strategies for career progression and engagement in advance. With the use of HR analytics, the company can allocate specific funds and budget for various functions by calculating cost per higher, expenses per employee, and revenue generated per employee.

HR analytics can provide answers out of the given data and can predict future related to anything whether recruitment or talent management. It focuses on the benefits of employees and the company such as how many hours employees work productively, who is coming in the range of succession planning, and what should be the chain of command. HR analytics helps companies to really work upon that bottom line where people are making a real impact. HR analytics can help professionals to identify that bunch of people who are better performers; the one hired 3 months back or the other one hired last year. On the basis of the performance, last quarter attainment, on the basis of time of productivity, etc., HR can understand the application of analytics in all facets of HR practices to be a successful department.

Analytics is a significant area of opportunities that HR can utilize to make an impact and show its value. HR needs to be much more aligned with the business strategies, goals, and objectives of the organization. HR can connect the business client and business executives to clearly articulate those goals and objectives and then can look into them and whatever business problems are they trying to solve. HR can help meet goals by enabling, training, and supporting the employees. HR can really use analytic to be much more purposeful, impactful, and really get aligned with the business more strongly.

HR professionals can leverage people's data to make better decisions. HR analytics uses various statistical techniques such as correlation metrics, recreation, and machine learning to systematically identify and quantify the HR drivers of business performance. These techniques are used to analyze people's business performance, engagement of employees, competencies of people, absenteeism, and their influence on the business outcomes. If HR can quantify the people's performance and the work they are doing is driving business or not then it is the use of HR analytics.

For example: A Dutch police company has a number of helicopters but the problem with the business is that their pilots left their jobs after a few years of services. The company is investing a lot of money into training them, paying for their licenses, and putting them on role as agents. But a common trend has been noted that after a few years of service, pilots are switching their jobs. This is costing a lot to the organization as preparing a pilot requires an investment of a good amount. So the company thought of using analytic to identify the real cause of the problem. They compared the pilots having more than 8 years of experience with the pilot having a few years of experience. After comparing both the groups, a commonality was noticed that this was the source of a selection of these pilots. Most of the pilots with high experience were hired from the civil aviation sector were already trained, had licensed to become agents with police and a much cheaper source than the other group. The other group agents were those who were looking for new carrier options and they were sent to aviation school to train them as police helicopter pilots. Out of the two sources, one channel is truly effective but the question is which one.

The data clearly presented that the people from civil aviation usually wanted to be in the police because it was fun for them, they get an opportunity to fly police helicopter. However, after a few years, they went back to join civil aviation. The available data on various civil aviation sites presented that these companies really like police helicopter pilots because of better training and good experience. Whereas the other groups of pilots hired from aviation school were able to make a career in the police and stayed there for long. The difference that analytic has presented is people who are already in the force as police agents, they were enthralled by the job and it did not matter to them that civil aviation can pay them more. They had a better determination to be a police agent and stayed there for a longer duration not likely to leave the police force. On the other hand, the civil aviation pilot got experience as a police helicopter pilot for a few years and went back to join civil aviation for big pay packages. With the help of analytics, the department learned that people who are already in the aviation school is a better option to hire. It may be costly at the initial stage but the ROI was better for these candidates as their retention was for longer duration in comparison to the cheaper source—already trained civil aviation pilots.

This insight impacted the decision of people selection and better utilization of learning and development budget. This example ensures that HR analytics can solve business problems to make better decisions and ensure success and saves money of the company. There are multiple other examples present where the use of HR analytics resolved multiple people and business-related issues. Researches present that analytics deliver approximately 13 X (X is money in any unit) return on every X spent and that clearly makes it a fantastic ROI.

The second real example is of a company that was facing the problem of retaining young professionals. The turnover rate of young professionals was about 22%–25% which was costing the business a lot of money. This business also used analytics and compared the difference between people who stayed longer with the people who stayed shorter in the company. The analytics presented the common characteristics of the professionals and predicted turnover intention and a flight risk. The outcome presented that the major reason of turnover was travel distance and the young professionals did not prefer it. Similarly, HR can make certain more profiles from the data and can device new meaningful insight such as women with high-pressure jobs may drop out after getting married in comparison to men in a similar situation. It was because of the demand of traditional roles from both such as men felt that they needed to take care of the family financially whereas females felt more responsible for family and children. So creating such profiles can help HR to identify the people who are likely to leave for the above-mentioned reasons. HR can calculate the cost-benefit analysis or retention versus hiring to make further action plans in the benefit of the company.

HR analytics is able to gauge the individual turnover probability that "A" has a 20% chance of leaving the organization and "B" has an 80% chance. Under such conditions, HR can identify whether "B" is a high performer or critical talent for the company, and if found so, they can make a strategic shift of him horizontally or vertically for improving the probability of retention. By using analytics, HR has not to make decisions based on intuitions, it can make data-based decisions that can be profitable for the company. By using HR predictive analytics for employee retention, Key Decision Company was able to save 70 million dollars annually.

Based on application and importance of HR analytics, here are some of the real examples practiced in real organizations that have been published by Harvard Business Review. Companies are changing their ways of doing business and altering the way they are creating value by application of HR analytics. BEST Buy, a big electronic chain in the USA started facing issues related to employee engagement. Employee engagement is a concern as it affects company profitability and productivity by impacting absenteeism rate, incidents, defects, and most important employee turnover. Best Buy wanted to know how the engagement score impacted the profits and business. They compared engagement rate per store with the profit per store. With the application of analytics, it was observed that when engagement score is changed by zero to one, that is, from 70% to 71%, the profit in the same store will increase by 1 million dollar. These statistics helped HR to incentivize managers to start increasing engagement score, eventually, improving the overall performance of the store with more engaged workforce. Best Buy started doing more engagement surveys that helped them to identify various levers that the company could turn in order to make better decisions.

These cases present that companies are changing their way of doing business by the application of HR analytics and metrics. However, at this point, companies need to identify the factors that make HR analytics very tangible and a practice for developing business acumen. In an organization, data related to people is usually stored in different systems such as data on HRIS, HR pay system, performance management system (PMS), engagement surveys, applicant tracking system, and data related to employee motivation and morale. The biggest challenge in front of HR function is to combine the data retrieved from different sources. The engagement score cannot combine with performance data, general demographic profile data of employees cannot be combined with absenteeism data, and so on. Department is aware of the absenteeism rate but they do not know in which department the absenteeism rate is highest or HR knows that which people company is attracting with the help of applicant tracking system but they do not know which applicants are performing well a year later. So HR professionals cannot combine PMS with an employee tracking system. So HR analytics involves a number of steps that HR professionals need to understand and implement for developing better financial astuteness and profitable decision-making.

Analytics helps people to bring meaningful interpretation from data, recognizing various patterns, and helps in building predictive models to take appropriate measures for profitable outcomes. HR analytics can produce astonishing results and predictive HR analytics can help saving a huge amount of business money by making better decisions. If used properly, it can become a game-changer which not only measure the performance but also predicts the impact of various policies and procedure and people and organization. Instead of making intuitive decisions or gut-feel based decisions, HR can produce data-driven predictive models with the application of

HR analytics on the information stored in HRIS of companies. That way, HR can really become the strategic business partner in true sense.

HP (Hewlett Packard) saved approximately 300 million dollars by taking preventive action for retaining the identified high flight-risk employees with the application of HR analytics. Similarly, Wikipedia is able to have an eagle eye on its 750,000 editors and their performance to take corrective and motivational measures to retain them and keep them engaged. There are multiple benefits of using HR analytics but not limited to the following:

1. Identification and utilization of opportunities
2. Backbone of company's strategic business plans
3. Deciding investment priorities
4. Effective decision-making and workforce management
5. Impact of HR practices on bottom line of organization
6. Alignment of HR strategies with overall business strategies
7. Data-driven predictive models for employee turnover, retention, performance, engagement, satisfaction, etc.
8. Data integration and transformation into meaningful information
9. Employee inclusion
10. Competitive edge.

According to Deloitte's 2018 People Analytics Maturity Model, benefits of using HR analytics are multifold; however, only 17% of companies worldwide are assessing and utilizing people's data. Only 2% of this 17% are using advanced artificial intelligence (AI) supported tools for collecting, integrating, and analyzing the data and 15% of this 17% are using HR analytics to build predictive models on ad hoc basis. This presents that organization and HR professionals have to really work hard and need to develop an interest in learning and integrating these statistical tools that can help them in developing financial acumen.

2.4 HR AND WORKFORCE SCORECARD FOR FINANCIAL ACUMEN

For the greater employee performance, HR professionals require a forward-thinking tool that can provide an actionable insight and HR scorecard can serve the purpose. HR scorecard provides an imperative insight to HR executives to understand the workforce implications of business decisions. The workforce scorecard approach was developed by Branin Becker and Dick Beatty in 2005. The workforce scorecard provides a framework that helps in measuring the behavior, mindset, competencies, outcomes, and culture required for the success of the workforce eventually revealing the impact of each dimension of the bottom line.

The competitive HR professionals develop a measuring system that links the capability and the capacity of the workforce with the organizational strategies and goals. HR scorecard is a metrics describing the HR activities that helps employees complete their jobs effectively and efficiently. HR scorecard emphasizes on five determinants or elements, that is, achievement of workforce objectives, appropriate investment on workforce development, alignment of HR functions with business strategies, support of HR policies to business and required and skill-set and experience of HR professional supporting workforce. The HR scorecard keeps a regular track of the progress made on the preidentified KPIs.

Business success can be fueled by the performance of its intangible resources, that is, human resources. However, in many business setups, HR is the most underperforming asset leading to organization's incapability to gain the competitive performance and position in the marketplace.

Workforce scorecard ensures that organizations are achieving their goals by measuring the work quality of employees. It is a tool that can be used in any department. Workforce scorecard is a term coined out of a balanced scorecard, a tool that assists managers in measuring the important variables of the company's performance. With the help of a scorecard, HR managers align the performance categories with the overall organization strategy. These performance categories are then translated into metrics to keep a track of performance. This tool helps HR not only to ensure the measurement of workforce talent, intellect, and contribution to the organization but also measures individual contribution, analyze scorecard, and keeps a track of employees' satisfaction, performance, capabilities, retention, and engagement. Huselid defined workforce scorecard as a tool to identify and measure the important aspects of workforce that contribute in achieving organizational goals such as skill-sets, behavior, mind-sets, and results. The workforce scorecard focuses on four elements provided below in a sequence:

1. Culture and mindset of workforce: It ensures that employees understand the business strategies and embrace them. Workforce scorecard focuses on providing a culture needed to support the execution of strategies.

2. Competencies of workforce: Workforce scorecard ensure the availability of critical talent/strategically important positions required to execute business strategies.
3. Leadership and behavior of workforce: HR scorecard ensures that proper direction and guidance are provided to workforce to attain the organization's strategic objectives. It keeps a check on the consistency of leadership and workforce behavior.
4. Success of workforce: It measure whether the organization strategic objectives are met by the workforce or not.

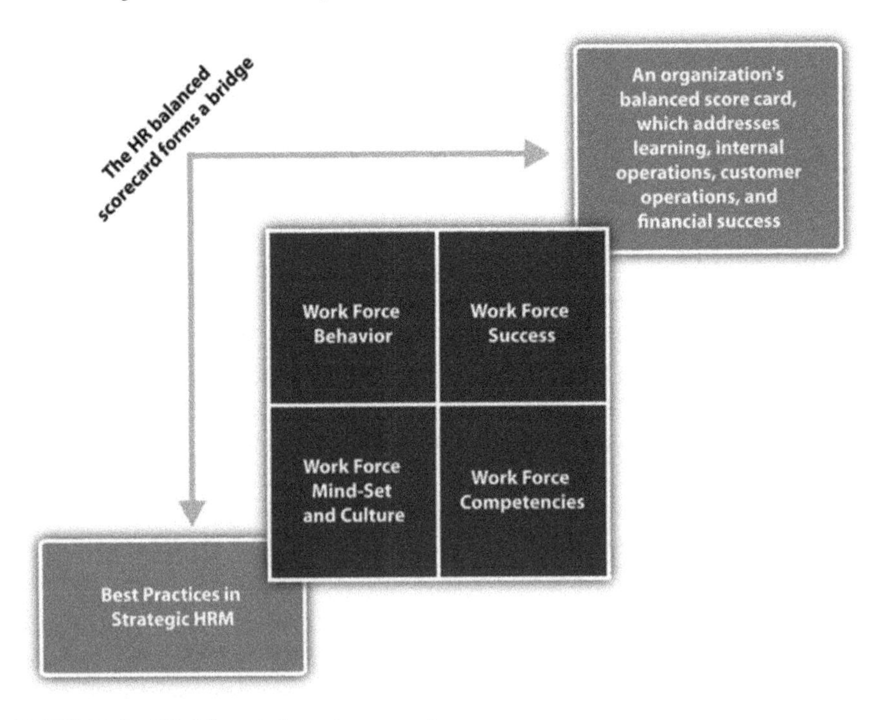

FIGURE 2.1 Workforce balanced scorecard.
Source: Mark A. Huselid, Brian E. Becker, and Richard W. Beatty (2005). "A Players" or "A Positions"? The Strategic Logic of Workforce Management. *Harvard Business Review.* https://saylordotorg.github.io/text_international-business/s16-05-tying-it-all-together-using-th.html (open access)

It is inevitable to understand the complexity of workforce scorecard and associate metrics to measure the intangibles of a company. Companies setting financial benchmarks to measure the intangibles associated with the workforce such as performance, change, innovation, engagement, etc. perform better due to their concerted efforts. It is important for HR executives to measure six workforce strategic areas, that is, financial performance, operating efficiency, client satisfaction, workforce performance, creativity, innovations and change, and social/environmental issues. The research presented those organizations who were keeping closer attention to these metrics and having the credible and reliable information were identified as industry leaders and reported a good financial performance than competitors.

Most companies have better control and feel more accountable for their raw material than their HR. However, HR is the most important asset of every organization to be taken care of and HR should feel more accountable and possess better control over workforce. It will enable a firm to get people with a positive mind-set, innovative, and compatible with organization strategic objectives. As per Figure 2.1, multiple benefits are associated with the use of workforce scorecard such as:

1. Linking the HR KOIs with corporate objectives through the application of a balanced scorecard and workforce scorecard approach.
2. Comparing the current HR practices with the company's historical action and external benchmarks to align and measure the HR ideas within the company's framework.
3. Determining the accuracy of workforce composition placement in the different function.
4. Creating an information reservoir where all the data is collected and compiled related to workforce.
5. Describes the strategic action plan to align the expected and actual performance of employees and find out areas of improvement.
6. Presents the status of achievement of objectives in term of appropriate investment on workforce, alignment of HR functions with business strategies, supportive HR policies and procedures, and listing of required skill-sets and competency in workforce to achieve endeavors.
7. Determine KPIs and keep a record of progress.
8. Emphasizes on considering HR as an asset instead of cost, introducing metrics to measure the strategic success at different levels and collaborative effort and joint accountability of line manager and HR managers for implementing the initiatives for uplifting workforce level.
9. Workforce scorecard measures skill-set, behavior, mind-set, competency, and culture required for achieving the organization objectives and impact of these dimensions on the bottom line.

10. Workforce scorecard is important for achieving organizational objectives by boosting human capital.

2.4.1 APPLICATION OF WORKFORCE SCORECARD FOR FINANCIAL DECISION-MAKING

Many organizations believe that employees are their most valuable asset but unaware of calculating the real worth of them and how HR practices can help people and organization in bringing their vision to reality. The reason is that it is very difficult to measure how HR functions impact organizational performance. Due to this reason, the tools and measures used by the HR department to measure their contribution in company's success fails to do so. The organizational performance and success depend upon the competitive advantage its workforce has, most of the time, HR department ends up measuring recruitment cost, replacement cost, training cost, etc. without valuing the competitive advantage of their most valuable assets, that is, people. To understand the application and calculation of it, companies have to measure workforce learning and commitment to identify competitive advantage. HR needs to develop an assessment center that is innovative and capable of measuring employee's contribution in organization profitability and improved shareholder value. The solution to this complexity is understanding and application of an HR scorecard that can ensure the strategic contribution of all employees and its impact on organization performance.

For Example: Sears decided to bring some strategy to turn around the current scenario of losing billions of dollars reason being customer loss. It identified that to attract investors, firstly, it is important to attract customers and if the company wants Sears to be an undeniable place to shop for customers, it needs to be an undeniable organization to work with, that is, its employees must love working with the company. So it decided to create strategies around building Sears an attractive workplace with the support of innovative ideas and inspiring policies. Once Sears become an enjoyable workplace for its people, it helped the company to attract customers to shop. Sears designed a set of competencies and positioned behavioral objectives to achieve each of its goals. These competencies and behavioral guidelines provided clarity to the organization that what foundation is required while building strategies around recruitment, training, compensation, and performance structures. The strategies and structures were so objectified and precise that HR personnel was able to measure the quantitative outcome of any initiative with confidence. For example: HR was able to calculate that if an employee's attitude is improved by 5 points, it will increase customer satisfaction by 1.5 points leading to a 6% increase in the company's revenue.

Deliverables of HR: By using a workforce scorecard, the HR department can become a valuable strategic partner than a traditionally seen head cost. For this, HR has to identify and measure its key deliverables, develop an HR system that can create these "deliverable" in supporting the organizational strategy. For example: at Sears, the key deliverable was turning the company into a compelling culture to work with. To bring this transformation, various functions such as recruitment, training, compensation, and appraisal structures were developed to hire people with creativity and innovativeness. Company also associated rewards with brilliant ideas to create a fun environment that compelled people to join there.

Architectonics of HR Strategy: HR can maximize its contribution in achieving the organizational objective of sustaining competitive advantage by focusing on system and architecture. To create a brilliant architectonics/structure in its value chain, HR has to focus on three dimensions: HR function must consist of professionals with strategic and financial competencies; HR system, its policies, practices, and procedures must be strategically aligned with organization high performance; and strategic focus on expected employee behavior in terms of motivation, key competencies, and expected behavior. A strong HR system is the backbone of its strategic influence on the overall system and among the known systems, high-performance work system (HPWS) is most effective and influential. The HPWS ensures the effective design and delivery of all elements in HR system. It validates its competency metrics by strategically associating it with talent acquisition and promotion decisions. It supports HR to develop strategies around the required skills for effective support to business goals. It also focuses on performance management and compensation policies to ensure the attraction, retention, and performance of critical and talented employees.

Creating the HR Scorecard: It is a strategic measurement tool helping businesses in managing and improving HR functions. The scorecard measures HR in form of—Deliverables; policies, procedures, and practices; and alignment of the overall system and efficiency of the workforce. The scorecard ensures cost minimization in HR activities by analyzing spreadsheets of data. The spreadsheets, dashboards, and metrics guide HR to link the organization's objectives and HR activities. It ensures that HR gets its due recognition by top executives after achieving the strategic goals of the firm. The scorecard helps HR professionals to frame SMART (specific, measurable, achievable, realistic, and time-bound) objectives. For example: a company wants to reduce per-employee medical expenses. Here HR has to identify alternatives present in the market that can provide mutual benefits, that is, neither the employee suffers nor the company loses. So they can provide health coverage to their employees to lower down the cost of the company.

With the available data, HR scorecard provides critical business indicators reflecting the present economic condition of the firm and the market. For example: HR can make profitable and strategic moves by confirming the market trends and evaluating the lag indicators such as unemployment rate. Through trend analysis, HR can hire and develop critical talent when the unemployment rate is high, can negotiate well and accomplish organizational futuristic needs and requirements at a low cost. With this move, they can spend the saved amount of training and development of hired critical talent to improve loyalty among new hirers and bringing innovativeness in business practices. HR scorecard helps practitioners to eliminate inefficiencies from its functions to reduce the cost and improve effectiveness. HR professionals are able to gauge the future requirements of the company based upon statistical analysis and presentation of the company's balanced scorecard. This will provide clarity of the future requirements and prepare HR to be ready with specific action plans to help organizations in achieving the goals better.

2.5 HR ACCOUNTING FOR FINANCIAL ACUMEN

HR accounting is the application of conventional accounting principles on the presently unaccounted HR practices. HR accounting identifies and reports the investments made on various HR activities such as talent acquisition, training and development or promotion, etc. According to the American Association Society, the HR accounting can be defined as "the process of identifying and measuring data about human resources and communicating this information to interested parties." According to Flamholtz, HR accounting measures the worth of employees to an organization and the cost associated with them to make better investment plans. According to Woodruff and Barry, "human resource accounting is an attempt to identify and report investments made in human resources of an organization that are presently not accounted for in conventional accounting practice."

Basically, HR accounting is an information system that tells the management what changes over time are occurring to the human resources of the business. HR accounting is a technique for systematically measuring both the asset value of labor and the amount of asset creation that can be attributed to employee's activities. It is generally an information system that shows management what changes usually are occurring overtime to typically the human resources in the enterprise. House Rent Allowance (HRA) is an information method able to assisting the supervision of effective selection and retention of people worthy to the organization. Therefore, HR accounting provides a comprehensive appearance of applied resource cost and beliefs information at one place for better decision-making process.

HR accounting keeps a check on the value produced by using the intellectual and knowledge of the people working at all levels in an organization. It is important to calculate the wealth of the organization in terms of human capital to calculate the real profit that can be earned by them. HR is an asset and it should be measured, accounted, and disclosed in the financial statement of the company. Although, HR is an intangible asset of the organization and difficult to value but this assets should be presented in balance sheet in some manner. In a more meaningful manner, HR accounting is a process where the worth of HR of an organization is properly valued, recorded and presented in the books of account. Although, it is difficult to value human resources, recording the value into book accounts and disclosing this information into financial statement, but after that also, the booked information can help HR and company executives in making effective strategies and better decisions.

Janet Hoffmann, president of HR Aligned Designs commented that "HR leaders should view themselves as business partners first with a specialty in human resources, an expertise in aligned people performance with business results and the ability to deliver profitable growth." According to her, HR professionals should know the company's financial statement, are able to calculate ROI, understand the business and its product and services, align with organization strategies, and develop financial and business astuteness for improving their decision and actions.

Unfortunately, HR accounting is an area where most of the HR practitioners lack and struggle. However, to develop a financial astuteness, it is important for HR professionals to study, understand, and interpret the various financial statements such as cash flow, profit and loss account, and balance sheet of the company. To develop financial acumen, it is vital for HR to understand factors affecting business profit, its cash flow, and be able to make effective decisions on the basis of available information. To be a true business partner, financial literacy is expected in every leader in any organization. According to Society of Human Resource Management's business terms glossary, "business acumen is the knowledge and understanding of finances, accounting, marketing, and operational factors of an organization which enables a leader to make good judgments and quick decisions."

Human resources are intangible assets that Sumantra Ghoshal classified in three categories to measure contribution and effectiveness, that is, intellectual capital, social capital, and emotional capital with an addition of spiritual capital. At macro level, HR is a combination of experience, knowledge, skills, creativity, imagination, intuition, and innovativeness possessed by all the employees in an organization. HR professional has to have financial acumen for the development of reliable and competitive management of human resources. It is important to present the value of the most important assets in the financial statement of the company. Sir Williams Petty presented workforce as "the father of wealth" in 1691 and made the attempt to value the labor in monetary terms to estimate organizational wealth. HR accounting presents relevant information about the value of workforce and the return of investment for the information of internal and external stakeholders. It also links the HR interventions with the financial results of the company.

To quantify the HR investments, it is important for HR professionals to understand the concept and application of various accounting methods and processes. This will help managers to find the cost and worth of an employee to the company. Two widely known approaches of HR accounting are historical cost approach and value approach. To book the value of HR into accounts, the cost approach uses four models, that is, historical cost model, replacement cost model, opportunity cost model, and standard cost model.

Historical cost approach is popularly known as acquisition cost model where five parameters are used to measure the investment made by an organization on its employees. These five parameters are recruitment, acquisition, training (formal and informal) and familiarization (formal and informal), development, and experience. The model emphasizes on capitalizing the HR investment into balance sheet instead of putting in profit and loss account. It is required to amortize the HR capitalized amount from time to time. In this method, human assets are considered similar to physical assets where age is considered as a factor of measurement at the time of recruitment and retirement from the company. The replacement cost method calculates the cost of replacing an employee. The replacement cost includes recruitment and selection cost, compensation and benefits cost, and training and development cost. This approach helps HR professionals to decide whether to retain the current staff or hire a new one. This is a flexible approach that allows for changes in the cost of various HR practices. Opportunity cost model focuses on determining the value of alternative use of employee and if employee does not have any alternate use then this opportunity cost value is empty for him. It helps HR to bid for critical, skilled, and scarce employees (current) which are difficult to replace. The opportunity cost of an employee is determined by the offer made by another department. This approach helps HR to optimally allocate human resources and it becomes a quantities method for appraising, evaluating, and further developing a personnel. The fourth approach under the historical cost is the standard cost method which determines year after year the standard cost associated with talent acquisition of per grade employee and their development. So the total cost calculated this way is the standard cost of all the employees in the organization and can be used for HR accounting purposes.

On the other hand, the value approach focuses on the worth of an employee, not on the associated cost. The value approach uses four models to estimate the value of an employee, that is, present value for future earning methods; reward valuation model, net benefit model, and certainty equivalent net benefit model. The present value for future earning method calculates the economic valuation of future earners based upon the present value of an employee. This method helps organization to find out critical talent by calculating the present worth of an employee's future contribution. This method determines the present value to adjust the chances of employee's separation, retirement, or death. The formula to calculate the present value of future earning is:

$$\Sigma\,(Vy) = \Sigma\, Py\,(t{+}1)\,\Sigma\, I\,(\mathrm{T})\,(1{+}R)\,t{-}y$$

where

$\Sigma\,(Vy)$ = Expected value of "y" year old employee's human capital

t = Employee's retirement age

$I(T)$ = Expected earning of that employee in that period

$Py\,(t + 1)$ = Probability of employee's retrenchment form organization

R = Discount rate

This model calculates the human value in various stages; first, categorization of all employees are done on the basis of age and skill; second, the average annual earning is calculated for all categories; third, the expected total earning determine for each group before their retirement age and this total earning is discounted as the rate of the cost of capital. This way, the net value of the human assets is calculated in the organization. Reward valuation model tries to fulfill the flaws of the present value of future earning method by two ways: first by considering the probability of person movement from one career to another for purpose of promotion and growth, and second, it also considered the option of employee leaving the organization earlier for other reasons than retirement or death. This model considered the expected realizable value of the employee estimated by calculating the current value of the employee based upon the set of future services they are expected to deliver during the probable tenure of him with the organization.

The third model, that is, net benefit model determines the net value of an employee by calculating the net benefits organization is getting from the job of that employee. It calculates the present worth of the person for the organization. The certainty equivalent net benefit model is an extension of "net benefit model" which calculates the worth of an employee by accruing both: the present value of the employee and the certainty of net benefits in the future. There are multiple other models present for HR accounting purpose and the focuses on keeping a record of every penny spent on HR activities whether recruitment, selection, placement, orientation, development, training, incentives, and other benefits provided to each employee.

HR professionals must be able to apply relevant model and associated formulae for the valuation of HR as assets and incorporating the cost and income in the balance sheet of the company. The objectives of HR accounting are not limited to calculating the net worth or associated cost of the employees. This not only helps the organization to effectively use HR but its scope is also extended in multiple directions such as:

1. For achieving cost-effective business goals and better decision-making about talent acquisition, development, and maintenance of human resources by furnishing cost value information.
2. Categorizing the human resources on the basis of depleted, conserved, or appreciated assets.
3. Optimum utilization of human assets.
4. To help in better decisions for change and development of management practice and principles by determining their financial outcomes.
5. To focus on human assets and keeping the record of information related to them.
6. To identify best fir for every job and improve HR as function.
7. Improving the image and good will of organization by attracting and retaining talent.

8. To provide information to share holder/investors/interested parties about the revenue and capital expenditure of HR assets.
9. To improve profitability of an organization and develop competitive talent.
10. Effective decision-making and HR assets control based upon furnished cost value information.

With the above-mentioned points, it is evident that the literacy of HR accounting is important for HR professionals in today's 4.0 revolution. HRA is not limited to measuring the cost of investment associated with various HR practices; it also quantifies the economic value of employees for the organization. The nonaccounting of employees and the related investments may provide a poor picture of the revenue and profitability of a company. HR accounting helps measures the potential and worth of an employee in monetary terms. It not only helps HR practitioners to examine the cost associated with employees such as expenditure incurred on acquisition, training, maintenance, etc. but also determines the value of employees in terms of worth the person will bring in future in return of the above investment.

A comprehensive HR accounting system addresses two additional areas, that is, budgeting and employee valuation along with the tracking of workforce-related investments. With the help of budgeting, an organization is able to concentrate on the cost incurred on employees across all function of organization. By focusing on cost information, HR professionals and senior executives can clearly analyze the impact of total employee cost on the organization. Employee valuation on the other hand, considers employees as organizational assets. The value of the assets can be determined by the efficiency and effectiveness of an employee based on his experience, qualification, leadership, innovativeness, etc.

This is very difficult to estimate the worth of an employee and quantify the variables mentioned above. However, if proper standards are fixed and scales are developed, it is possible to quantify the worth up to an extent, if not completely. And that is why; it is important for HR professionals to have knowledge of accounting principles to apply the same in making the profit and loss account of the most valuable assets of an organization. The same can be presented in the form of a balance sheet for attracting the interest of investors and generating capital. HR accounting has multiple benefits. It keeps a check on the corporate plan looking for technological advancement, expansion, or diversification, etc. to estimate the requirement of various skill sets and availability of the same for the key position. If the required skill set is not available inside and outside the organization, the HR can suggest leaders to modify or change the corporate plan. It ensures that human assets are properly utilized in the organization and not get wasted and ensure high returns and profit for the organization. It assists organization to take necessary steps to improve employee performance and contribution. HR accounting curbs cost of various functions and activities associated with maintenance of human assets such as recruitment cost, training cost, placement cost, etc. by improving the methods and techniques of procuring human resources. The motive of HR accounting is to make HR executives trusted business advisors for the company.

2.6 CONCLUSION

Indeed HR has witnessed a lot many changes in the last four decades. Before 1980, it was known as personnel management department used to engage mostly with hiring, training, administrating, and firing of people. After almost 40 years, today, it performs a range of roles from managing talent to becoming strategic business partner to shared operational services being a strategic and tactical department. HR is more critical in today's era of AI and Internet of things. It acts as a differentiating factor, more focused on measuring the impact of business decisions on employees, their engagement, and motivation.

HR function is capable of optimizing its impact with the adoption of HR practices that help professionals to make effective financial decisions. The contribution to the bottom line can be increased with the understanding and application of accounting and statistical principles in the area of HR. Employee engagement and motivation are quite critical for contribution in the organizational efforts and success. However, according to a trend analysis conducted by Gallup in 2017 of past 20 years, it was noticed that almost 70% of the employees are not appropriately engaged and contributing very minimal in achieving organizational goals. The report supported the famous Pareto principle as only 30% of the employees were actively engaged and contributing in organizational success. This is an indicator that organizations are unable to optimally utilize their human capital and returns are little on the investment made on an employee by the organization.

According to Culture Amp's 2018 report, companies with exponential growth are focusing and maintaining a high level of employee retention and engagement with the use of technology, AI, and application of latest HR practices such as big data analytics and benchmarking. Organizations are looking for HR professionals who are well equipped with latest technologies and HR practices that can help company to achieve overall objectives. Growth Force Company developed SMART core values employers expect in potential HR employees where S = snap (quick problem-solving capability); M = meaning (understand the concepts and application of accounting and love with numbers); A = accountable (ready to take responsibility of their actions and plan in advance); R = resourceful (knowledge hub and keen to share their learning), and T = team player (believes in collaboration).

Financial acumen is a requirement of business from years and people were holding that skill set but it is a buzzword nowadays in the area of HR. To be strategic business partners, HR professionals have to understand that how business makes money, where the finances come from, about the company's products and services, and study of competitors and market at large. This will help them to device HR policies and practices according to the changing need of the organization. It will not just help them to save money; they will be able to plan their strategies around the critical needs of people and organization. To develop financial acumen, financial literacy is important. They have to learn about the internal environment, external environment, competitor's analysis, and functioning of their own business to have an insight that how organization functions, what is required, and what HR strategies need to be implemented to have a competitive edge.

To create a conducive and competitive environment within the company, HR has to have a close look at the political game of the business, how people treat each other and what is the impact of internal politics on the overall business. Overall, it can be concluded that financial and business acumen is required in an HR practitioner. HR professionals have to break the traditional approach of using qualitative aspects, to be a strategic business partner; they need to quantify HR functions to really understand the impact of their activities on the organization success in terms of profitability and sustainability.

KEYWORDS

- **financial acumen**
- **HR practices**
- **HR metrics and analytics**
- **HR accounting**
- **workforce balanced scorecard**

REFERENCES

Abubaka, S. (2006). *A Critique of the Concept of Human Resource Accounting*. Ahmedu Bello University, Zaria, Nigeria.

Acito, F., & Khatri, V. (2014). Business analytics: Why now and what next? *Business Horizons*, 57(5): 565–570.

Agarwal, D., Bersin, J., Lahiri, G.; Schwartz, J., & Volini, E. (2018). People data: How far is too far. https://www2.deloitte.com/us/en/insights/focus/human-capital-trends/2018/people-data-analytics-risks-opportunities.html

Barends, E., Rousseau, D. M., & Briner, R. B. (2014). *Evidence based management: The basic principles*. Amsterdam, Netherlands: Center for Evidence-Based Management.

Becker, B. E., Huselid, M. A., & Ulrich, D. (2001). *The HR Scorecard: Linking People, Strategy and Performance*. Boston: Harvard Business School Press.

Becker, B. & Huselid, M. (2006). Strategic human resources management: Where do we go from here? *Journal of Management*, *32*, 898–925.

Becker, B., Huselid, M., & Beatty, R. (2009). *The Differentiated Workforce: Transforming Talent into Strategic Impact.* Boston, MA: Harvard Business Press.

Brummet, R. L., Flamholtz, E., & Pyle, W. C. (1970). Human resources measurement—A challenge for accountamnts. *The Accounting Review*, 43(2), 217–224.

Caplan, E. H. & Landekich, S. (1974). *Human Resources Accounting: Past, Present and Future*. National Association of Accountants 1974.

Chen, R. R., Ravichandar, R., & Proctor, D. (2016). Managing the transition to the new agile business and product development model: Lessons from Cisco Systems. *Business Horizons*, 59 (6): 635–644.

Cronin, G. (2007). Measuring strategic progress. Choosing and using KPIs. *Accountancy Ireland*, 39 (4): 30–31.

Earley, C. E. (2015). Data analytics in auditing: Opportunities and challenges. *Business Horizons*, 58 (5): 493–500.

Eckerson, W. W. (2009). Performance management strategies. *Business Intelligence Journal*, 14(1): 24–27

Fitz-enz, J. (2002). *How to Measure Human Resource Management,* 3rd ed, McGraw Hill.

Flamholtz, E. (1973). Human resources accounting: Measuring positional replacement. *Human Resources Management*, 12(1): 8–16.

Flamholtz, E. G. (1999). *Human Resources Accounting: Advances in Concepts,* 3rd ed., San Fransisco, USA: Jossey – bass.

Flamholtz, E, Bullen, M., & Hua, W. (2002). Human resources accounting: a historical perspective and future implication. *Management Decision*, 40 (10): 947–954.

Griffin, J. (2004). Developing strategic KPIs for your BPM system. *DM Review*, 14 (10): 70.

Guenole, N. & Ferrar, J. (2014). Active employee participation in workforce analytics. IBM Smarter Workforce Institute. Available at https://hosteddocs.ittoolbox.com/employeeparticipation.pdf

Guenole, N., Feinzig, S., Ferrar, J., & Allden, J. (2015). Starting the workforce analytics journey: The first of 100 days: IBM smarter workforce institute report. Available at http://www-01.ibm.com/common/ssi/cgi-bin/ssialias?htmlfid=LOL14045USEN

Gupta, D. K (1991). Human resource accounting in India: A perspective. Administrative Staff College of India. *Journal of Management*, 20 (1): 9–10.

Harvey, J. (2005). KPIs—The Broader strategic context. *Credit Control*, 26 (4): 65–66.

Hendricks, J. A. (1976). The impact of human resource accounting information on stock investment decision: An empirical study. *The Accounting Review*, 51 (2): 292–305.

Hirsch, W., Sachs, D., & Toryfter, M. (2015). Getting started with predictive workforce analytics. *Workforce Solutions Review*, 6 (6): 7–9.

HR metrics and their impact on business *by Leon Teeboom; reviewed by Michelle Seidel, B.Sc., LL.B., MBA; Updated May 13, 2019, Academy of Innovative HR,* https://smallbusiness.chron.com/hr-metrics-impact-business-62267.html

Huselid, M. A., Becker, B. E., & Beatty, R. W. (2005). *The Workforce Scorecard: Managing Human Capital to Execute Strategy*. Harvard Business Review Press.

Corbin, J. (March 7, 2017). Surprising results from the 2017 Gallup Employee Engagement Report. https://www.theemployeeapp.com/gallup-2017-employee-engagement-report-results-nothing-changed/

Jones, J. (March 11, 2016). Business acumen: More than just business knowledge. Retrieved from: https://www.shrm.org/resourcesandtools/hr-topics/organizational-and-employee-development/pages/business-acumen-more-than-business-knowledge.aspx

Kuppapally, J. J. (2008). *Accounting for Managers*. Prentice-Hall of India Private Limited. New Delhi.

Ferrar, J. (February 26, 2019). How can HR professionals build business acumen? https://www.myhrfuture.com/blog/2018/12/7/how-can-hr-professionals-build-business-acumen

Kaplan, R. S., & Norton, D. P. (1996). *The Balanced Scorecard.* Harvard Business School Press.

Kashive, N. (2013). Importance of human resource accounting practices and implications of measuring value of human capital: Case study of successful PSUs in India. *Journal of Case Research*, 4 (2): 113–144.

Kaur, S., Raman, A., & Singhania, M. (2014). Human resource accounting disclosure practices in indian companies. *Vision: The Journal of Business Perspective*. 18: 217–235. 10.1177/0972262914540227.

Kavanagh, M. J. & Thite, M. (2009). Human Resource Information Systems: Basics, Applications, and Future Directions. Thousand Oaks: SAGE Publications, Inc.

Kogut, B. & Zander, U. (1992). Knowledge of the firm, combinative, capabilities and the replication of technology. *Organization Science*, 3: 383–397.

Kouhy R, Vedd R, Yoshikawa T, Innes J (2009). Human resource policies, management accounting and organisational performance. Journal of Human Resource Costing and Accounting, 13(3): 245–263.

Lawler III, E. E., Levenson, A., & Boudreau, J. W. (2004). HR metrics and analytics: Use and impact. *Human Resource Planning*, 27(3): 27–35.

Lawson, R., Hatch, T., & Desroches, D. (2008). *Scorecard Best Practices*, John Wiley & Sons.

Weatherly, L. A. (January 2011). *The value of people: The challenges and opportunities of human capital measurement and reporting. HR Magazine.* http://findarticles.com/p/articles/mi_m3495/is_9_48/ai_108315188.

Lev, B. & Schwatz, A. (1997). The use of Economic concept of human capital in financial statement. *Accounting Review*, 46(1): 103–112.

Likert, R. M. (1967). *The Human Organization: Its Management and Value*. New York: McGraw-Hill Book Company.

Likert, R. M. (1961). *New Patterns of Management*. New York: McGraw-Hill Book Company.

Lockwood, N. (2006). Maximizing human capital: Demonstrating HR value with key performance indicator. *HR Magazine*, 51 (9): 1–10.

Marcia Moore, MSSW, AIHR, https://smallbusiness.chron.com/components-balanced-scorecard-approach-strategic-hr-67048.html

Huselid, M. A., Becker, B. E., & Beatty, R. W. (2005). 'A Players' or 'A Positions'? The strategic logic of workforce management. *Harvard Business Review.*

Mark Feffer (September 21, 2017). 9 tips for using HR metrics strategically. https://www.shrm.org/hr-today/news/hr-magazine/1017/pages/9-tips-for-using-hr-metrics-strategically.aspx

Mayo, A. (2018). Applying HR analytics to talent management. *Strategic HR Review*, 17 (5): 247–254. https://doi.org/10.1108/SHR-08-2018-0072

McIver, D., Lengnick-Hall, M. L., & Lengnick-Hall, C. A. (2018). A strategic approach to workforce analytics: Integrating science and agility. *Business Horizons*, 61 (3): 397–407. doi:10.1016/j.bushor.2018.01.005

McKenzie, J. L. & Melling G. L. (2001). Skills-based human capital budgeting: A strategic initiative, not a financial exercise. *Cost Management*, 15 (3): 30.

Okpala, P. & Chidi, O. (2010). Human capital accounting and its relevance to stock investment decisions in Nigeria, University of Lagos, Nigeria. *The European Journal of Economics, Finance and Administrative Sciences*, 21: 64–76.

O'Regan, P., O'Donell, D., Kennedy, T., Bontis, N., & Cleary, P. (2001). Perceptions of Intellectual Capital: Irish Evidence. *Journal of Human Resource Costing & Accounting*, 6 (2): 29–38.

Patra, R. & Khatik, S. K. (2003). Human resource accounting policies and practices: A case study of Bharat Heavy Electronics Limited, Bhopal, India. *International Journal of Human Resource Development and Management*, 3 (4): 285–295.

Porwal, L. (1998). *Accounting Theory an Introduction.* New Delhi: Tata McGraw Hill Company.

Rasmussen, T., & Ulrich, D. (2015). Learning from practice: How HR analytics avoids being a management fad. *Organizational Dynamics*, 44 (3): 236–242.

Roslender, R. (2004). Accounting for intellectual capital: Rethinking its theoretical underpinnings. *Measuring Business Excellence*, 8 (1): 1–38.

Savino, D., McGuire, K., & Kelly, M. (2012). Human asset measurement: A comparison of the artifact-based approach versus input methods. *Journal of Management Policy and Practice*, 13 (1): 39–45.

Sandervang, A. (2000). From learning to practical use & visible returns: A case in competence development from a Norwegian firm. *Journal of Human Resource Costing & Accounting*, 1 (2): 82–100.

SHRM Foundation (2016). Use of workforce analytics for competitive advantage. Available at https://www.shrm.org/foundation/ourwork/initiatives/preparing-for-futurehr-trends/Documents/Workforce%20Analytics%20Report.pdf

Siegel, E. (2013). *Predictive Analytics: The Power to Predict Who Will Click, Buy, Lie, or Die.* Hoboken, NJ: John Wiley & Sons.

Sonara, C. K. (2009). Valuation and reporting practices of human resource accounting in India. *Journal of Indian Management and Strategy*, 14 (3): 47–52.

Souza, G. C. (2014). Supply chain analytics. *Business Horizons*, 57(5): 595–605.

Srimannarayana, M. (2010). Status of HR measurement in India. *VISION—The Journal of Business Perspective*, 14 (4): 295–307.

Tootell, B., Blackler, M., Toulson, P., & Dewe, P. (2009). Metrics: HRM´s holy grail? A New Zealand case study. *Human Resources Management Journal*, 19 (4): 375–392. http://dx.doi. org/10.1111/j.1748-8583.2009.00108.x

Toulson, P., & Dewe, P. (2004). HR accounting as a measurement tool. *Human Resource Management*, 14 (2): 75–90. http://dx.doi.org/10.1111/j.1748-8583.2004.tb00120.x

Ulrich, D. (1998). *Delivering Results: A New Mandate for Human Resource Professionals*. Boston, MA: Harvard Business Press.

Ulrich, D. & Brockbank, W. (2005). *The HR Value Proposition*. Massachusetts: Harvard Business School Publishing.

Ulrich, D. & Dulebohn, J. H. (2015). Are we there yet? What's next for HR? *Human Resource Management Review*, 25(2): 188–204.

Ward, M. J., Marsolo, K. A., & Froehle, C. M. (2014). Applications of business analytics in healthcare. *Business Horizons*, 57(5): 571–582.

Weatherly, L. (2003). The value of people: The challenges and opportunities of human capital measurement and reporting. *SHRM Research Quarterly, 3*: 14–25.

Wright, P. M. (2008). Human resource strategy: Adapting to the age of globalization. SHRM Foundation. Available at https://www.shrm.org/hr-today/trends-and-forecasting/specialreports-and-expert-views/Documents/HR-StrategyGlobalization.pdf

Yoon, S. W. (2018). Innovative data analytic methods in human resource development: Recommendations for research design. *Human Resource Development Quarterly*, 29 (4): 299–306.

CHAPTER 3

Challenges to HR 4.0 in the Global Business Scenario

RICHA GOEL,[1*] SEEMA SAHAI,[2] ANITA VENAIK,[3] and MEHDI BENFADEL[4]

[1,2]*Amity International Business School, Amity University, Uttar Pradesh 201313, India*
[3]*Amity Business School, Amity University, Uttar Pradesh 201313, India*
[4]*Iveco France, 14 Avenue du 24 Août, 69960 Corbas, France*

[*]*Corresponding author. E-mail: rgoel@amity.edu*

ABSTRACT

The digital age has brought about a revolution in the entire world. Companies have started investing in tools and machine which support the latest technologies. In this chapter, we will discuss how artificial intelligence as a tool with robots and bots has affected the business processes and has brought about the HR 4.0 revolution. The challenges that an organization faces is mostly reflected in its HR policies and reforms. This chapter will give an insight into how the challenges are overcome and the change in technology is managed by an organization to sustain itself.

3.1 INTRODUCTION

Internet age and digitization has completely revolutionized the International Business Sector. According to some researches, it is the transition to 3rd industrial revolution and many believe that it is the beginning of the 4th industrial revolution—Industry 4.0 basically the emerging and integration of new technology with the globalized world. With the new wave of technological innovation, there has been emergence of institutional settings which has led to changes across the countries in the mechanisms responsible for standardization (Armstrong, 2009).

This rapid spread of digitization across the globe has raised the competitive and regulatory stakes in the hands of few multinational enterprises (MNEs) like Amazon, Alibaba, Google, Apple, Facebook, and many more which have led to debate that whether this transition is positive or negative. Many economists argue that technology as a tool is meant to make the world a decentralized place but the reverse is happening. Whether this integration of digital age with International Business is a boon or bane, it will be influenced by the strategies used by the global organizations. Thus, information age including use of artificial intelligence (AI), robotics, big data, algorithms, and neural networks with MNE strategies has become an important area of study for International Business Scholars.

Secondly, such industrial transformation and rapid changes in technology, economic, and social environment are creating new opportunities for SME and startups. The new digital world or Industry 4.0 is changing the sectoral boundaries, deconstructing traditional industries, and stimulating the emergence of new sectors, thereby increasing flexibility impact upon labor market and employment practices. The advancements in technologies have led to succession in our life, thereby building it hassle-free and contented. Today, automation is all the buzzword. The day-to-day activities of the businesses are automated and then easily monitored as well as managed from devices like smart phones, laptops, etc. The automation or digital revolution has taken the manufacturing sector by storm. It has streamlined the processes, reduced both time and human efforts, minimized errors, and enhanced the overall productivity (Brynjolfsson, 2018).

Recently adopted trends like business analytics, block chain, AI, and machine learning will surely help in the development of society, in detection of defects, and production failures; it will enhance the productivity and connectivity. The integrity of information with authenticity will be at higher pace.

Industry 4.0 was first proposed by the Government of Germany in 2013. It includes a rich amalgamation of traditional manufacturing processes with state-of-the-art technology. This revolution extends and elaborates the impact of digitalization in many ways.

At present, the significant concern is adoption of the change and welcoming the new revolution which is Industry 4.0. Undoubtedly, human capital is the biggest asset of any organization so it is mandatory to protect and nourish this asset for leading and having competitive edge.

Organizations can barely contend without exceedingly talented workforce and without the persistent interest in the human capital. Moreover, the success of any organizations depends firmly on having capable people. Thus, the capacity to manage talent adequately has been and will be one of the greatest difficulties confronting organizations today.

It has been noted from ancient times that changes in the habits of the workforce global organizations know the disturbing fact that significant population, economic, social, and technological changes also alter the workforce considerably. Changes in birth rates, migration patterns, and cultural, economic, and social standards have undoubtedly absorbed the workers of all ages and backgrounds (Brynjolfsson, 2014).

We are conscious of the theory that is often theoretically divided into two distinct methods known as "soft" and "tough" methods in the field of human resource management (HRM). The difficult strategy derived from Michigan College is linked to a more rational philosophy of leadership, where management is based on a series of logical thinking to action. Employees in this strategy are viewed like any other significant resource and should be managed rationally. According to this strategy, the function of executives is to efficiently handle staff and keep them in sync with certain organizational demands. In general, as in the difficult strategy, HRM is more worried with closing policies, systems, and operations on human resources.

Contrary to the difficult strategy, the soft model is also known as HRM's Harvard model which underlines the significance of incorporating HR strategies with organizational goals. This strategy is more worried with valuing individuals as an organization's critical asset and a source of competitive advantage. The soft model, which is more influential, is more concerned with coping with individuals as critical assets and emphasizes the significance of employees' engagement, adaptability, and high skills. In this perspective, staff is viewed in productive procedures as proactive rather than passive inputs. According to Armstrong (2009), HRM is described as a consistent approach to the management of the most valued resources of an organization—individuals employed there who contribute separately and collectively to the accomplishment of its objectives. We can deduce from this definition that HRM or merely HR is a feature in organizations intended to maximize employee performance in serving the strategic goals of their employers. HR is concerned mainly with managing people within organizations, concentrating on policies and systems. HR also deals with industrial relations, that is, the balance between organizational procedures and regulations.

3.2 STRATEGIC HUMAN RESOURCE MANAGEMENT AND ITS ROLE

Strategic human resource management (SHRM) which plays a vital role in creating strategy is intended to assist businesses best satisfy their employees' requirements while supporting business objectives, vision, and mission. Management of human resources deals with all elements of a company as well as its staff, business that impacts staff, such as hiring and firing, pay, rewards and recognitions, and training and performance management. Strategic management of human resources is proactive people's management. It involves thinking ahead, and planning methods for a business to better fulfill its employees' requirements, and for staff to better satisfy the company's requirements. This can influence the way things are accomplished on a company site, from recruiting methods and training programs for employees to evaluation methods (Crawford, 2019).

SHRM helped top managers understand that if HR is strategically managed, staff can enhance efficiency and productivity. The real SHRM came into being when businesses implemented employees' programs "to win hearts and minds." On the other side, SHRM does not have a single definition. However, well described are the SHRM methodology and strategic frame works. Furthermore, SHRM is always related to human resources strategic thinking. The high-performance organization always creates the HR structure that is created using strategic management's fundamental principles and includes HR professionals' strategic thinking. Strategic thinking is an elaboration of a business plan that divides the strategy into actionable HR initiatives and projects. Not only is it a strictly technical discipline, but each organization should use a specific set of distinct strategic methods; strategic thinking and planning should be applied. Rethinking and revisiting all processes and procedures is a challenge for human resources; they need easy, user-friendly, streamlined, and renovated HR solutions that can obviously define SHRM.

HR needs to create a conducive environment and work culture where teams operate in a cross-functional manner, where SHRM plays a significant role in bringing human resources to think about initiatives in cross-functional projects and organizations to work together as a team. Therefore, HRM must recognize possibilities that it brings to the team. HR must recognize initiatives-related hazards and put proposals on the table. It becomes a part of strategic debates and has an impact on strategic plan design. SHRM brings HR organization to the next level. It integrates HR into a company, identifies fresh circumstances for development, and aligns human resources staff with business (Danneels, 2004).

When a human resource department strategically develops its recruitment, training, and compensation plans based on the objectives of the organization, it guarantees a greater opportunity of achievement in the organization. Let us think about this strategy in relation to a basketball team where player A is the SHRM and players B–E are the other departments within the organization. The whole team wants to win the ball game, and they may all be phenomenal players alone, but one great player does not always win the match. If you have watched a lot of sports, you understand that five great players will not win the game if each of the five great players is the most valued player.

HR departments are accountable for evaluating and assisting each player or department to strengthen any faults that need to happen. Strategic management of human resources is then the method of using HR methods such as coaching, recruitment, compensation, and staff interactions to build a better organization, one employee at a moment (Figure 3.1).

Difference b/w SHRM & HR Strategy

SHRM	HR Strategies
• A general approach to strategic management to HR • Aligned with the organizational intention with future directions • Focus on long term people issue • Defines the areas in which specific HR strategies need to be developed • Focus on macro concern such as structure & culture • Strategic HRM decisions are built into strategic business plans	• Outcome of the general SHRM approach. • Focus on specific organizational intentions about what needs to be done. • Focus on specific issues that facilitate the achievement of corporate strategy. • Human resource strategy derived from SHRM

FIGURE 3.1 Difference between SHRM and HR strategy.
Source: Reprinted from Sharma. Open access

3.3 TRANSFORMATION OF HR TO HR 4.0—A CHALLENGE

Smart HR 4.0 strategy to muddle with Industry 4.0 transformation challenges. Today, emerging technologies, such as Internet-of-Things, big data analytics, and AI will automate most HR processes, resulting in skilled and leaner HR teams. For effective application of smart HR 4.0 to enable HR agencies to engage in a more tactical role in the general development of the organization, both organizational structure and leadership style modifications would be needed (Dirican, 2015).

A few HR digital transformation examples are as follows:

1. Use social media, AI, and online gaming to set an example of a company like Unilever where changes in the recruitment and selection processes are made.
2. By organizing "Hackathons" such as the You Belong @Cisco app and Ask Alex CISCO builds fresh HR products.
3. IBM's tests to drive fresh digital HR solutions are well known.

3.4 ROLE OF HR 4.0 IN TODAY'S GLOBAL BUSINESS SCENARIO-LEADING CASE STUDIES

Volatility, uncertainty, complexity, and ambiguous universe of business today demands high productivity from employees. It calls for highly skilled people to embrace the change, which enable change management, transformation, and strategy enablement for the employer. With technology advancing to a "sentient" stage, the workforce needs to readjust the change which is radical to enhance productivity, and create interoperability.

Due to globalization, changes are witnessed in every form such as: in technology; in organization structures and culture; in business models (going from brick and mortar to brick and click or pure play models); in people demographics [business process outsourcings (BPOs) and knowledge process outsourcings]. Also, most importantly, it can be seen in the way they perceive work and want to work nowadays. Organizations are technology centric and so are the employees who prefer to work in flexible modes (The Royal Society & British Academy, 2018).

The responsibility and role of HR is to carry vision for the future as much as they do insights from the past. HR responsibility as an organization is to retrain and rescale people to make them more employable for the future. The gaps that are social, technological, economical, and environmental have to be filled by HR 4.0 to sustain and remain competitive. HR 4.0 calls for the development of a diverse and individualized set of career options too. It is a re-engineering process of "Old school in New school" that requires more dedicated application of knowledge management to handle fluctuation within the industry. Organizations need to change from "tech-reluctance" to "tech-advocacy." They have to focus on data-driven insights for better decision-making. Role has changed from specialists to strategists to drive the business.

3.5 WHAT EXPERTS SAY

Relevant is for today—being ready is for tomorrow. If you are not relevant today, you can never be ready for tomorrow. All of us have to engage ourselves in something, something to learn and share something to do differently, something to participate, which is more "transformational."

Cyrus Jalnawala, Dow Chemicals: In 2010, Unilever created a "Sustainable Living Plan," which included programs to decrease the business' environmental effect and enhance the health and well-being of employees. This plan involves nine specific obligations, including the purchase of all agricultural products from renewable sources by 2020. Netflix guarantees the consistency of the recruitment, evaluation, and recognition scheme with corporate values, saying "Values are what we value." As a consequence, a great deal of time and attention is dedicated to evaluation; HR requires delivering the correct abilities and capacities (Ernst and Young, 2018).

We are looking for reverse mentoring where we will have the younger workforce assigned to senior leaders for sharing their knowledge on technology, automation, and on what they are strong at to have HR 4.0 successfully implemented.

Lakshmi Nadkarni: Give employees freedom. Give them the responsibility, the framework, and the choice. Educate them about the choices; help them understand it better so that they can make good decisions on their own. Often times, we underestimate their capability and tend to spoon feed them far too longer than they need to be.

Rajan Sethuraman, Latent View: Tomorrow's HR must be strategically deliberative in defining their ownership to develop future leaders. They will need to draw up future leadership capabilities based on forward-looking business trends and knowledge-based systems and "software-as-a-service" platforms to simplify and enhance efficiency for all stakeholders.

3.6 PRESENT CHALLENGE

Uniqueness and value is something that employees have to create and then tailor HR policies and frameworks to suit different combinations of uniqueness and value.

As per a survey report by Mckinsey, man and machines will be working alongside each other, complementing each other. While most occupations will be impacted by automation, a very few occupations will actually be replaced. About 30% of the activities in 60% of all occupations could be automated. But, the nature of these occupations will undergo changes and emergence of possibilities of newer opportunities in learning and progressing in capabilities.

We can cite examples of education industry where most of the work and assignments given to the students will be through massive open online course, another example can be in the healthcare, where AI algorithms will read diagnostic scans to facilitate medical professionals accurately diagnose and prescribe the right treatment at right time, which will save many lives. At online retail store like Walmart employees who earlier manually stacked merchandise are now becoming robot operators to monitor automated operations. But, professionals need to smartly adapt to reap the benefits of this change. There will be no dearth of work but people will need to undergo transitions, acquire new skills and align themselves to the machines accompanying them at work. From redundant occupations to emerging ones, adaptive change is demanded in HR 4.0 era where changes in workflows, workspaces, and workplaces will take place (Frank, 2019).

New and emerging technologies will integrate the physical, digital, and biological worlds to impact all economies and industries, even as it will push the boundaries of human capabilities. In looking at future of HR, the impact of the 4th industrial revolution cannot be underestimated. Driverless cars, bots, augmented, and virtual reality have set in motion unimaginable facets of robotization that directly or indirectly will impact people, work, and work culture.

Undoubtedly, Industry 4.0 has the powerful potential of transforming organizations into real-time enterprises. Connecting the dots in real time becomes a reality and this enables speed and agility for faster "go-to-market" innovation. In addressing the challenges of digitization to move ahead, neither status quo nor a roll-back is an option but to sustain one has to embrace the change. Continue to leverage automation and AI for sustained performance and productivity enhancement is the only mantra for success. Over a period of time, these technologies create economic pluses to enhance RoI. Hence, in its reinvented form, HR needs to be the prime mover of such collaboration to shape a future that puts people first and empowers them. The 4th industrial revolution may seem like a dehumanizing development that could robotize people out of jobs. It is here that HR will act as an influencer to assert that there is, and there will be work for everyone despite automation but for getting that work, one needs to have right skills and right attitude, the work will be different and will require newer skills and adaptability to master those skills.

HR's strategic capacities need to set up the correct teaching programs—it will be essential for HR to train and retrain mid-career employees as well as fresh generations for future problems. This needs the joint efforts of governments, leaders in the government and private sectors, and entrepreneurs—they need to behave as a change agent and coordinate government and private projects, generate the correct incentives, and allow greater human capital investment. There is no doubt that the future with automation and AI will be difficult, but a much richer one if we use aplomb technology-and mitigate the negative. Jobs may not become redundant but jobs will change so the nature of jobs (Gries & Naudé, 2005).

Suhas Kadlaskar, Mercedes Benz: It is essential for HR to restructure themselves for this seamless flow and filter of work. This includes the right vision, right strategy, right approach, right knowledge, right infrastructure, and right skills talent—plus the agility to scale in scope, capabilities, and volumes as HR emerges as reinvented business leaders.

As a change adopted by Cognizant of the four causal forces, there are five capabilities one need to focus on to build a robust internal organization.

Technology no doubt would make certain forms of labor obsolete, but it would also allow humans to get more involved in tasks which require greater knowledge, critical thinking, and conscience. The economic opportunity created by HR 4.0 would in turn create new jobs and new opportunities for growth.

Creating opportunities for collaboration between humans and AI in the future 4.0 is initiating automation and implementing AI in organizations cannot be completely rejected from employment-related issues.

3.7 HUBS OF DATA CREATION WITH THE HELP OF SKILLED EMPLOYEES

We can say that human beings are the most significant asset to any organization they are innately able to recognize pictures, interpret languages, draw inferences, and differentiate items, computers need an exhaustive dataset to learn and mimic such abilities. These hubs could hire primitive computer literacy individuals to produce AI system training content.

3.8 ADVISORY SOLUTIONS COMPRESSING HUMAN EXPERTISE INTO MACHINES

Therefore, it is a profitable chance to digitize their field knowledge by constructing consultative alternatives through AI to switch from human learning to machine interactive stage. This could considerably enhance the accessibility of expert advice to a big population across the agricultural, rural health and finance sectors, the fashion industry, and the education sector.

3.9 GREATER INVOLVEMENT OF HEALTHCARE PROFESSIONALS WILL BE IN DEMAND

Healthcare industry will make the greatest use of AI, as it will assist doctors, radiologists, nurses, and other healthcare providers to devote their time and knowledge to critical instances, as AI assistants would perform routine duties. Healthcare suppliers could also concentrate on the humane and empathic side of care delivery, eliminating repetition in the workplace, and comforting the sick. It will increase the worker's positivity and help improve the workers' ill mental health.

3.10 CREATION OF NEW ROLES WITHIN IT SERVICES

In the coming days, jobs such as research analysts, data entry operators, system engineers, and test engineers are now gone by professionals such as research scientists, language processing specialists, developers of robotic process automation, and man–machine team directors. Organizations, therefore, need to channel their employees' mindset and provide them with ample chance to learn and adapt changes. Given the widespread employment of the Indian workforce in IT services and BPOs, the short-term impact of automation is expected to be high. Overseas customers were of paramount significance to many Indian IT businesses generating income (Grosz, 2016).

3.11 RESKILLING THE WORKFORCE FOR NEW AGE EMPLOYABILITY

Many significant steps need to be taken now to retrain staff around human–AI cooperation for new age employability. The skill sets that the future workforce is anticipated to be well versed with a mundane technology, algorithmic, and programmatic know-how. HR 4.0 has a strong demand for qualified staff in the field of AI, robotics, machine learning, and data sciences as businesses are working to implement this technology for a broad variety of apps. The abilities are no longer restricted to the technology sector alone, but are also widely used across the fashion, healthcare, banking, retail, and other industries.

It is compulsory for experts to work closely with the creation of multiple AI-driven alternatives to promote critical and analytical thinking, as well as their duty to enhance scientific literacy and enable innovation and creativity. This is needed in order to design diverse apps by borrowing expertise from multidisciplinary areas where offering qualified workforce will be a challenge for HR. Therefore, cross-functional teams need to work together to achieve a correct job culture equilibrium (Jarrahi, 2018).

3.12 STRATEGIC AND MANAGEMENT DECISION-MAKING

Making the strategy and work on data which is both unstructured and semistructured as well trained skill is in demand who would be required to take major decisions for eliminating redundancy, cutting cost, designing competitive strategy, making improvement in performance, and make investment keeping in mind the actionable insights gained from AI applications.

These will raise a real challenge for human resources department in the coming new generation of HR. The best way to predict the future is to create it.

3.13 IMPORTANCE OF EMERGING TECHNOLOGIES ON HR 4.0 AND RISING CHALLENGES

The constant evolution of technology around us has created a fear of loss of jobs and chances of inequality. There is no doubt about the fact that AI reduces the human efforts and further help in increasing the productivity. For developing nations like India, China, and Brazil, it helps in achieving economies of scale. But, one needs to understand how much is too much?

With India riding forward on the back of rapid digitalization and industrialization, it is important to understand that incremental use of technology is important rather than rapid use of the same. With the skilled and educated unemployed ones are increasing in India, the rapid digitalization could have strong implications on the workforce and at workplace. The role of humans has changed now. The role is not to just produce but to create things in future. Innovations like cloud computing, machine learning, and AI will have certain degree of impact on businesses, various jobs, and at workplaces (Kolb & Jones, 2012).

Evolution of artificial technology is so uncertain that it is hard to understand the impact of the same. There are claims of robots surpassing us the humans and replacing us at workplaces. One of the biggest revivals of many sectors is because of the advancements in AI. The major buzz around AI is that it is a "disruptive" technology. In past few years, there has been projection of the amount of job losses or jobs gains that has or will happen. It has been observed that manufacturing sector has experienced the job losses due to the rapid automation.

It is said that many jobs will be faded off after the advent of AI as machines will have the capability to do the same task in more efficient and better manner than humans. We use technology all around us all the time from working at office to ordering food or shopping online, but it has never crossed are minds that this rapid automation may have caused so many losses of potential jobs. There is no doubt that the technology has the ability to make faster calculations to solve difficult algorithms and adapt quickly with the environment but it lacks the emotions. The decisions which involve human emotions cannot be achieved by the machines.

In technologically advanced economies, it has been observed that growth of productivity is continuously declining. Analysts and many observers have pointed out that these continuous advances in AI may lead to job losses and widening of gap between the rich and the poor. Due to increment in technology, it has been pointed out that there will be increasing inequality. A report has shown that about 30% of workers in United Kingdom and about

40% of workers of United States are at major risk of being replaced by the technology. Jobs are at major risk in India and China according to the World Bank Report in 2017. It is hard to reach to a conclusion at such an early stage because AI is still at a very nascent stage.

Cultural and social changes are linked with economic growth and technological changes. Modern researches have shown that in different countries same technology can be implemented in a very different way from each other. It has also been seen in automobile industry in 20th century. Organized work accompanied the technological change and which are further linked to other cultural and social factors. Many new start-ups are crowd funded which means that individuals are developing faith and are willing to increase the technological intervention in their lives. In United States, increasing inequality between the highly educated and lesser educated workers is also because of the increase in immigrations or import of new talent (Kolbjornsrud, 2016).

The import of such skilled workers in US is reason for the decreasing growth of income levels for middle and lower educated workforce in US. Dyer in 2012 pointed at the researchers in field of geography, economics, and sociology who have proven that there is a shift in balance of power between workforce and the capital. It is important to have laws and policies at place to protect our workforce. There are clusters of highly skilled workforces in some parts of the world and not in the rest part which has further increased the gaps. Many publications have pointed at the impact of technology on production capacities and employment levels.

Variations in findings exist among the various studies pointed at the negative effects of automation on levels of employment and economic growth. The other school of thought suggested that technology, automation, and robotics adoption in manufacturing sector would increase the productivity of workers and provide businesses the benefits of economies of scale. In Germany, adoption of robotics actually increased the level of employment for low skill workforce as manufacturing unemployment was replaced by employment in service industry.

It was pointed out that a single automated robot is the reason for unemployment of approximately two employees in manufacturing sector. It meant that there will be approximately 250,000 jobs losses from 1995 to 2015. With rapid liberalization, globalization, and privatization in the developing and least developed countries, technology is the backbone of this change. The increase in technology growth also helps in bringing boost in earnings for the workforce.

All jobs may not require the same level of automation as compared to the other jobs, although, every job has some component of automation in it. As a result, the significance of automation varies from job to job. A study pointed at the following (Sahai, 2018):

- AI may affect jobs at various levels of education and compensation.
- There may not be 100% automation of a job but there will certainly be transformation of the jobs into new automated job roles.
- It was also suggested that for jobs from now to 2030, the workforce needs to constantly work on skill upgradation and adapt quickly with the fast pace of digitalization.
- Whether an AI technology replaces or compliments human for a task still, it is an extension to human capabilities. AI provides various solutions to our economic and social problems. The world has seen reduction in friction between humans after the advent of AI developed technologies.
- Human is considered to be the most important resource for a company. Therefore, AI technology is used in human resource analytics to analyze the employee's behavior and how they react to different changes in the work environment.

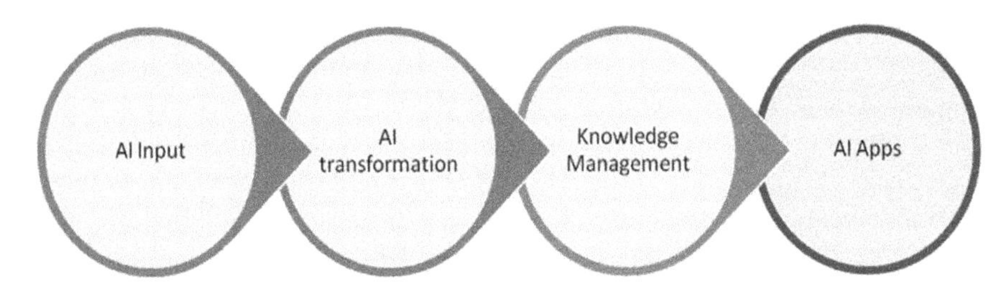

FIGURE 3.2 Development of AI applications from AI inputs.

AI input are the voice, image sensor, and video that further gets transformed through neural networks and image recognition in addition to speech to text further, the AI is mixed with knowledge management systems to develop AI-enabled technologies and applications, commonly called chart assists, virtual scribes and it helps human resource professionals in avoiding burnout in analyzing the behavior of its employees (Figure 3.2).

The impression that automation of the workforce can help in reducing the costs of an organization is a myth. It is very important to have a responsive work force which can adapt to the changes in the technology. This also needs a very optimistic leadership which can groom the workforce to accept such transitions. Such a workforce can be termed to be an agile workforce, which has been trained to adapt to the technological changes. "An agile workforce can enhance an organization's ability to survive in a volatile global business environment." *The World Economic Forum Report of 2015* said that industry 4.0 would mean a loss of 7.1 million jobs worldwide, which is a very alarming feature (Shaw, 2018).

For organizations, the first challenge would be to recruit people with the correct skill set. According to Boston Consulting Group, transformation in technology will lead to a very different approach in recruiting which will focus more on the capabilities of the workforce as compared to the qualifications which are determined by degrees and roles.

Once the recruitment process is over and the company has hired people who are well versed with AI, the question comes up whether they are well versed with the business processes. This again becomes a challenge to train the newcomers to make them familiar with the organization's work process and culture. The McKinsey Global Institute report said the by the year 2030 almost 14% of the workforce across the world will need to change occupational categories as a result of technology advancement like AI and digitization.

Besides training, retention of workers has become an important feature of modern organizations. In the day of AI and digitization, the challenge of retention can be overcome to some extent by allowing the workforce opportunities for personal growth. This means constant upgradation of skills and making them dynamic along with the dynamic changes in technology.

Knowing the various ways in which the industry 4.0 is going to affect HR is very necessary to sustain. It may require a major revamp of the entire sector. Leaders who are proactive and futuristic will survive this revolution. Leaders have to make themselves well versed with all the tools that are used in industry 4.0 so as to make the employees ready for it. Figure 3.3 shows all tools used in smart HR 4.0.

Daily tasks in an organization include very routine work like timesheets, expenses, etc. These routine jobs can be taken by IT tools, which are available and humans can work more on tasks which require deep thinking and creativity (The British Academy & The Royal Society, 2018).

The cognitive technologies that allow for AI in the organizations and are now available, actually enhance the prospects of enriching the various jobs that exist and do not prove to be a threat to the jobs. Here, we need to say that the machines which behave like humans and tend to interact with people like humans are still going to do. Only repetitive task and all work that requires thinking and creativity will still be handled by humans and probably give them more time and rewarding roles. Here, effects can only seem to be positive.

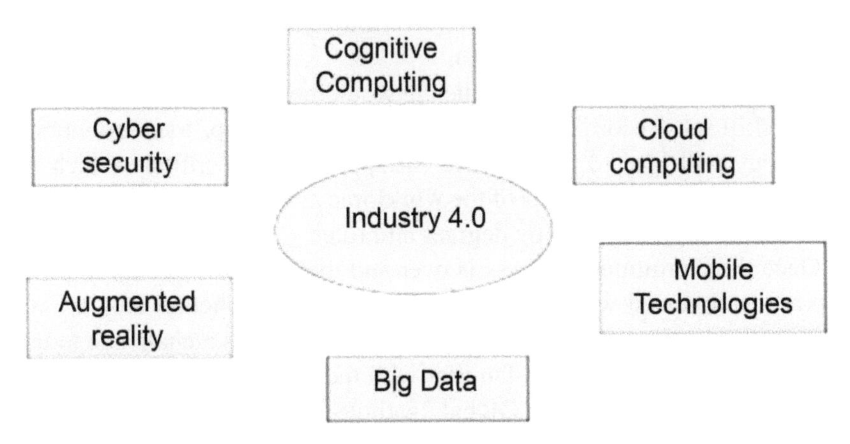

FIGURE 3.3 Tools of smart HR.
Source: Reprinted from Shaw and Verghese, 2018. Open access.

A study by Deloitte said that 42% of people feel that within 2–3 years organizations will be taken over by AI. This, however, did not mean that people will not be working or that only machines will be there in organizations. All this meant was that people will be taking more help of technology in doing their tasks. Many leading companies like IBM, Facebook, and a lot of other technological giants are moving toward implementing AI (Upchurch, 2018).

There is a need for human involvement which cannot be forgone. This involvement is also required in developing the AI solutions. It is not only during the development stage, but also in implementing and testing the solutions. The biggest irony today is that organizations are galloping to recruit and train people in latest technologies but technology cannot work in isolation. Humans have to always be there (Wired Insider, 2019).

3.14 CONCLUSION

Digitization is all oriented toward inserting the feeling of being effective and efficient in this rapidly evolving age than using traditional company ideas. In the digitization age, the use of technology such as AI has become one such paradigm. The primary objective of this section is to shed light on the continuing trends in international business, IT integration with HR and global companies, and understand its role in value cocreation, resource inclusion, and service exchange services.

While addressing the issues in these areas, the researcher has found more prevalent problem areas that are known and cannot be prevented, such as jobs and ethical considerations, energy development, and much more. In addition, countries like India have an enormous economy with young aspiring worldwide executives, so exploring and studying the above elements for this nation can bring fresh theories, perceptions, job possibilities, creative thoughts, and so on. How strategies are made and enforced is very obvious. Digitization and fresh techniques allow worldwide executives to interact directly with the outside world, regardless of their place, via digital channels.

While digital channels/medium are considered the most suitable approach to reach large clients, the latest growth of infrastructural equipment in terms of Internet connectivity and telecommunications has helped a lot. In this age of highly efficient and effective economies, it is essential to have adequate strategies to implement such as adequate infrastructural growth, rehabilitation of HR strategies, and public policies to help integrate emerging technologies with worldwide companies.

KEYWORDS

- **HR 4.0**
- **digital age**
- **artificial intelligence (AI)**
- **robots**

REFERENCES

Armstrong, M. (2009). *Armstrong's Handbook of Human Resource Management Practice.* 11th Edition, Kogan Page Limited, London

Brynjolfsson, E., Liu, M., & Westerman, G. F. (2018). When Do Computers Reduce the Value of Worker Persistence? *SSRN Electronic Journal.* https://doi.org/10.2139/ssrn.3286084

Brynjolfsson, E., Mcafee, A., & Manyika, J. (2014). Will Your Job Disappear? *New Perspectives Quarterly, 31*(2), 74–77. https://doi.org/10.1111/npqu.11457

Crawford, K., West, S. M., & Whittaker, M. (2019). *Discriminating Systems: Gender, Race and Power in AI.* Retrieved from http://cdn.aiindex.org/2018/AI Index 2018 Annual Report. pdf.

Danneels, E. (2004). Disruptive Technology Reconsidered: A Critique and Research Agenda. *Journal of Product Innovation Management*, *21*(4), 246–258. https://doi.org/10.1111/j.0737-6782.2004.00076.x

Dirican, C. (2015). The Impacts of Robotics, Artificial Intelligence on Business and Economics. *Procedia–Social and Behavioral Sciences*, *195*, 564–573. https://doi.org/10.1016/j.sbspro.2015.06.134

Ernst and Young. (2018). The New Age : Artificial Intelligence for Human Resource Opportunities and Functions e. Retrieved from https://www.ey.com/Publication/vwLUAssets/EY-the-new-age-artificial-intelligence-for-human-resource-opportunities-and-functions/$FILE/EY-the-new-age-artificial-intelligence-for-human-resource-opportunities-and-functions.pdf

Frank, M. R., Autor, D., Bessen, J. E., Brynjolfsson, E., Cebrian, M., Deming, D. J., … Rahwan, I. (2019). Toward Understanding The Impact of Artificial Intelligence on Labor. *Proceedings of the National Academy of Sciences*, *116*(14), 6531–6539. https://doi.org/10.1073/pnas.1900949116

Gries, T., & Naudé, W. (2005). Artificial Intelligence, Jobs, Inequality and Productivity: Does Aggregate Demand Matter? *IZA Discussion Papers*. Retrieved from www.iza.org

Grosz, B. J., Altman, R., Horvitz, E., Mackworth, A., Mitchell, T., Mulligan, D., … Shah, J. (2016). *Standing Committee of the One Hundred Year Study of Artificial Intelligence*. Retrieved from https://ai100.stanford.edu.

Jarrahi, M. H. (2018). Artificial intelligence and the future of work: Human-AI symbiosis in organizational decision-making. *Business Horizons*, *61*(4), 577–586. https://doi.org/10.1016/j.bushor.2018.03.007

Kolb, R., & Jones, P. C. (2012). International Labour Organization (ILO). In *Encyclopedia of Business Ethics and Society*. https://doi.org/10.4135/9781412956260.n438

Kolbjornsrud, V., Amico, R., & Thomas, R. (2016). *The Promise of Artificial Intelligence: Redefining Management in the Workforce of the Future. Accenture Institute for High Performance and Accenture Strategy*. Retrieved from https://www.accenture.com/_acnmedia/PDF-32/AI_in_Management_Report.pdf

Sahai, S., Goel, R., Malik, P., Krishnan, C., Singh, G., & Bajpai, C. (2018). Role of Social Media Optimization in Digital Marketing with Special Reference to Trupay. *International Journal of Engineering and Technology (UAE)*. https://doi.org/10.14419/ijet.v7i2.11.11007

Shaw, P. Verghese R. M. (2018). Industry 4.0 and Future of HR. *Journal of Management, 5(6)*. http://www.iaeme.com/JOM/issues.asp?JType=JOM&VType=5&IType=6

The British Academy, & The Royal Society. (2018). The Impact of Artificial Intelligence on Work. Retrieved from https://www.thebritishacademy.ac.uk/sites/default/files/AI-and-work-evidence-synthesis.pdf

The Royal Society & British Academy. (2018). The Impact Of Artificial Intelligence on Work. Retrieved from https://royalsociety.org/~/media/policy/projects/ai-and-work/frontier-review-the-impact-of-AI-on-work.pdf

Upchurch, M. (2018). Robots and AI at Work: The Prospects for Singularity. *New Technology, Work and Employment*, *33*(3), 205–218. https://doi.org/10.1111/ntwe.12124

Wired Insider. (n.d.). AI and the Future of Work | WIRED. Retrieved June 29, 2019, from https://d1ri6y1vinkzt0.cloudfront.net/media/documents/AI and the Future of Work_FIPP_VDZ.pdf

CHAPTER 4

Global HR Challenges in Industry 4.0

LALIT KUMAR SHARMA

Jaipuria Institute of Management, Ghaziabad 201014, Uttar Pradesh, India

**Corresponding author. E-mail: lalitks4@yahoo.com*

ABSTRACT

It is never easy to predict the future. Many experts have been mistaken in their estimates many times. It is important for a business to know about socioeconomic changes and technological advancement that will have an impact on business in coming years. The new era of digitalization, robotics, and artificial intelligence is going to revolutionize the business world. It creates the need for companies to invest in the solutions that integrate their machines, processes, employees, and products in a network of data collection, data analysis and performance improvement and company development. Opportunities provided by computers and robotics have been revolutionized now, case was not same earlier. The base for revolution is data, mechanism of its collection, analysis, and use of information for right decision-making. Right decision-making is an edge for many companies over to others in Industry 4.0 scenario. This fourth Industrial revolution demands the help of robots and AI to ensure the performance in business. This chapter tries to look upon the effect of automation in managing human resources as per new scenario of HR 4.0. Challenges like employee skill development, workspace complexities, retention of jobs, and manpower in changing scenario of Industry.

4.1 INTRODUCTION

According to the estimates of the Economic Policy Committee and the European Commission, by 2050 the population in the 24–60 age group will decline by 16% to 48 million while the elderly population will increase to 58 million (Qina et al., 2016). Western countries, which anticipate that this change in population structure will force them to compete in the future, have attempted to redesign their industries in such a way as to require the least labor force. Challenges are high in the changing world where the businesses are finding difficulties in coping with them. Innovations and speed to success in shortest possible time are getting essential. Various developments were at center in all industrial revolutions like first industrial revolution with steam power and mechanization in the late 18th century, second industrial revolution in the late 19th century with electricity. The Second Industrial Revolution has emerged with changes in basic raw materials and energy sources. Steam, coal, and iron, as well as steel, electricity, petroleum, and chemical materials have been used in the production process (Alçın, 2016). Third started in the mid-20th century with the power of computer and Internet and fourth embarked as Industry 4.0 is witnessing the power of new information technology and automation. New technologies and innovations like virtual reality (VR) for digital world, robotics, driverless cars, Internet of Things (IoT), AI, and 3D printing are the biggest reasons for Industry 4.0. In the industry, the virtual environment is used at every point such as production planning, design, production, service, maintenance, testing, and quality control (Bayraktar and Kaleli, 2007). The customer needs are changing in the new business world. VUCA (volatility, uncertainty, complexity, ambiguity) is the current acronym that has to be embraced by the companies. Smart production systems are used to create the flexibility and capacity to meet the customer expectations in market. High automation is substituting conventional and monotonous processes. As the new processes are becoming complex, new qualifications for current workforce are needed that makes HR job more critical. The workforce needs to focus more on creative activities. The competencies in management are vital to meet the arising challenges of fourth industrial revolution. The difficulties of the diversity can be summed up in four areas: economic, social, technical, and environmental (Hecklaua et al., 2016).

Every industrial sector is impacted because of influence of Industry 4.0 as it will bring a lot of changes in their businesses. The fourth industrial revolution can be analyzed in two manners. Some says that it is extension of the changes under the third industrial revolution, and others feel that it is a shift to a new world. Changes like digitalization, robotics, and AI are going to revolutionize almost all industrial sectors across the world. Industry 4.0 refers to the shift toward rising data exchange and automation in every industry that result in the high speed and better outcome.

Cloud computing, cognitive computing, IoT, and cyber-physical systems are part of Industry 4.0. Fast changes and developments in technologies as well as automation are making a new world for industry that is highly network oriented.

This revolution is bringing a promising change with more disruption than ever. Sooner rather than later global business to get ready for opportunities and challenges generated by this paradigm shift everywhere.

4.2 CHALLENGES FOR TRADITIONAL ORGANIZATIONS

The major challenges for traditional organizations can be summarized as "continued and accelerating disruption." Leaders have understood the need for constant transformation in all industrial sectors, but it is challenging to transform as per new dynamics of Industry 4.0 in future.

The revolution is not about simply introducing digital technologies to the business—it requires changes that in turn include following:

- Helping employees to accept the continuous disruption.
- Attracting the talent from different demographics in the world.
- Ensuring the development and engagement of employees in organization.
- Focusing on digital customer experience for business growth.
- Engaging the customers in business processing on regular basis.
- Moving in the direction to be digital leader.
- Developing capacity of digital transformation.
- Achieving the sustainability in digitalization of business.

In driving such changes, balancing of human capital is very important. Continuous interconnection of employees within the organization is very important for sustainable change and development in business. Organizations are looking for the employees who can work in team to develop more relevant products and services.

4.3 BIG DATA AND ANALYTICS IN HR

As data exchange in business is a key in Industry 4.0, there is a priority of many organizations to empower the right professionals for the changes that lie ahead in future. Now, the challenge is bigger for HR managers to hire the potential talent who can lead the organization in the era of Industry 4.0. Situation is more complex now than previous. HR team is looking for the answers of questions now which did not exist earlier.

Conventionally, the functions of human resource have been more oriented on recruitment, performance appraisal, employee engagement, and compensation designing. Now, the potential of current and new talent is evaluated on more analytical situations about performance, innovation, engagement, ethics, and remuneration as per the global strategies of organization. New dynamics of concerns is prevailing in the schedule of HR as per changing scenario of Industry 4.0.

Use of data analytics is very helpful in recruiting the talent and developing the right kind of leadership with cost efficiency in the organizations across the globe. It is a big tool in ensuring right people in right jobs. The use of data analytics in HR increases productivity and engagements, activate innovation, empower talent, and transform leadership. Talent acquisition and retention practices are more effective with the use of HR analytics. HR analytics can serve as an "extra brain" for HR leaders to answer the unanswered questions in changing dynamics. Proactive approach can be developed for sustainability with the help of data analytics in HR.

Approach for handling the data generated through reward and benefits, interviews, performance, assessments, engagement, and attrition data should be developed with a new angle of evaluation to help in decision-making for the problems faced by organizations in scenario of Industry 4.0.

New data sets like social media interactions are widely used for understanding expectations of potential employees and employers by HR. These types of practices should create the mechanism for any team to employ either basic or complex analytical tools, which are as follows:

- Invest in a well-integrated system that leads for better business understanding as well as efficiency and functioning of system. It is to be understood that systems are not the substitute for understanding business situations.
- Utilize the organizational knowledge for sustainable business growth with the use of the right HR analytics talent.
- Ensure learning from data of your own environment as well as outside environment in the world. This will provide better understanding to work out the upcoming challenges in HR.
- Create unique selling propositions for your organization along with following some benchmarks.
- Learn to deal with complexities in business in order to keep outputs simple, effective, and efficient.
- Maximize the use of data to predict the future, but always keep in mind that there is no single, predetermined future!

4.3.1 INDUSTRY 4.0—HOW TO LEAD

An agile workforce can enhance an organization's ability to survive in a volatile global business environment (Katayama and Bennett, 1999). As per the PwC report 2017, automation and thinking machines are replacing human tasks and jobs raising huge organizational, talent, and HR challenges. First mooted in 2011, Industry 4.0 promotes the computerization of manufacturing, bringing about what is known as "smart factories."

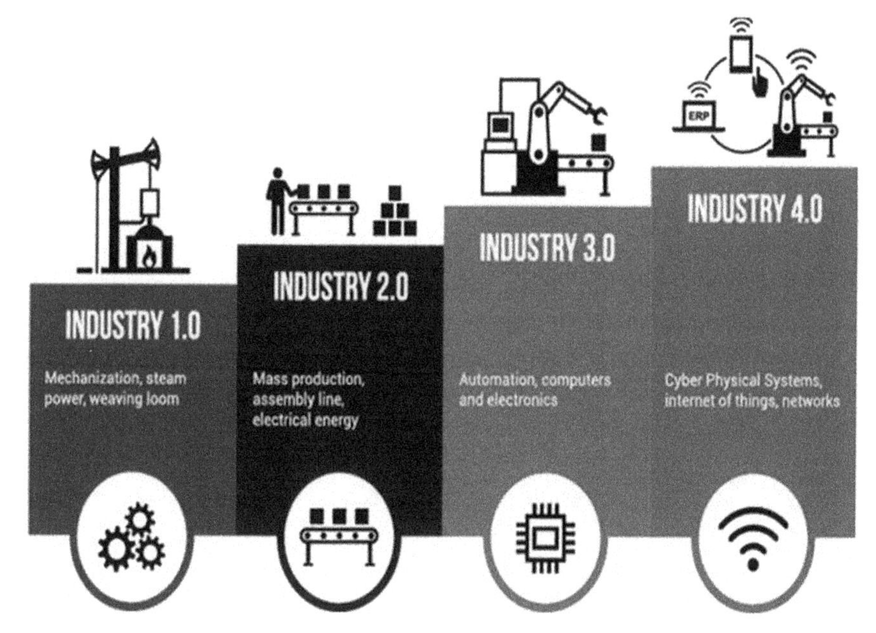

FIGURE 4.1 Concept of changes from Industry 1.0 to 4.0.

All types of organizations including small and medium enterprises across various sectors are influenced with the impact of digitization. Impact is different one in each sector, so it is important for the companies to have right understanding of digitization and its effects on the industrial sector. Proper understanding about opportunities and threats will help the companies making effective decisions for the growth (Figure 4.1).

There are dynamic operating conditions under which a company has to deal with the challenges and preparing them for future in the world of digital impact. It is a major issue with leaders to take care of the task of restoring respect and confidence in business and leadership as Industry 4.0 is creating performance pressure on old establishments. They have been expected to lead the organization through the times of uncertainty and turbulence. They need to ensure the pathway for success under new dynamics of business conditions like increasing use of automation across the globe.

Leadership in the world of digitalization is all about ensuring the achievement of organizational objectives with your ability to influence and impact your stakeholders and followers. Relationship management, empathy, critical thinking, and emotional intelligence are next generation leadership competencies required for achieving organizational goals under digital foot prints.

4.4 THE FOURTH INDUSTRIAL REVOLUTION (INDUSTRY 4.0)

Professor Klaus Schwab (founder and executive chairman of the World Economic Forum) published a book entitled *The Fourth Industrial Revolution* in which he describes how this fourth revolution is fundamentally different from the previous three industrial revolutions. These were mainly characterized by advances in technology.

Schwab defines the first-three industrial revolutions as the transport and mechanical production revolution of the late 18th century; the mass production revolution of the late 19th century, and the computer revolution of the 1960s.

Integration of IoT, cyber-physical systems, and Internet of Systems leads for the development of Industry 4.0 or fourth industrial revolution. In other words, it is the thought in business units in which machines are augmented with web connectivity and associated with a system so that the entire production process can be visualized and monitored and makes decisions on its own.

A range of new technologies will combine the physical, biological, and digital worlds in Industry 4.0 (Figure 4.2). Our economies, industries, and even our thought processes that mean to be human will be challenged by these new technologies.

Technological innovation is the key driver for bringing changes in the global economy under Industry 4.0 revolution. These technologies will generate more benefits and challenges in the entire world. In such scenario, Schwab mentions that public–private research collaborations should increase, new sphere of knowledge, human capital should be created to benefit all. This impact of technology and disruption will create many managerial leadership challenges that may result in external factors on which leaders may not be having any control or little control.

However, the sustainable development of organizations will be guided by the leaders only. They need to ensure the readiness of organization for adopting the changes required as per scenario of Industry 4.0. They have to develop the right combination of technology and human capital for sustainable growth of organization. How to bridge the performance gap will be an important decision for all leaders in changing scenario. For this, they need to shape the fourth industrial revolution by grasping power and opportunities and ensure the organizational values in the success.

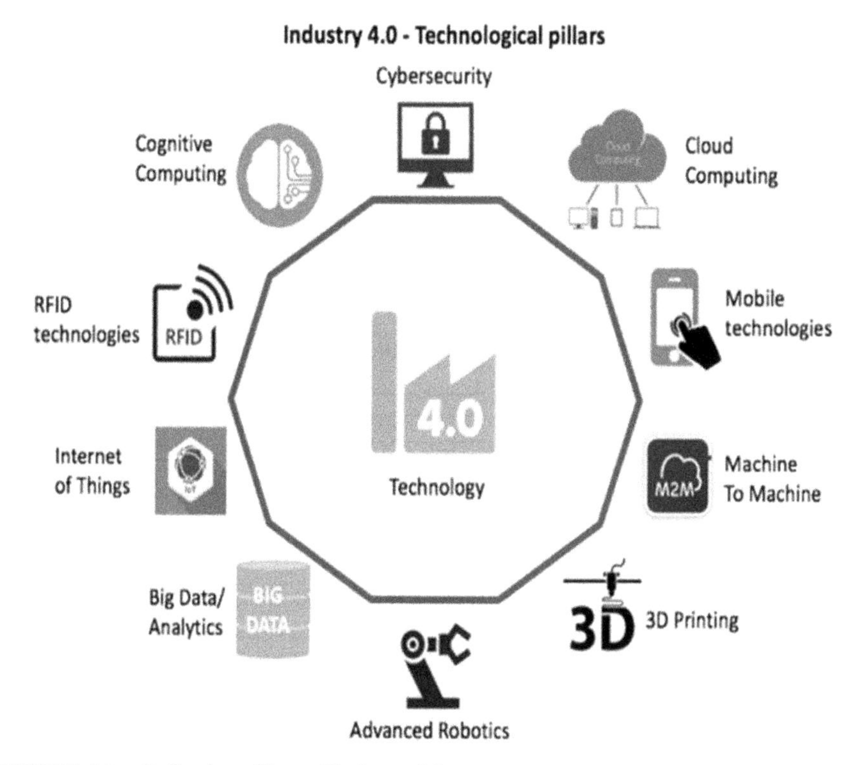

FIGURE 4.2 Reflecting pillars of Industry 4.0.

Development of a comprehensive view about the use of technology and its impact on human capital in organizations at micro- and macro-level how it is changing various environmental factors, is important for leaders to take care of organizational benefits.

It is very difficult for many leaders to come out of conventional thinking approach and learn the high use of technology and making the required changes and substituting the old standards of performance. Still, linear thinking is very promising in many organizations among their decision makers. This may result in shocking conditions for a business as per recent developments. Use of technology and disruption is vital for shaping the future of organization.

4.5 IMPORTANCE OF COGNITIVE READINESS

Cognitive readiness is the mental preparation developed by leaders and their teams to prepare them to face the ongoing dynamics, undefined and unpredictable business conditions under impact of digitization in industry 4.0. Advanced thinking can be developed with the help of cognitive skills development.

The Executive Development Associates has identified the seven key cognitive readiness skills collectively known as Paragon7. These cognitive skills help in the development and enhancement of leader's ability to navigate successfully in new conditions.

These seven cognitive readiness competencies are as follows:

1. *Mental cognition:* Recognize and regulate your thoughts and emotions.
2. *Attention control:* Manage and focus your attention.
3. *Sense making:* Connect the dots and see the bigger picture.
4. *Intuition:* Check your gut, but do not let it rule your mind.
5. *Problem solving:* Use analytical and creative methods to resolve a challenge.
6. *Adaptability:* Be willing and able to change, with shifting conditions.
7. *Communication:* Inspire others to action; create fluid communication pathways.

Overall, heightened cognitive readiness allows leaders to maintain a better sense of self-control in stressful situations.

4.6 EFFECTIVE LEADERSHIP AT ALL LEVELS

Digital impact does not influence only at a particular level; it affects all levels across the organization. Thus, organization is in need of effective and efficient leadership at all levels to drive the digital strategies and achieve organizational goals. Expansion of digital transformation across the organization leads for the need to consider the approach for building a potential leadership pipeline with all capabilities to lead the organization in digital era and achieving success.

Positions stretching the leaders beyond their competencies and skills set are helpful in development of the pipeline of effective leaders. This requires coaching and mentoring to build digital capabilities.

Some conventional leadership capabilities like effective communication, motivating, and empowering others still remain important in any organization. As there are new requirements of leadership capabilities at all levels of organization, it demands a combination of new mindset and behaviors, digital skills and knowledge to lead the teams successfully in the era of Industry 4.0.

4.7 HR PROBLEM IN AN ORGANIZATION

Organizations are gradually becoming innovative and global because of globalization process and fourth industrial revolution worldwide. New practices are required to build competent and productive workforces at workplace. Managing workforce diversity derives the questions of values, traditions, norms, culture, and subculture. These questions should be addressed in order to remain competitive and sustain the position of organization. Effectiveness of HR strategy will be measured on the basis of the performance of screening, training, pay, and other human resource policies and practices, by using technology and other resources to encounter the challenges due to fourth industrial revolution Industry 4.0.

4.8 CHALLENGES FOR HUMAN RESOURCE EXPERTS

There are numerous challenges for HR experts in the scenario of Industry 4.0. Some of the key challenges for HR experts are shift in role of HR, talent acquisition, outsourcing, working environment, workforce diversity, managing technology, leadership development, and organizational culture.

4.8.1 SHIFT IN ROLE OF HR

The conventional role of HR is going to be changed in terms of procedures and responsibility. New dimensions of responsibility are emerging under Industry 4.0. Now, the data exchange and management are becoming important to success for HR functions. Analytics is becoming important for strategic approaches in HR. Now, the role will be more focused on accountability as compared to responsibility.

Measurement of performance is priority in changing role of HR. This will ensure the growth and sustainability of business under new dynamics. Now, the job of lip service is not important for HR professionals, they need to find the metrics for growth and performance in the organization.

4.8.2 TALENT ACQUISITION AND DEVELOPMENT

Focus on talent acquisition and development is required in order to meet with the changing scenario of business across the world. New technologies and innovations like VR for digital world, robotics, driverless cars, IoT, AI, and 3D printing are looking for the competent talent for achieving success in business. VUCA is to be managed by the emerging talent in new conditions of business. There need to be a focus on development of existing manpower as per developments and changes in business world. Some effective recruitment strategies can be as follows:

- Referrals.
- Cloud recruitment.
- Employment branding.
- Developing relationship with strong candidates for future positions in business.

4.8.3 NETWORKING AND OUTSOURCING

Outsourcing will be significant to more companies as it will help them to deal with the challenges of VUCA in their business. For attaining economical benefits in global market, companies are looking for better and efficient HR practices and it can be done with the help of networking of organizations very well. So, networking is important. New business locations will be creating advantages for many companies. A few reasons for outsourcing their HR services are as follows:

- Cost reduction
- High focus
- Compliances
- Accessibility to best technologies
- Unavailability of internal resources

4.8.4 WORKING ENVIRONMENT

Competitive advantage cannot be achieved, if the working environment in organization is sick, exhausted, and stressed. There is a growing importance of relationship between work environment and health as well as well-being of employees. It is a big motivation and attraction to the new generation of workforce.

4.8.5 WORKFORCE DIVERSITY

Sustainable success can be achieved only when organization is capable in managing workforce diversity. Industry 4.0 will create new equations of success to those organizations which are able to mange workforce diversity effectively. The level of diversity of manpower will increase in future across the globe in digital environment. Automation, AI, robotics, etc. will need talent in every industry from different parts of the world. Their need and expectations will be different from local workforce. Managing the demographic and psychographic characteristics of the emerging workforce will be one big challenge for HR experts.

4.8.6 MANAGING TECHNOLOGY

Automated procedures will eliminate many jobs that exist today. This makes HR jobs more strategic and performance oriented. AI, robotics, IoT, VR, and automated production processes are realities in Industry 4.0 which have to be incorporated by HR experts successfully in their organizations. The companies which are not able to meet with new technological challenges, possibility of doing business will be eliminated for them. Use of technology in HR functions also becomes important for future sustainability.

4.8.7 LEADERSHIP DEVELOPMENT

Leaders have understood the need for constant transformation in all industrial sectors but they are finding difficulties in the ability to develop as per the requirements of growing industries. Leaders have to focus on developing the competencies of their employees according to the need of changing scenario in an organization. Moreover, taking care of the personal benefits and welfare of workers is equally an important task to all HR leaders in the world of automation and robotics. Performance data of the real and the virtual world is important and should be merged for a new insight in business. Use of data analytics is very helpful in recruiting the talent and developing the right kind of leadership with cost efficiency in the organizations across the globe. It is a big tool in ensuring right potential as per the demand of jobs. Leaders, who are having high level of interpersonal skills and understanding of the integration between advanced technologies and people, will be more dynamic and result oriented.

A new approach is required by the leaders in HR to handle various data types so that effective decision-making get possible to manage the changes in scenario of Industry 4.0. The importance of using technology and data cannot be ignored. Therefore, development of critical thinking, interpersonal skills, and socialization is gaining momentum with time. Those leaders, who can develop intelligence quotient along with emotional intelligence in their employees, will be better prepared for future needs. Leaders need to know about automation, AI, etc., and their implications in organizations very well. They need to learn how to make best use of data but more important is to enhance work performance and employee engagement.

Training and continuous development, standardization, dealing with system complexities, regulatory framework, safety and security, organizational change and development, and communication structure are some important areas to be considered by leaders to confirm the organizational success. Industry-related leadership style is the need of hour for the companies to be the leader in Industry 4.0 era. Better real-time control measures have to be developed to ensure the accountability of workforce for their performances. Dynamic leadership styles have to be developed by these leaders which can design the long-term sustainable approaches for organizational success. Possible steps should to be taken to ensure motivation of employees and flexibility for necessary changes for organizational growth and development.

Development of organizational culture which facilitates the creativity and new values is not an easy job. It requires changing the fundamental values of the managers in the organization many times. New practices to evaluate performance management are required in Industry 4.0 world. It will be important for the organizations to evaluate the readiness of managers and their teams to perform in an unknown future. Only assessment of current delivery will not be sufficient.

4.8.8 ORGANIZATIONAL CULTURE

Impact of Industry 4.0 will make several changes in the value system of an organization. This will create new dimensions in the organizational culture, which should be handled by HR heads to manage success and people under changing situations and roles and responsibilities.

4.9 TRENDS AFFECTING HUMAN RESOURCES

Oliver Wyman has done a research on fundamental trends or mega trends that will affect business over the next 20 years. On the basis of his study, some major trends affecting HR in years to come are given below.

4.9.1 THE BIG GENERATION GAP

The generation Y (between 20 and 35 years of age) or the "Millennials" will represent half of the workforce by 2020 and three quarters by 2025. All companies need to understand the expectations of new generation as their aspirations are different from the previous generations. They enjoy having fun in work, love to work in team, keen on self-development during work, and loyalty quotient is reducing.

Some obvious differences between old and new generation are as follows:

1. *Factors of engagement:* Old generation is a strong believer of hard working for recognition, while younger generation is able to promote its own contributions. New generation looks for more freedom in working and they have their aspirations on priority.
2. *Work-life balance:* Improper work-life balance has created the problems for many employees always. New generation is highly concern with their personal life along with professional life. They want organizations to take care of both. They do not want to compromise. They want "Balanced Life." This is why some of the organizations started giving better facilities to their employees within the organization and now it will be done by more number of organizations.
3. *The concept of career path:* New generation wants more clarity about their career path in the beginning of their career. Old generation used to think in terms of predefined concepts of career. Performance management is a critical task for HR managers as expectations

of new generation related to career are very vibrant. HR mangers were concerned with performance management, training programs, compensation, etc., but HR managers have to consider multichannel thinking, multidimensional identity, and need for agility for new generation.

4. *The impact of national culture:* As new generation is gaining the momentum in scenario, the impact of national culture is disappearing. As young generation has their own set of values and expectations, so the importance and impact of national culture is reducing. It creates the necessity for traditional organizations to bring changes in their value system so that workforce diversity can be managed by them easily.

4.9.2 TERMS OF ENGAGEMENT

Bringing happiness in work ensures better employee involvement. Now, happiness parameters have been changed under Industry 4.0, where companies need to think more about well-being of their employees. Employee engagement cannot be reduced only to economic activities in the jobs; they want recognition and contribution in the vision of organization. Engagement of employees in the digital era will need different social aspects to be covered by HR managers as employees are more vocal and expressive now on many social platforms. Interesting to see that current employees are seeking meaning of life and have different thinking compared to older generation employees. Gradually, society is becoming stronger stakeholder in business than owners.

4.9.3 PERSONAL DEVELOPMENT

Personal growth and development opportunities are an important criterion for selecting an employer by the new generation employees. Continuous enhancement of employability and professional growth is essential for upcoming generation of workforce. Realization of professional spheres and personal aspirations is important in changing trends. Developing workforce for the opportunities in other domain is equally important.

Seniority and qualifications will not be an important standard for pay rise in organizations. Task orientation and accountability are going to be strong measures of performance to the people. There will be more value for expertise and technical skills. Talent will be managed as per individual competencies rather than job descriptions. Simplification of HR models will be required to bring satisfaction among employees.

4.9.4 EMPLOYEES AS CUSTOMERS

Growth will be more in those organizations who value their employees like their customers. Taking care of the needs of employees is important in changing scenario of Industry 4.0. Employees cannot be treated under the influence of hierarchy. Better exchange of information at all level is required to ensure workforce engagement. Employees need to feel special within the organization. It is important to make your employees happy so that company can deliver the promises made to their customers. Rights of the employees have to be protected and claims should be settled to ensure their satisfaction.

Some important strategies which can be adopted by companies are as follows:

- Differentiating the value proposition based on employee profiles.
- Modularizing the value proposition according to the lifecycle stage of individual employees.
- Developing new nonmonetary compensation models.
- Expanding the range of services and fostering employee wellbeing.
- Establishing tailored on-boarding systems to maximize the chances of success of new recruits.

4.10 CONCLUSION

As various developments like steam power and mechanization in the late 18th century, electricity in the late 19th century, power of computer and Internet in the mid-20th century, and the power of new information technology and automation in Industry 4.0 have taken place in the global market in different era, industries have been transformed accordingly. New technologies and innovations like VR, robotics, driverless cars, IoT, AI, and 3D printing are making revolutions worldwide and responsible for changing scenario in industry. Acceptance of disruption, talent identification and attracting them, employee development and engagement, focus on digital customer experience, customer involvement in business processes, digital leadership, capacity development for digital transformation, and making business sustainable are major challenges for traditional organizations.

Use of data analytics provides the support in talent recruitment and leadership development and ensures cost effectiveness in any industry. It helps in aligning right people in right jobs. Data analytics in HR helps in increasing productivity and engagements of employees, increasing innovation, and transforming leadership. HR practices become more effective with the help of data analytics. HR analytics acts as an extra brain for HR leaders in the corporate. A proactive approach can be developed by HR managers with the use of HR analytics to make their organizations more sustainable.

Ambiguity in operating business conditions is a big task for any HR leader as it demands better integration of people and advanced technology to create a success. Leaders are expected to build the confidence among employees in the time of uncertainty and turbulence. Leadership need to focus on relationship management, empathy, critical thinking, and emotional intelligence as next generation leadership competencies under digital foot prints.

The cognitive readiness competencies are mental cognition, attention control, sense making, intuition, problem solving, and adaptability and communication for performance and accountability of an employee. Development of digital capabilities among workers is essential to meet with digital challenges.

Innovations and growth, assessing leadership gaps, identifying skills, managing barriers to change, and creating future goals are major tasks for a leader in Industry 4.0. Nurturing of people, reskilling, and adaptability is important in dynamic conditions of business.

KEYWORDS

- **challenges**
- **artificial intelligence (AI)**
- **workforce**
- **automation**
- **performance**
- **leadership**

REFERENCES

Alçın, Sinan: "Üretim İçin Yeni Bir İzlek Sanayi 4.0". *Journal of Life Economics* 8. 2016, pp. 19–30.

Bayraktar, E., Kaleli, F.: Sanal Gerçeklik Uygulama Alanları. Akademik Bilişim, 31 *Ocak-2 Şubat: Afyon.* 2007.

Chang, S., Gong, Y., & Shum, C. (2018), "Promoting innovation in hospitality companies through human resource management practices." *International Journal of Hospitality Management*, 30, 812–818.

Hecklau, F., Galeitzke, M., Kohl, H., & Flachs, S. (2016), "Holistic approach for human resource management in Industry 4.0." *Procedia CIRP*, 54, 1–6.

Katayama, H. & Bennett, D. (1999), "Agility, adaptability and leanness: A comparison of concepts and a study of practice." *International Journal of Production Economics*, 60–61, 43–51.

Qina, J., Liu, Y., Grosvenora, R. (2016), "A Categorical framework of manufacturing for industry 4.0 and beyond". *Procedia CIRP* 52, 173–178.

CHAPTER 5

Management Approaches for Industry 4.0

RENU PAISAL

Gautam Buddha University, Greater Noida, Uttar Pradesh 201308, India

**Corresponding author. E-mail: renu_paisal@hotmail.com*

ABSTRACT

Industry 4.0 is portrayed by savvy fabrication and execution of cyber-physical systems for generation, that is, implanted a device for moving mechanisms and trigger, systems of microcomputers, and connecting the machines to the merit chain. It accelerates thinking about the redesigning and promoting development of items. These governing policies can assume a significant job in the improvement of vigorous capacities, and feasible learning and improved atmosphere. This nominal target to offer a perspective on best reasonable administration rehearses which can advance the atmosphere of development and learning in the association, and consequently encourage the business to coordinate the rate of Industry 4.0. This nominal is one of the underlying endeavors to draw the consideration toward the significant job of the executives rehearses in Industry 4.0, as the greater part of the ongoing investigations are examining the innovative viewpoint. This nominal likewise proposes exact and quantitative examination on these administration approaches with regard to Industry 4.0.

5.1 INTRODUCTION

Industry is a vital piece of any economy. Since the development of industrialization, industry encountered the changes in outlook because of the mechanical changes and advancements. These changes in outlook are known as "modern upsets," for instance, motorization (first, mechanical transformation), high utilization of electrical vitality (second, modern unrest), gadgets, and computerization (third, industrial revolution). The present economy is going to confront the fourth mechanical insurgency, activated by social, monetary, innovative, and political changes (Lasi et al., 2014). his fourth mechanical insurgency is otherwise called Industry 4.0 (Burmeister et al., 2015), which is a subclass of advanced change in existing business and current procedures, supplanting the manual business activities by computerized PC structures (Lansiti & Lakhani, 2014). Monetary triggers of Industry 4.0 are shown in Figure 5.1. As demonstrated by Figure 5.1, there are number of components fundamental for Industry 4.0 including nonappearance of capable workforce, developing society, resource viability and clean urban age, mass customization, growing thing variability, shorter thing life cycle, dynamic worth chain, flighty markets, and cost abatement weight. Everyone of these components needs particular administration to adapt up to difficulties. Indeed, even the world's tremendous economies are standing up to these challenges like China is going up against the trial of clean urban age on account of unreliable high defilement. Japan and China are going up against the issue of developing work control. Additionally, cost diminishing and customization should work in parallel, as customization causes additional expense (Jones et al., 2003). The wheel of item life is likewise compressed because of evolving patterns, and now organizations need to concentrate on transient development (Lasi et al., 2014).

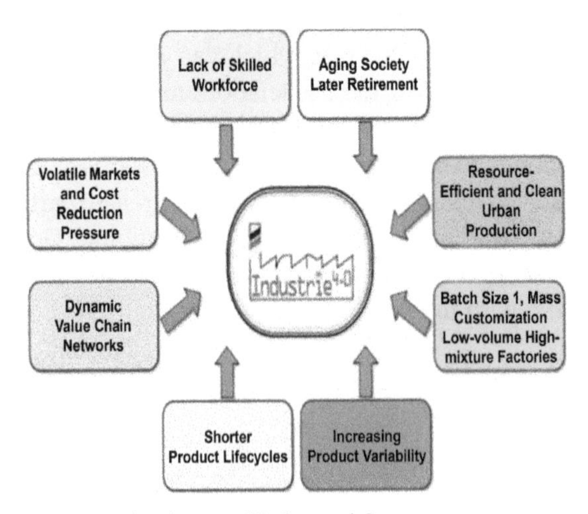

FIGURE 5.1 Socio-economic triggers of Industry 4.0.

Source: Thttp://www.slideshare.net/SPRICOMUNICA/basque-industry-40-the-fourth-industrial-revolution-based-on-smart-factories

The focal thought of Industry 4.0 is to actualize the cyber–physical systems (CPS) for generation, for example, inserted actuators and sensors, systems of microcomputers, connecting the machines to the value chain (Porter & Heppelmann, 2014). It further contemplates the propelled update and reengineering of things. Figure 5.2 is displaying a condensed thought of Industry 4.0 generation style. It is likewise portrayed by profoundly separated redid items, and well-organized blend of item and administrations, and furthermore the worth included administrations with the genuine item or administration (Lansiti & Lakhani, 2014). In basic words, Industry 4.0 should have shrewd machines, stockpiling framework, and generation office. It limits the human mediations and increment profitability. It accentuates on decentralized and profoundly robotized generation, as appeared in Figure 5.2.

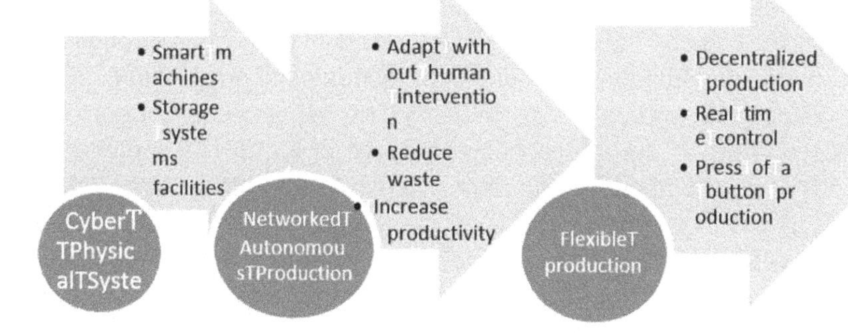

FIGURE 5.2 Idea of Industry 4.0 production style.

Store system structures in Industry 4.0 are depicted by versatile methods and high efficiency which should be cost saving just as are relied upon to offer focal points like improved organization for complex things, diminished time to market, and age on premium (Burmeister et al., 2015). Regular separation and charge authority are measured as opposing challenge techniques, yet Industry 4.0 has the test to empower them at the same time (Fleisch et al., 2014; Olschewski & Weber, 2014)T

In such an unsure business condition, there are numerous difficulties with respect to the administration as it draws near, for instance, to the plan of action development (Burmeister et al., 2015), as in the time of Industry 4.0 key accomplishment factor for certain endeavors is the headway capacity (Lasi et al., 2014). In such a space, the activity of delegates is critical who are convinced to contribute in legitimate learning, and imaginative strategies in the affiliation. Since in condition, for example, Industry 4.0, where pace of progress is quickening with more noteworthy power and recurrence, firm should be touchy to innovative requirements of clients and new kinds of contenders. To empower the representatives to work as indicated by Industry 4.0 needs pace; it is fundamental to give a climate of advancement and learning, as it is a critical facilitator of learning and innovative practices on occupation (Van der Sluis, 2004).

That is the explanation this paper offers a viewpoint about organization practices fitting for giving a climate of learning and progression to the delegates, which can empower the laborers and relationship to meet the necessities of Industry 4.0. As fitting administration practices can possibly improve dynamic abilities, which can prompt advancements (Saban et al., 2000). The greater parts of the investigations on Industry 4.0 are talking about the innovative parts of the business and endeavor. This assessment is a fundamental undertaking to draw the thought toward the best organization approaches for Industry 4.0, in order to facilitate with the pace of mechanical bounces by overhauling the environment of learning and progression.

The executive approach for Industry 4.0 is dependent upon the progression limit of enormous business (Lasi et al., 2014); either it is about CPS (e.g., embedded actuators, sensors, and PC frameworks), hing reengineering, partition, or some generation system issues. In the event that alliance should be shrewd, they need sharp operators, and condition for learning and headway which requires appropriate association rehearses. The board for Industry 4.0 is without a doubt a basic issue which relatively few individuals have inspected into. Industry 4.0 needs to make limits crosswise over various estimations in the relationship. There is a need to make capacities to effectively oversee methodologies and consider portfolios to get to potential market and clients, to improve worth chain strategies and frameworks, chance the authorities and legitimate issues, and social association by uprightness of globalization.

It is particularly certain that in Industry 4.0, associations will confront numerous monetary, social, and mechanical difficulties, which require dynamic capacities and inventive work power. Along these lines, it is critical to talk about that in what capacities organizations would be able to upgrade their abilities which prompt developments to coordinate the prerequisites of Industry 4.0. This investigation proposes a few administration practices to make the association perfect with Industry 4.0 by building up an atmosphere of learning and development, which can at last improve the hierarchical abilities. These practices are examined as follows.

5.2 ORGANIZATIONAL STRUCTURE

In the earth of quickening paces of changes, authoritative structures can assume a noteworthy job in the improvement of an atmosphere appropriate for learning and development (Van der Sluis, 2004). On a more extensive range associations can run from robotic structure to natural plan (Daft, 2015). Robotic plan is portrayed by a brought together structure, specific assignments, numerous standards and conventions, vertical interchanges, and exacting chain of command of power and it is reasonable in a steady domain, and an unbending society (Burn & Stalker, 1961), which is not the situation with Industry 4.0. Industry 4.0 is described by an insecure evolving condition, and is good with the natural structure of association which is portrayed by decentralization, strengthening, few guidelines and customs, level correspondence, and synergistic cooperation (Burn, & Stalker, 1961). This sort of configuration is progressively reasonable for advancement methodology and changing conditions (Daft, 2015). So, in Industry 4.0, while making hierarchical structure, a chief ought to stay in the natural worldview of plan. It is not sensible to suggest one single structure for Industry 4.0, as affiliations need to design the versatile structures according to their needs and conditions, no single system is suitable for every affiliation, each one has its focal points and hindrances (Daft, 2015). Anyway following are a few reasonable alternatives for Industry 4.0 condition.

Matrix structure: It implies a fundamental structure in the affiliation, where activities are balanced among more than one position line (Grant, 1996b). The system structure accumulates people and resources by limit and thing shrewd at the same time using a twofold reporting system (Jones et al., 2003; Grant, 1996b). The best thing of the system structure is that it is genuinely versatile and can rapidly respond to the need of progress (Grant, 1996b). In the system structure, each specialist needs to work with two administrators, one thing executive, and other helpful boss (Jones et al., 2003; Grant, 1996b). Cross-section structures can in like manner empower the formal interfacing parts by joining basic thinking from the thing and helpful chairman (Van der Sluis, 2004). The lattice structure of affiliations can be a better than average wellspring of coordination with the pace of Industry 4.0.

Project teams: A gathering-based structure places different techniques and limits in a solitary social event to investigate a regular objective (Grant, 1996b). It isolates the helpful and departmental tangles, quickens the essential authority process, updates generalist capacities, and supports the learning in the affiliation (Nonaka & Takeuchi, 1995; Shah & Mulla, 2013). In an uncertain space like Industry 4.0, where changes are typical from time to time and headway is a key to advance, adventure-based gatherings are a nice decision to energize learning and improvement (Van der Sluis, 2004; Aubry and Lièvre, 2010). To improve the progressions, new data and new calendars are required. Technical upgrades require new specific courses of action and the reuse of existing plans (Keller, 1992). Undertaking gatherings, especially in a learning circumstance, can be a facilitator of data sourcing, and reuse (Khedhaouria & Jamal, 2015), which is essential for advancements (Gray & Meister, 2006). Regardless, learning source and reuse for improvement in adventure gatherings is liable to accumulate part's target course (Khedhaouria & Jamal, 2015). So, adventure boss should stir the social occasion people to learn by bearing the slips up, or by grasping a capacity of arranged supervision (Fleisch et al., 2014; Lasi et al., 2014).

Flat hierarchy: Dynamic framework is the affiliation's chain of importance of initiative; it shows the authority of executives at different levels in the affiliation (Jones et al., 2003). Level structures are depicted by less level of hierarchy of leadership, and the scope of control is wide (Jones et al., 2003). This suggests that in a level structure, there are less authoritative/staff levels in chain of significance anyway number of laborers offering an explanation to one boss is ordinarily higher than a tall structure, where the quantity of levels is high. A level structure energizes speedier correspondence, and lessens the partition among laborers and the top organization (Jones et al., 2003). Thusly, the level structure fabricates the chances of delegate support in talks and essential initiative, which grows the chances of laborer learning and besides quick and noise charge contribution to top organization, by virtue of level correspondence (Daft, 2015). So, it is not unexpected to battle that a compliment various leveled structure can be immaculate with Industry 4.0, as it energizes definitive learning and advancements by growing specialist collaboration, and speedier analysis to top organization.

Decentralization: With decentralization, the situation to take decision is moved to the lower levels of affiliations. In decentralized structures, the power and learning of activities lies with the agents as opposed to supervisors or top organization (Daft, 2015). In the decentralized system, lower chiefs and nonregulatory staff have the situation to take their decisions, for example, how to use definitive resources (Jones et al., 2003). They need not mess with support from top organization. In sketchy condition where situation changes routinely, decentralization is perfect for certain affiliations. It empowers the specialist to take the propitious decision to change the course with the modification in the business condition. This kind of system supports quick decision and learning. So, it might be fought that decentralization can urge the affiliation's comparability to Industry 4.0.

5.3 LEADERSHIP STYLE

Activity bowed to impress others, awaken, push and to achieve an objective, we should targets our activities (Jones et al., 2003). Leaders fulfill their perfect targets with their partners by getting the best possible activity style as demonstrated by the condition. It is suggested by the way target speculation of activity too (House, 1971). Like as, in the once-over of most creative associations of the world, Apple Inc. is between top ones. As showed by far most of the logical examinations it is not a result of the specific aptitudes of Apple's CEO Steve Jobs, it is a result of his power capacities (Li et al., 2016), for instance, he understood how his hard work best yields from his laborers. Also, achievement of Microsoft Corporation is every now and again credited to the position style of Bill Gates. Therefore, to stimulate the strategy of progression and learning in Industry 4.0, there should be a specific activity style to be acquired. The most generally discussed power style for advancement and learning is the transformational organization style (Aryee et al., 2012; Slåtten, 2014; Afsar et al., 2014; Birasnav, 2014). Some further position styles as genuine activity (Muceldili et al., 2013) and esteem-based organization (Birasnav, 2014; Politis, 2001), Industry 4.0 wants something more than the transformational authority, which ought to be progressively express to learning and headway. As the change organization is confined to romanticizing sway, moving motivation, insightful actuation, and giving vision (Bass, 1985), Industry 4.0 needs to focus on data, learning and improvement. Thusly, effort has been made by displaying the learning centered power work, by uniting the transformational and worth-based style of organization (Donate & de Pablo, 2015). Data masterminded authority is logically express to learning and advancement; yet, simultaneously there is potential to extend the work of data arranged organization to be used in Industry 4.0., for instance, by testing and after that including innovative activity showing, invigorating data scattering, unfaltering behavior, arrangement, directing, and mentoring to the work of learning centered position. As this widely inclusive form of data organized, authority can urge the relationship, animate the pace of headway, and learning in the relationship to be great with Industry 4.0.

5.4 HUMAN FUNDS (HF) PRACTICES

HF practices are contemplate as one of the necessary sources by which associations can form the aptitudes, abilities, practices, and disposition of its workers to complete association objectives (Collins & Clark, 2003). Administrators can upgrade the creativity, information the executives limit, and learning among representatives by structuring the HF rehearses as needs be (Chen & Huang, 2009). As HF practices are basic for upper hand in a learning-based economy (Chen & Huang, 2009), HR rehearses which should be structured appropriately for development and learning are preparing, staffing, execution evaluation, pay, and occupation plan (Chen & Huang, 2009). In Industry 4.0, chiefs need to plan these HF rehearses with the expectation to advance ingenuity and learning in the association.

Training: Associations in Industry 4.0 way to structure their arrangement programs by which it can improve the creative limit and learning. Associations may offer diverse kind of preparation to the workers to empower them for performing multiple tasks. It is not fundamental that these trainings ought to be straightforwardly significant to representative employment; yet, to expand the assortment of aptitudes (Chang et al., 2011) these instructional courses ought to be progressive. Trainings ought to likewise concentrate on group building and collaboration abilities, and coaching ought to be the normal movement of directors, particularly to the new contracts (Ma Prieto & Pilar Perez-Santana, 2014). Moreover, they must prepare sessions to upgrade the critical thinking abilities of the workers (Chen & Huang, 2009).

Staffing: In Industry 4.0, enrollment must to be based on assortment of aptitudes, and mixed learning and these should be tried in the screening procedure beforehand choosing the competitor (Chang et al., 2011). Associations should devote significant exertion in choosing the correct possibility for each activity by utilizing broad enlistment and determination techniques (Ma Prieto et al., 2014). For instance, to procure imaginative workers, selection representatives should concentrate on recognizing the properties fundamental for inventive conduct, for example, receptiveness to encounter, which can be assessed via psychometric testing in the determination procedure as receptiveness to new knowledge is described by dynamic creative mind, internal inclination mindfulness, assortment inclinations, scholarly interest, innovativeness, and adaptable reasoning (Costa & McCrae, 1992; Barrick & Mount, 1991). Besides, individuals who are profoundly exposed to new knowledge demonstrate progressively inspirational frame of mind toward learning (Barrick & Mount, 1991). During the time spent enlistment and choice, associations ought to likewise assess the objective direction of the applicant, which can be learning direction and execution direction. For advance development and learning in the association, spotters ought to incline toward up-and-comers with high learning direction as workers with learning objective direction like to take part in testing undertakings and are anxious to improve themselves (Button et al., 1996), to build up another arrangement of abilities, and will in general accomplish dominance (Kim & Lee, 2013).

Compensation: The remuneration framework in Industry 4.0 mirrors the commitment of representatives to the organization. Representatives should get the pay dependent on individual, gathering, and hierarchical execution (Ma Prieto & Pilar Perez-Santana, 2014). There must be a connection among execution and the reward, for example, benefit sharing and extra motivator pay (Chen & Huang, 2009). Such a remuneration framework can possibly encourage the atmosphere of development and learning in the associations (Chen & Huang, 2009; Ma Prieto & Pilar Perez-Santana, 2014).

Performance evaluation: A presentation examination framework which can suit Industry 4.0 should be centered around worker improvements, outcome-based methodology, and conduct-based methodology, as these methodologies can encourage knowledge and advancement (Chen & Huang, 2009). Representatives should get the input on their exhibition on repetitive premise. Moreover, the exhibition evaluation should be progressively objective, for example, there must exist frameworks to assess the presentation quantitatively. A perfect examination procedure incorporates the foundation of execution guidelines, conveying the desires, estimating the genuine presentation, contrasting the real execution and the principles, talking about the evaluation with the worker, and starting the remedial activity where vital (Lee & Kelley, 2008). MBO can be clarified as "A presentation examination technique that incorporates common target setting and assessment dependent on the fulfillment of the particular destinations" (Lee & Kelley, 2008). A commonplace MBO program is described by explicit objectives where goals are succinct articulations of anticipated results. For participative basic leadership, directors do not dole out the targets to the workers singularly. Objectives are not forced in the MBO program, supervisors and workers set the objectives and the approaches to accomplish the objectives by shared talk and accord. Timetable is additionally characterized for every objective, and there is progressive input in the MBO program. Progressive input enables directors and representatives to screen the exercises and make the remedial move as need be (Lee & Kelley, 2008). MBO is a decent method of execution evaluation to be perfect with Industry 4.0.

Job design: It is depicted as "the manner in which the position and the undertakings inside that position are sorted out, including how and when the assignments are done and any elements that influence the work, for example, in what request the errands are finished and the conditions under which the assignments are finished" (Lee & Kelley, 2008). Occupation configuration to advance the atmosphere of development in knowledge ought to be portrayed by occupation revolution, adaptable projects in numerous territories, the broad exchange of errands and obligations to the workers. Moreover, employment configuration ought to encourage cooperation and joint effort, and requiring aptitudes assortment (Ma Prieto & Pilar Perez-Santana, 2014). In the business 4.0, where conditions are portrayed by change and advancement, such work configuration can assist the association with adjustments as indicated by the business condition.

5.5 CENTERING MOMENTRY DEVELOPMENTS, YET LONG HAUL ABILITIES

Nature of the ventures in Industry 4.0 is described by short formative periods (Lasi et al., 2014). It does not imply that associations should not think about the more extended term viewpoints. As the pace of progress in mechanical, social, financial, and world of politics is high in the Industry 4.0 (Lasi et al., 2014), the advancements in them will not continue over a more drawn out period. Association should make the advancement procedure a piece of schedule, by building up the long-haul capacities in workers, for example, by building up the imaginative work conduct and upgrading the learning the executives rehearses in the association, which can possibly decidedly impact ingenuity (Donate & de Pablo, 2015). By embracing the correct administration rehearses, associations can build up the dynamic capacities for development (Saban et al., 2000). In straightforward words, associations and representatives should be skilled enough to alter their course as indicated by the evolving circumstances.

5.6 READINESS TO RELINQUISH SPECULATION AND LEARNING

As talked about that in the faulty business 4.0 condition, formative periods and progress periods are consolidated (Lasi et al., 2014), so there is a need to evacuate the regular style of endeavor (Herrmann et al., 2006). To empower the headway techniques, affiliations should be glad to surrender their present hypothesis and data, if required. Instead of using out of date learning and other resources, affiliation should increase new data, make an understanding of the obtained data into focus capacity, and a short time later develop new things reliant on the middle wellness (Hermann et al., 2006). Learning and progress are the basic achievement factor in Industry 4.0, and once in a while it foresees that excitement should leave information, experience, and speculation to suit new advancement (Hermann et al., 2006). For instance, there are chances that to spare the present learning and experience, alliance may overlook the new frameworks for working or new headway, which can incite an increasingly significant calamity.

5.7 CONCLUSION

The fundamental objective of this paper is to offer a viewpoint and propose the best organization practices for the associations preparing for the fourth current change. The business condition of Industry 4.0 is talked about, which is dubious and unsteady. At such a juncture, the significant objectives and difficulties of Industry 4.0 are examined such as keen assembling, usage of CPS for creation, inserted actuators and sensors, systems of microcomputers, connecting the machines to the worth chain (Porter & Heppelmann, 2014), advanced improvements and reengineering of items, profoundly separated tweaked items, well-organized mix of item and administrations, and furthermore the value included administrations with the real item or administration (Lansiti & Lakhani, 2014), proficient store network (Burmeister et al., 2015), and empowering cost initiative and separation at the same time (Fleisch et al., 2014; Olschewski & Weber, 2014). Based on the contention that accomplishment in Industry 4.0 is reliant on the advancement ability of big business (Lasi et al., 2014), this investigation offers perspective on reasonable administration works that include hierarchical structure, authority, and HR rehearses. In addition, his paper likewise stresses on the need of transient development, long haul abilities, and the readiness to surrender speculation and learning, whenever essential.

This paper additionally offers course for future research on the administrators' practices with respect to Industry 4.0. Definite and quantitative research later on can support the conflicts made in this assessment subject to the compromise of composing and bases. For future research, converse with essential examination to endorse the dispute is required concerning Industry 4.0 trailed by a survey-based audit to test the revelations by quantitative methodology. Unit of examination ought to cut edge adventures drew in with sharp gathering, and the execution of CPS.

All in all, this examination offers suggestions for chiefs and endeavors to receive suitable administrative ways to deal with and endure to develop in the fourth mechanical upset. This investigation additionally gives suggestions to the analysts by offering a hypothetical structure for the upcoming research. Industry 4.0 requires savvy assembling and keen business tasks, which require developments. Development is subject to individuals' ability which is encouraged by learning and information. In the event of inconsistency, associations need to reconsider and upgrade their administration draws. That is the reason, it is similarly critical to talk about the administration approaches for Industry 4.0 alongside the innovative and profoundly logical examinations. The marvel how fitting administration practices can prompt learning, upgrade abilities, develop the abilities to meet the difficulties of brilliant assembling and business activities, and similarity with Industry 4.0, which likewise speaks profoundly on the subject and principle thought of this paper. These practices give a domain and atmosphere that are reasonable for adapting new abilities to address the necessities and difficulties of Industry 4.0. Learning and information enables the executives build the ability of the representatives by making them increasingly inventive and creative (Chen & Huang, 2009). Progressively innovative and inventive specialists will be in better position to contribute in keen gathering and business exercises which are the rule characteristics of Industry 4.0, and advancement limit is one of the essential thought required for progress Industry 4.0 (Lasi et al., 2014). Along these lines, suitable administration practices can make the association good with Industry 4.0 by encouraging picking up, improving ability, advancement, and brilliant assembling and business activities.

KEYWORDS

- **challenges**
- **Industry 4.0**
- **management practices**
- **organizational structure**
- **leadership style**
- **HR practices**

REFERENCES

Afsar, B., Badir, F.Y., & Saeed, B.B. (2014). Transformational leadership and innovative work behaviour. *Industrial Management & Data Systems, 114*(8), 1270–1300.

Aryee, S., Walumbwa, F.O., Zhou, Q., & Hartnell, C.A. (2012). Transformational leadership, innovative behaviour, and task performance: Test of mediation and moderation processes. *Human Performance, 25*(1), 1–25.

Aubry, M., & Lièvre, P. (2010). Ambidexterity as a competence of project leaders: a case study from two polar expeditions. *Project Management Journal, 41*(3), 32–44.

Barrick, M.R., & Mount, M.K. (1991). The big five personality dimensions and job performance: A meta-analysis. *Personnel Psychology, 44*, 1–26.

Bass, B.M. *Leadership and Performance Beyond Expectations*; The Free Press: New York, 1985.

Birasnav, M. (2014). Knowledge management and organizational performance in the service industry: The role of transformational leadership beyond the effects of transactional leadership. *Journal of Business Research, 67*(8), 1622–1629.

Burmeister, C., Luettgens, D., & Piller, F.T. (2015). Business model innovation for Industry 4.0: Why the industrial internet mandates a new perspective on innovation. *Die Unternehmung: Swiss Journal of Business Research and Practice, 70*(2), 124–152.

Button, S.B., Mathieu, J.E., & Zajac, D.M. (1996). Goal orientation in organizational research: a conceptual and empirical foundation. *Organizational Behaviour and Human Decision Processes, 67*(1), 26–48.

Chang, S., Gong, Y., & Shum, C. (2011). Promoting innovation in hospitality companies through human resource management practices. *International Journal of Hospitality Management, 30*(4), 812–818.

Chen, C.J., & Huang, J.W. (2009). Strategic human resource practices and innovation performance—The mediating role of knowledge management capacity. *Journal of Business Research, 62*(1), 104–114.

Collins C.J., & Clark K.D. (2003). Strategic human resource practices, top management team social networks, and firm performance: The role of human resource in creating organizational competitive advantage. *Academy of Management Journal, 46*(6), 740–751.

Costa, P.T.J., & McCrae, R.R. *Revised NEO Personality Inventory and NEO Five-Factor Inventory Professional Manual*; Psychological Assessment Resources: Odessa, FL, 1992.

Daft, R. *Organization Theory and Design*; CENGAGE: Boston, MA, 2015.

Donate, M.J., & de Pablo, J.D.S. (2015). The role of knowledge-oriented leadership in knowledge management practices and innovation. *Journal of Business Research, 68*(2), 360–370.

Fleisch, E., Weinberger, M., & Wortmann, F. Business Models and the Internet of Things, Ph.D. Thesis, Bosch Internet of Things & Services Lab Universität St. Gallen, Aug. 2014.

Grant, R.M. (1996b). Toward a knowledge-based theory of the firm. *Strategic Management Journal, 17*(10), 109–122.

Gray, P.H., & Meister, D.B. (2006). Knowledge sourcing methods. *Information & Management, 43*(2), 142–156.

Herrmann, A., Tomczak, T., & Befurt, R. (2006). Determinants of radical product innovations. *European Journal of Innovation Management, 9*(1), 20–43.

House, R.J. (1971). A path goal theory of leader effectiveness. *Administrative Science Quarterly, 16*(3),321–329.

Jones, G.R., George, J.M., & Hill, C.W. *Contemporary Management;* McGraw-Hill/Irwin: New York, 2003.

Keller, R.T. (1992). Transformational leadership and the performance of research and development project groups. *Journal of Management, 18*(3), 489–501.

Khedhaouria, A., & Jamal, A. (2015). Sourcing knowledge for innovation: Knowledge reuse and creation in project teams. *Journal of Knowledge Management, 19*(5), 932–948.

Kim, T.T., & Lee, G. (2013). Hospitality employee knowledge-sharing behaviours in the relationship between goal orientations and innovative work behaviour. *International Journal of Hospitality Management, 34*, 324–337.

Lansiti, M., & Lakhani, K. Digital Ubiquity: How Connections, Sensors, and Data Are Revolutionizing Business. In *Harvard Business Review*, Harvard Business Publishing: Boston, MA, 2014.

Lasi, H., Fettke, P.D.P., Kemper, H.G., Feld, D.I.T., & Hoffmann, D.H.M. (2014). Industry 4.0. *Business & Information Systems Engineering, 6*(4), 239–242.

Lee, H., & Kelley, D. (2008). Building dynamic capabilities for innovation: An exploratory study of key management practices. *R&D Management, 38*(2), 155–168.

Li, N., Long, X., Tie, X., Cao, J., Huang, R., Zhang, R., ... & Li, G. (2016). Urban dust in the Guanzhong basin of China, part II: A case study of urban dust pollution using the WRF-Dust model. *Science of the Total Environment, 541*, 1614–1624.

Ma Prieto, I., & Pilar Perez-Santana, M. (2014). Managing innovative work behavior: The role of human resource practices. *Personnel Review, 43*(2), 184–208.

Muceldili, B., Turan, H., & Erdil, O. (2013). The influence of authentic leadership on creativity and innovativeness. *Procedia-Social and Behavioral Sciences, 99*, 673–681.

Nonaka, I., & Takeuchi, H. *The Knowledge-Creating company: How Japanese Companies Create the Dynamics of Innovation*; Oxford University Press: New York, 1995.

Olschewski, F., & Weber, M. (2014). Geschäftsmodelle der Industry 4.0, *In: Inspect, 5*(1), 535–547.

Politis, J.D. (2001). The relationship of various leadership styles to knowledge management. *Leadership & Organization Development Journal, 22*(8), 354–364.

Porter, M., & Heppelmann, J.E. How Smart, Connected Products are Transforming Competition, In *Harvard Business Review;* Harvard Business Publishing: Boston, MA, 2014.

Saban, K., Lanasa, J., Lackman, C., & Peace, G. (2000). Organizational learning: A critical component to new product development. *Journal of Product & Brand Management*, 9, 99–119.

Shah, T., & Mulla, Z.R. (2013). Leader motives, impression management, and charisma: A comparison of Steve Jobs and Bill Gates. *Management and Labour Studies*, *38*(3), 155–184.

Slåtten, T. (2014). Determinants and effects of employee's creative self-efficacy on innovative activities. *International Journal of Quality and Service Sciences*, *6*(4), 326–347.

Burn, T., & Stalker, G.M. *The Management of Innovation*; Tavistock Publications: London, 1961.

Van der Sluis, L.E. (2004). Designing the workplace for learning and innovation: Organizational factors affecting learning and innovation. *Development and Learning in Organizations: An International Journal*, *18*(5), 10–13.

CHAPTER 6

Recent Reforms in the HR 4.0 Industry

GUNEET KAUR MANN

Jaipuria Institute of Management, Ghaziabad 201014, Uttar Pradesh, India

**Corresponding author. E-mail: guneet@jaipuria.edu.in*

ABSTRACT

Rapid response to employee analytics involves taking a call into the employee requirement for concentration, time-off, and career development initiatives using HR software. Companies are trying to transform employee engagement into employee experience to create an ecosystem to inculcate engagement, culture, and performance management. Artificial intelligence tools have taken over the workplace activities related to recruitment and other functions. Digital solutions are the latest resource for changing HR thinking strategies. Another significant technical discipline is to develop interactive dashboards that bridge the gap between middle and senior managers. The simple format of gamification has come out to be a great source for talent motivation and retention. Virtual teams at different geographical locations improve productivity as well as employee satisfaction. Flexible work arrangements like managing the "gig economy" helps to identify demand hiring and lowering the costs associated with human resources. Talent sourcing practices decouple the limitation of location and its association with productivity. In this era, where we breathe technology, how can any organization defy the colossal existence of a digital transformation reform in the coming industrial setup?

6.1 INTRODUCTION

In today's era, when the world is upsurged with technology and catalyzing on the digital revolution why would human resource management (HRM) be left behind. The central role of HR is to reshape the traditional standard operating procedures and formulate innovative systems to cater to the ever-changing workforce requirements. Referring this change as reforms in the new-age industries, it is a productive approach toward result-oriented organizations. These reforms are instrumental in effectively bringing out change and also managing ambiguities, stress, and problems inherent within the processes of change. The workable areas under HR reforms are contractual and service conditions, workforce planning, talent management, performance management directed toward a fully integrated people strategy, and a new HR delivery model. Apart from these function-led activities, an enhanced cultural development will have to support the creation of an engaged workforce leading to an inclusive work environment. The leadership in such a scenario would be passionate about making a difference in the organization with their integrity, fair, and impartial practices in addition to being results-oriented, proactive, adaptable, and resilient. Focusing on the multifunctional role of HR, the cooperation of all subsystems working within the organization would require an overhaul. Improved management will definitely endeavor to create a positive difference in effectiveness, efficiency, and sustainability. The idea is to target management of strategic resources, infrastructure, change, and employee management. Hence, today's human resource function delves into various realms like digitalization of processes, optimizing data-based policies, and not to mention the gamification framework. Talent trends have to be carefully studied to enhance the penetration of artificial intelligence (AI) and machine learning. These recent practices have led to an advanced form of HRM which goes beyond the deterministic capabilities with a great implication on economics, culture, and policies.

6.2 THE PEOPLE ANALYTICS

Post the establishment of the psychometric era, today the main area of focus for human capital is the *Psychology of People Analytics*. Over a world, significant change has occurred in employment trends with increased emphasis on digitization and the penetration of technology across sectors. This has resulted in creation of a demand in skilled workforce and added on to the pressure of recruiters. The objective of HR analytics is to provide an organization with appropriate insights for organizational management and strategic management (Levenson, 2017). In an age, where products can be imitated and evolved, the main focus of differentiation is the human factor.

The other ancillary agents being vendors, business associates, supply chain partners, and many others. HR would need to equip itself with the tools and resources that empower them with in-depth comprehension of behavior patterns, opportunities, and challenges (Kumar, 2016). HR analytics or people analytics is formally defined as the mathematical, statistical, and data mining approach toward HRM (Kumar, 2016). It is highly aligned with the business data and corresponds to every change that happens in the organization. All inclusive, people analytics may be used to look at the manpower quality, in terms of its education, experience, knowledge, skills, and everything that leads to a developed organization along with social harmony. It will help in weeding out fake applicants enabling the HR to take informed decisions related to recruitment. It enables managers to comprehend more about the complexities of people factor like their performance which creates value to the organizations. But, one of the prime concerns of people analytics is which data is pertinent, which should be captured, and which should not be. And, further how this data can be modeled to predict capabilities which lead to optimal returns (Kumar, 2016). The prolonged debate on the quantification of HR functions, finds its answers in people analytics. According to data researched by first advantage across India in Q3 from background verifications and reference checks, highest discrepancy between skilled and unskilled workforce was found in banking, financial services, and insurance sector at 35% followed by IT at 29%. Another survey by Kelly Services reported that across all professional/technical sectors, 60% of managers globally, lack the right combination of hard and soft skills (KGWI, 2016). Following the pattern, 60% of millennials are concentrating on upskilling themselves due to their upbringing within technological purview and disruptive business models (KGWI, 2016). Rather than using self-report measures and ratings from peers and supervisors, the use of technology-driven psychometric assessment has induced more scientific insights into people analytics (Chadha, 2018). Today's tests allow immersive and realistic simulations and gaming providing a platform to exhibit knowledge, attitudes as well as behaviors relevant for the job scenario (Chadha, 2018). It has assisted HR not only in selection of apt candidates, but also in personnel development, team building and development, and a systematic progression in career. But, HR analytics is not sufficient; the data horizon goes beyond and covers financial and other business data. Today, the people data can be used maximum for predicting employee churn, performance, return on investment, or long-term workforce planning (Chadha, 2018). This means ensuring that right candidates are selected, top performers are rewarded and retention is strong. For example, an airline can capture the number of repeat customers and their flying pattern to develop a model which can predict and diagnose problems, pinpoint training solution, and also aggravate sales. It supports the top managers in understanding, attrition, recruitment metrics, labor cost, and employee engagement by area, business unit, or across levels. Endless permutations and combinations of data lead to better results and strategy designing (Guzzo, 2015). There are four categories under people analytics—descriptive, diagnostic, predictive, and prescriptive according to the utility of data.

Descriptive analytics looks into the present scenario, summarizing the data into understandable information through charts and reports and also representing trends over a period of time (Chadha, 2018). Using data aggregation and data mining, it is used to convert data into meaningful findings. Trends can be studied in customer behaviors and their habitual actions to be used for marketing strategies and customer service. Diagnostic analytics depicts the reasons (why factor) behind the occurring of descriptive analytics. If the root cause is known, it becomes subtle to make efforts in the right direction and lessen the problematic effect. Few techniques that use diagnostic analytics are attribute importance, principal components analysis, sensitivity analysis, and conjoint analysis. Predictive analytics emphasizes on the futuristic view of a business. It tries to predict on the basis of past events. The techniques used in predictive analytics are data modeling, machine learning, and AI. Moreover, the common used tools for this type of analytics are Python, R, RapidMiner, etc. Prescriptive analytics is the most powerful decision options among all these as it suggests favorable outcomes and options to be taken on the basis of predictions. Because of aspects like AI, this system uses a strong feedback that constantly learns and updates the relationship between actions and outcomes. The scientific method of predicting and prescribing data allows organizations to create more opportunities for themselves and choose the best possible actions required. These decisions are relatively linked to organizational goals like increased revenue, cost optimization, and justifications of departmental decisions.

6.2.1 HR FUNCTIONS CAPITALIZING ON PEOPLE ANALYTICS

The core functions within an organization related to HR are talent acquisition, employee engagement, employee development, and compensation. Some of the important by-products of people analytics are skill investments, training interventions, and employee assessments (Fitz-Enz, 2010). Let us see how people analytics assists in gathering relevant information to support the issues and problems affecting managers.

(I) *Employee management:* Common HR metrics include voluminous data on demographics, performance, job analysis, compensation, and training. It can be bifurcated as follows:

- Employee data: background, demographics, skills, and engagement.
- Systematic data: attendance, adoption, training programs, development programs, leadership programs, and project outcome assessments.
- Performance data: ratings, performance evaluation scores, and succession programs.

(II) *Talent management:* Starting from recruitment, to onboarding to engagement of employees, trackers can be the key to minimizing costs and maximizing output. Even, inputs can be extracted from fresh incumbents on the organizational culture at the time of induction and converted into metrics. The information can be used for preventive measures to control attrition. Employee satisfaction can be achieved even with less budget and high engagement. But the ways need to be innovative. Inclusions in talent management trackers would be the following:

- Hiring cost per hire
- Offer acceptance ratio
- Department wise attrition rates
- Causes of attrition and absenteeism
- Employee engagement level
- Training cost per employee
- Learning and development opportunities.

With the help of these HR metrics, it would become easy to tap efficient recruitment sources, plan for future successions, and identifying absenteeism patterns and leaves to control costs and improve productivity (Anderson, 2004).

(III) *Compensation and benefits:* In this area of HRM, reward and compensation is often the least analyzed expense for an organization. Most of the times, it is beyond numbers and secretive nature of remuneration makes it a bone for contention. The key areas to be manager in this are as follows:

- Cost per full-time-equivalent (FTE)
- Average remuneration
- Satisfaction with performance management
- Over paid employees
- High performers who leave after appraisal.

Analyzing these parameters will diagnose the compensation system of the organization and keep a track of its impact on retention, detect, and prevent compensation problems.

(IV) *Employee engagement and productivity:* An engaged employee is a happy employee. He puts his heart and soul into the work that he performs, which in turn improves the effectiveness as well as efficiency. HR needs to chase the versatile factors that impact employee engagement like organizational culture, team dynamics, management styles, and work life balance. The key metrics for this parameter are as follows:

- Target achievement
- Profit per FTE
- Employee productivity index
- Performance differential rate
- Healthcare cost per employee
- Benefits cost per employee

These kinds of people analytics rather than HR analytics serve as a better source of alignment with company's goals and objectives. Diagnostic and predictive analytics can bring out the amount of work outside of normal working hours, percentage participation in ad-hoc meetings and initiatives, hours per week in meetings, and manager's time spent with his team.

(V) *Cognitive biases:* The working of human mind can sometimes be affected by the situation or perspective toward the problem. The true picture is hazy at times and in no sense draws the required concentration. But, a machine will never fail to deliver its performance and will not base the conclusion on perceived character, personality, intentions, or efforts rather focus on context, opportunities, or constraints.

Some of the major cognitive biases that limit the authenticity of brain are as follows:

- *Algorithm aversion*: People tend to follow their intuitions more than trusting algorithms.
- *Assuming a normal distribution*: Taking the fact that human behavior is also bell shaped can delimit the judgment about human capabilities. People analytics can help create an unbiased view of the world by churning data in the most insightful manner.
- *Hindsight bias*: This bias is related to the "story bias" and the "overconfidence effect" as we all tends to like good coherent stories and believe that we are too good at making predictions.
- *The fallacy of one cause*: Thinking that one reason is holistically important for understanding a cause leaves room for undetected and ignored aspects of a situation.
- *Implications for people analytics*: To understand the reasons of employee behavior and attitude, it is required that causes have to be studied. Like what a good leader should be like and how and why people leave organizations. People analytics team needs to introspect and find out the complex reality in a simple way.

6.2.2 CHALLENGES FACED IN USAGE OF PEOPLE ANALYTICS

(i) *Ethical issues*: The data collected through people analytics would be put to right use is an apprehension with the employee. Like information about an employee's health would not be put against his employment and he would be assured protection against any harm is a major concern of the employee. HR analytics needs to be used with a high level of clarity and transparency in order to attain valid responses from employees and diffuse their fear of breaking trust.

(ii) *Analyzing immaterial data*: Employers may be tempted to check and monitor each and every piece of data whether it can be mined into useful information or not. This could lead to wastage in time and resources and losing out on the larger picture.

(iii) *Data usage*: There is a possibility that employers might use data for a specific purpose or motive. Predictive performance analytics uses internal data to help assess potential employee turnover. Moreover, the same set of data may be utilized for decisions related to retrenchment and promotion. But that could be a concern, as the real purpose of analytics will fail if it is put to every situation or circumstances. Managers may deviate from the exact objective while consulting data for several purposes which may prove to be reverse in outcome.

6.3 GAMIFICATION OF HR

Trying to understand how organizations keep up the enthusiasm and fun-visualization for the millennial workforce, gamification of processes has touched the lives of employees' competitiveness. For example, LinkedIn when asks its members to complete their profile details, they give a percentage which shows the profile strength, skill endorsements in a gamified way. Gamification is a process of exploiting gaming elements and activities like winning badges, earning points, and topping leaderboards, in a nongame environment. It is a new way to appeal to the individual's sense of competition and desire for recognition (Mekler, 2017). Just as Facebook satisfies the psychological satisfaction of an individual to be liked and endorsed, gamification techniques tend to promote the competency acumen in an employee.

6.3.1 OCTALYSIS

The basic framework of gamification is based around a game mechanics theory of octalysis. According to the one and only gamification guru and internationally acclaimed keynote speaker Yu-kai Chou, people have feelings, insecurities, and reasons why they want to do some things and to get the job done quickly the old-aged "function-focused" systems are too old school for them. Gamification optimizes their feelings, motivations, and even engagement. As per the right-brain theory, people do not necessarily require a goal or reward to use their creativity, have fun with friends as these activities are self-rewarding. Organizations like Facebook, LinkedIn use a combination of left- and right-brain drivers. The second level of octalysis optimizes the experience of an employee through four phases of a player's journey, which are as follows:

- Discovery: The reason behind any one even starting something.
- Onboarding: Making the users learn the rules and tools of the game.
- Scaffolding: Repetitive actions toward achieving a goal.
- Endgame: How to make your experts remain with you.

The third level onward, octalysis starts factoring in different types of players using different motivational tactics for each. In total, there are five levels; however, many organizations usually suffice with level 1 for a better HR experience (Economou, 2015). Gamification has enhanced the organization's effectiveness, employee engagement, and strengthens their employer brand. The area is yet to be researched, but HR and gaming have started to get together for some time now. As per the Society for Human Resource Management, there are two types of gamification, which are as follows:

- Structural gamification: Applying gaming rewards such as badges, levels, leader boards, etc., to job-related activities.
- Serious games: Where a simulation is created for specific purposes such as training or sales simulation.

Some areas where gamification has been tried and tested are:

1. *Administrative work:* Documentation work done at the time of induction seems to be tedious and boring. But with gamifying the process, like reward points for timely completion of these activities encourage efficiency and compliance. For example, in Google when employees were late in submission of their travel bills, it gamified the expense process by letting employees who did not spend their whole allowances to decide the fate of the remaining money to be paid in their next paycheck, saving it toward a future trip or donating it to charity of their own choice. This initiative brought forward a 100% compliance within 6 months of launching the program.
2. *Recruitment and selection:* There is a term named as "recruitainment" for integrating gamification into the process of recruitment. It may include quizzes related to industries, challenges, company related quests, and behavioral quizzes. This enhances a greater pool of candidates for selection as it involves a stimulated work environment. Organizations can comprehensively test their candidates' abilities, competencies, creative thinking skills, aptitude, and cognitive skills. This results in well-managed costs and time. This happened in Marriott International Inc., which developed a hotel-themed online game similar to Farmville to match prospective employees with Marriott as an organization and its culture.
3. *Training and development:* Inculcating levels to gain badges or points through games can take employees to the leaderboard competition,

and get trained on the job and cross functionality. And, then recognizing the leaders at award functions will motivate them toward a self-learning culture in the organization. For example, the Deloitte Leadership Academy has trained around 10,000 + executives from all over the world through gamification to increase the knowledge level and sharing. These games allow creation of a real-life environment with built-in networking features, enabling the development of massively multiplayer online role playing games. It results in greater collaboration and cooperation and reduces dreariness involved in just training.

4. *Engagement and retention:* Some of the areas of employee engagement where gamification can be implemented is employee wellness. Like there is, Mindbloom's Life Game, which is used by Aetna, a free online social game aimed at improving employee health and wellness by encouraging them to interact with a metaphorical "self." Through this, employees can keep a track of their health by choosing and developing plans to foster wellness. Sometimes, a team game is created to encourage participation and enhance team spirit. Some companies gamify fitness goals to motivate them toward optimal health choices, decreasing insurance costs, enriching organizational culture, and lead to employee welfare.

6.3.2 ADVANTAGES AND DISADVANTAGES OF GAMIFICATION

Gamification though not a complete solution to HR problems, but can act as an interesting method of implementing training and other systems within the organization. It cannot give for sure solutions but can support in enhancing employee satisfaction and communication. The gamification has to be thoughtfully programed so that the main objective remains learning and not false incentives. Also, implementing gamification techniques should not mean that other techniques become useless. A lot of employees may not find it effective and hence like other ways of storytelling, training, etc., gamification also has a demotivating effect if it is based on money alone. All we can conclude is that gamification has been able to deliver a lot of effectiveness when it comes to learning and developing employees, even recruiting and engaging employees, but it has to be implemented with a lot of care and caution as there are some downsides to it.

6.4 DIGITAL LEADERSHIP

Wearing the leadership hat in a digital world, needs quite a lot of business acumen these days. It requires the way of understanding not only money, but HR, operations, business, and law (Sheninger, 2019). Taking into account today's erratic changes, organizations are coping up with the digitalization in order to become data-strong and focus on transforming their business processes. Some of the key skills needed in this digital age are discussed below.

1. *Proactive:* To adjust the business processes with the changing environment, leaders of today are expected to keep pace with the remodeling and stay in market. Today's slogan is "evolve or evaporate." So, the leaders also need to be digital-literate in the sense of technology to be used and what will be the roadmap to adopt it. They should be ready to experiment and also face the challenges with regard to failures. The interest of personnel or workforce should be the priority when it comes to adapting any new technology as the success rate of usability of technology lies with the end users.
2. *Proficient:* The new age manager should be well acknowledged with the latest digital processes which are expected to be of the highest level. He should be a change agent for the organization to build a technology-based platform.

6.5 AI AND HR

AI is referred to the use of an array of technologies that allow a machine like computer to act in a highly cognitive manner including decision-making. The subsequent option after data analytics is applying AI to the processes of an organization. Although, in marketing the application has been very much relevant, it follows a slow pace in the area of HRM (Rana, 2018). Today, many of the menial tasks of HR function are done by software and AI can contribute to massive changes in managing the workforce, HR planning, and employee engagement. Some of the ways in which AI can redesign the domain and take it to a higher level are as follows:

1. *Recruitment:* Talent acquisition software can eliminate monotonous work in the process of shortlisting candidates by scanning, screening, and filter appropriate applications in a very less time. This facilitates

recruiter to focus more on the quality of applicants, leading to a good hiring decision as well as a low cost per hire.

2. *Onboarding:* The process of inducting employees into the organizational system has been the centre of emphasis for many managers. A lot of efforts are needed to pay attention to the new hire and his/her anticipations. AI takes over when managers lack the time to be given to every new joining. A systematic machine based onboarding procedure can position the organizational policies and guidelines in the mind of the new worker. This has led to increase in retention rates and low absenteeism and employee turnover.
3. *Training:* With the advent of technological changes occurring at a higher pace, improving employee's skills and knowledge is a challenge for the HR department. While they are adjusting and coping up with the environmental changes, there is an equal and urgent requirement of continuous learning and development of employees by the HR. AI can very efficiently plan, organize, and execute training programs for workers. Techniques like online courses and digitized material-based learning have enabled organizations to sufficiently provide solutions for performance or competency gaps. AI helps employees to learn at their own pace and fit classes or lessons according to the preferences of individual employees.
4. *Performance analysis:* Every organization strives to possess employees who are engaged and committed toward organizational strategies and values. They are constantly trying to assess and monitor the behaviors, results, and attitude of employees toward their jobs and goals. Using AI, managers can implement an effective performance appraisal system, wherein it is easy to measure key performance indicators. AI tools help set objectives and standards for employee performance and allow them to be measurable. It gives a complete and holistic performance review of every staff member and department. Such a system will be able to identify performance lacunae and detect team members who lack in skills or efforts required for a job.
5. *Retention:* Considering the volatile market environment and everchanging organization policies, keeping best employees is an uphill task. Herein also, AI plays an important role by analyzing and predicting employee needs. It assists in diagnosing individual aspirations and desires and tries to provide appropriate motivation to employees. It could be monetary or nonmonetary benefits which

are to be bestowed upon personnel to keep them running in the organization. This is a proactive as well as preventive approach toward managing people and solving problems even before they arrive.

KEYWORDS

- **digital transformation**
- **employee experience**
- **gamification**
- **HR analytics**
- **people analytics**
- **talent management**

REFERENCES

Anderson, M. W. (2004). The metrics of workforce planning. *Public Personnel Management*, *33*(4), 363–378.

Angrave, D. & Charlwood, A. (2016). HR and analytics: Why HR is set to fail the big data challenge. *Human Resource Management Journal*, *26*(1), 1–11. https://doi.org/10.1111/1748-8583.12090

Chadha, S. (2017). Psychometrics: The science of measuring the human mind, *20*(9).

Chadha, S. (2018). The psychology of people analytics, human capital, *22*(2).

https://www.academia.edu-gamification-making-work-fun-or-making-fun-of-work

Economou, D., Doumanis, I., Pedersen, F., Kathrani, P., Mentzelopoulos, M., & Bouki, V. (2015, July). Evaluation of a dynamic role-playing platform for simulations based on Octalysis gamification framework. In Intelligent Environments (Workshops) (pp. 388–395).

Fitz-Enz, J. (2010). The New HR Analytic Predicting the Economic Value of Your Company's Human Capital Investments.

Guzzo, R. A., Fink, A. A., King, E., Tonidandel, S., & Landis, R. S. (2015). Big data recommendations for industrial–organizational psychology. *Industrial and Organizational Psychology*, *8*(4), 491–508.

Kumar, R. (2016). HR analytics: Where do we stand?, *20*(5).

Levenson, A., & Fink, A. (2017). Human capital analytics: too much data and analysis, not enough models and business insights. *Journal of Organizational Effectiveness: People and Performance*.

Marler, J. H. & Boudreau, J. W. (2016). An evidence–based review of HR analytics. *The International Journal of Human Resource Management*, *28*(1), 3–16. https://doi.org/10.1080/09585192.2016.1244699

Mekler, E. D., Brühlmann, F., Tuch, A. N., & Opwis, K. (2017). Towards understanding the effects of individual gamification elements on intrinsic motivation and performance. *Computers in Human Behavior*, *71*, 525–534

Rana, D. (2018). The Future of HR in the Presence of AI: A Conceptual Study. The Future of HR in the Presence of AI: A Conceptual Study (November 24, 2018).

Sheninger, E. (2019). *Digital Leadership: Changing Paradigms for Changing Times*. Corwin Press.

CHAPTER 7

Behavioral Finance for Financial Acumen

DINESH KUMAR SHARMA and SHIV RANJAN*

Gautam Buddha University, Greater Noida, 201308, Uttar Pradesh, India

**Corresponding author. E-mail: sranjan@gn.amity.edu*

ABSTRACT

In behavioral finance, we deal with repeated biases, heuristics, and pricing inefficiencies present in financial market with limited, uncertain, risky information having time constrains, and strategic in nature. The psychological biases, emotions, stress, and individual differences influence financial decisions which further impact on financial acumen. So we need to understand how behavioral finance impact on decision-making and can we control these impacts. This study will analyze how individuals process financial information in their brain and how and what decisions arise within the brain. In fact, we will able to develop such a training module which will minimizes the impact of biases and improves the decision-making of an individual, whether investor or advisor. Technology advances especially data analysis and artificial intelligence can further play an important role for decision-making on financial analytics based on machine learning but final decision would be taken by human being. For that such a training module will help to make rational financial decision. Further analyses of impact of financial market anomalies and personality traits on financial decision-making, so that financial acumen will improve and new methods of financial acumen will develop for the betterment of the individuals and the society.

7.1 INTRODUCTION

A basic financial analysis is to understand what drives your company's earnings and expenses and to use key financial indicators. You do not have to be a financial genius, but you need financial literacy to:

- Understand the financial objectives of your business.
- Recognize why management takes certain actions and what needs improvement
- Determine how to help your company earn more money, save money and improve the key indicators supervised by management.

You need financial knowledge to contribute more to meetings, negotiate strategically, and manage your own career. Marina of the American Management Association says that "strengthening your financial acumen is a must in today's economy."

The financial vision and prophecy is complex. To create economic value for your organization, you need to understand and effectively manage multiple financial metrics related to goals, products, stakeholders, platforms, resources, rules, geography, time zones, and markets. To be able to understand not only the figures in your income statement, but also to use and manage your human resources, time, and finances, while respecting government requirements and regulations, risk management and compliance, staying in step with the competition, recalibrating new technologies, and delivering to customers with a higher level of quality can be very difficult. To refine these skills, inspire and motivate your team and advance your vision, you need to take a two-pronged approach to leadership: mastering people's skills and fully understanding their financial acumen.

The financial success depends both on uncertain cost and uncertain revenues and in order to be successful, one must assess these uncertainties to a high degree. Proper investment analysis is crucial to deal with these uncertainties, to reduce and anticipate them whenever possible. But due to behavioral finance, people behave irrational and are not able to take right financial decisions on right time even though they are financially literate. Getting knowledge of finance and interpretation of financial statement can be easily acquire, but behavioral finance impacts on their financial acumen.

7.2 REVIEW OF LITERATURE

A study by Belsky and Gilovich (1999) focuses solely on the determinants of financial education and the preparation of young people for financial decision-making. School outcomes and cognitive abilities have been found to be important determinants of financial education, but they are not the only determinants. In fact, many variables continued to be important predictors of financial education, aside from education and cognitive abilities. In addition, education and cognitive skills alone cannot explain the great differences in financial literacy among young people.

Another study by Dehnad (2011) emphasizes the link between financial literacy, financial literacy, and financial behavior. The meta-analysis method was used to analyze the data. The meta-analysis establishes clear rules for inclusion and exclusion criteria, as well as coding methods to characterize similarities and differences between studies. It has been found that financial education, like other types of education, collapses over time; Even an excellent educational program with a large number of teaching hours has a slight effect on behavior. Correlation studies that measure financial education show a closer relationship with financial behavior. Using three empirical studies, it has been found that the consequences of financial education are considerably reduced when it is necessary to use them.

Another study by Huberman (2001) is related to the influence of various socio-demographic factors on various aspects of financial education among urban youth in India. The study also examined the relationship between aspects of financial education. Hastings et al. studied the problems faced by consumers when making financial decisions using two variables: financial illiteracy and current impatience/prejudice.

Further, Ritter (2003) elaborated the main objective of the study was to study the nature of the term financial education. This document begins with a reference to existing problems regarding a person's ability to effectively use financial information and its consequences. It has been argued that financial literacy and financial literacy are not synonymous and that financial education is a complex phenomenon that requires adequate conceptualization.

Statman (1995) stated the importance of financial education in developing countries, especially given the global financial crisis is unspoken. He also referred to the efforts of major international organizations to strengthen financial education in developing countries. It has been stated that financial education is an active process in which the transmission of information is only a beginning. The ultimate goal is to empower individuals to make efforts to improve their financial well-being.

Jureviciene and Jermakova (2012) proposed study which was based on two aspects. The first was a methodological study, in which a randomized control experiment was conducted to assess the impact of financial training on the individual outcomes of micro-entrepreneurs and businesses in the Dominican Republic. Also the second was conceptual, where tests were done to determine if the type of program determined the effectiveness of the training. The results of the study showed that a thorough knowledge of finance and financial accounting certainly had a positive impact on the practice of managing small businesses in an emerging market such as the Dominican Republic. But it has been shown that the effectiveness of this training depends largely on the organization of financial education training.

Shefrin (2000) argues that behavioral finance seeks to understand and predict the systematic effects of the financial market on psychological decision-making. Brabazon (2000) stated behavioral finance, as part of a behavioral economy, is a branch of finance that, using theories of other behavioral sciences, particularly psychology and sociology, attempts to detect and explain phenomena that are incompatible with the paradigm of the expected utility of wealth. Behavioral economics is essentially experimental and uses research methods rarely used in traditional financial literature.

Tversky and Kahneman (1973) argue that behavioral finance is a behavioral economy. Behavioral economics is a combination of two disciplines, namely psychology and economics. This combination should explain why and how people make irrational or illogical financial decisions when they save, invest, spend and borrow money.

Fama (1998) "Behavioral finance is a study of the influence of psychology on the behavior of a finance specialist and its subsequent impact on the market." He studied the theory of behavioral finance and the market's efficiency with a different view. He focused on how and why the market can be ineffective due to irrational behavior of people.

Thaler (1999) stated that behavioral finance is no longer as controversial as it was before. While financial economists are used to thinking about the role of human behavior in raising stock prices, readers will look at articles published in the past 15 years and wonder why such a fuss. He predicts that in the near future, the term "behavioral finance" will be correctly perceived as a useless sentence. What other type of financing is there? In their enlightenment, economists generally include in their models as much "behavior" as they observe in the real world. In the end, doing the opposite would not be wise. These definitions provide clear information about behavioral finance and its importance for financial decision-making.

The Nobel Prize in Economics was awarded to Kahneman (2003), himself a professor of psychology, which shows that the contribution of behavioral finance to the financial economy is accepted worldwide. Professor Daniel Kahneman has studied and simplified the heuristics and prejudices resulting from financial decisions in the face of uncertainty. Thus, work related to behavioral finance has grown considerably thanks to several experiments in which it was concluded that prejudices affected the decision-making process and the process of forming a person's preferences. Financial education is very important for making financial decisions.

In addition, some studies have shown that financial literacy is very important to perceive investment risk. Shefrin (2001) has highlighted significant differences between financial experts and the uninitiated with less financial knowledge than financial experts. As a result, people as a general rule, had a higher risk appetite than financial experts and had a preference for membership (i.e., to find more reliable suppliers and sellers than nonmembers professional). In addition, these people felt less financially that financial products were more likely to be too complex. In another study (Tversky & Kahneman, 1986), the authors of a study on risk perception in established a strong correlation between knowledge and risk scales. Further they concluded that participants saw several more understandable investment products, perceived as the same less risky products. Thus, financial education will be more important in the financial environment than it is envisaged.

However, there is an increasing body of literature that shows that people do not have enough financial education to help them make informed financial decisions (Bernerd & Thomas 1988; Gustavo, 2010; Uzar & Akkaya, 2013). In an OECD study conducted in 14 countries, Solt and Statman (1989) concluded that financial illiteracy was widespread in many countries like Albania, Poland, Malaysia, United Kingdom, and South Africa. It seems that people around the world suffer from a lack of financial literacy, no matter what country they live in. In addition, recent data show that lack of understanding of financial problems may be the main reason for the diversification sub-portfolio (Shiller, 2000) and the decrease in equity holdings (Ariel et al., 2005), insufficient preparation for retirement (Kahneman et al., 1990), lack of wealth accumulation (Pompian, 2006). Despite for the presence of a meaningful and useful preliminary study describing the role of financial education in financial behavior, Oba (2018) explained how financial education can influence financial behavior. In addition, although these studies appeared at one time or another while studying the impact of financial education on any behavior or financial outcome, they did not include any other explanatory variables. Limited studies have examined several variables in addition to financial education, such as risk aversion (Pompian, 2006), self-regulation, future direction (Tseng, 2006), at risk (Novemsky & Kahneman, 2005), negative emotions, past behaviors and attitudes (Kaneko, 2004), the role of self-perception that is the locus of internal control relative to the external (Herbert, 1979), trust (Lin, 2011). The vast majority of studies dealing with complex financial decision-making or complex financial behavior in a pure financial literacy assessment imply that there is a need to expand the relevant literature to better understand these complex and complex processes.

Although the literature on behavioral finance is very broad, it is suggested to present some empirical case studies to understand behavioral finance and its application to decision-making. Kahneman and Tversky (1973), who have been recognized as the fathers of behavioral finance, are better explained by their work at different stages. In 1979, they presented an article criticizing the theory of expected utility, which empirically concluded that low-weight individuals had only probable results compared to the results obtained with certainty; In addition, shared components are generally excluded from all perspectives considered. According to the theory of perspective, value is added to gains and losses instead of final assets: probabilities are also replaced by decisional weights. The theory that they have been confirmed by experience predicts a pronounced fourfold risk-taking trend, a risk aversion of moderate to high probability gains and low probability losses, as well as risk-seeking gains.

Thaler (1980) in "Behavioral Finance, Investment Budget and Other Investment Decisions" conducted a review of the literature on the impact of behavioral distortions on the investment budget by biased managers over-investing cash flows Creating and sustaining new businesses and projects Low-productivity investment strategies generally result in: longer contractual incentives, but their effectiveness in curbing investment seems limited. Behavioral finance is a part of the literature and shows how irrational fissile behavior affects investment decisions.

The behavior that ignores the basic economic principles of buying, offering, and buying stocks due to price momentum is called behavioral behavior in finance and leads to a wrong decision. In the late 1990s, venture capitalists and private investors invested heavily in Internet-related companies, although most did not have a strong business model. Adverse reactions to news, excessive, and inappropriate responses were often positive or negative on the financial market. The principle of repentance states that a person evaluates expected reactions to events or future situations. Psychologists have found that resulting decision-makers have more regret if this decision is less conventional. This principle can also be applied to the field of investor psychology in the stock market. Regardless of whether an investor has considered buying a stock or a mutual fund has declined, buying the expected collateral will provide an emotional response to the investor; avoid selling stocks whose value has fallen to a bad investment decision and reporting losses. In addition, it is sometimes easier for investors to buy "popular or popular stocks of the week." In short, the investor only follows the "crowd." As a result, if the value of a share or mutual fund drops significantly, investors can more easily rationalize their investment options. The investor may reduce sentiments because a group of individual investors have also lost money on the same wrong investment.

7.3 FINANCIAL LEARNING

For more than 10 years now real estate prices have been rising continuously and one might be inclined to think that this trend will continue for another 10 years. But among the many things we can learn from history, one lesson is that the future is uncertain: just a few years ago the global financial system was close to a meltdown and few would have predicted then the strong growth in real estate prices that we observed in the past few years. Quo Vadis?

Project developers need to cope with uncertainty, think about it and find strategies to deal with it. One way of doing so has always been flexibility. If demand is low a developer can wait and build when markets have recovered, or he can change the project to another use. When looking into practice, however, the employed financial analysis tools neither properly account for flexibility, nor uncertainty when dealing with development of projects.

One way of financial analysis that deals with flexibility and uncertainty is real options analysis. This approach is closely linked to the valuation of financial options and is thus based on assumptions that are not necessarily applicable in the world of project development. Also we argue that due to a lack of accessibility and applicability, real options analysis has not yet gained much attention in the project development practice. We thus introduce a more intuitive, simulation-based real options model grounded on the works of de Neufville, Scholtes, and Geltner. This so-called Engineering Approach works with "classic" net present value calculations and combines them with simulations of future market scenarios. Based on these scenarios, decision rules are implemented and optimized to find the optimal behavior for the simulated future. Thereby we make flexibility and uncertainty more accessible in the analysis of development of any projects.

The correct assessment and management of risk is therefore indispensable and belongs to the main capabilities of successful developers. Some of the main risks of project development as:

Development risk: The risk of not planning an adequate use for a specific location and the risk of planning financially nonfeasible projects.

Time risk: Due to the financial leverage of most projects time risk is among the most important risk factors. Delays can harm the profit of developers substantially.

Approval risk: All development projects need to be approved by the authorities. Neighbors can raise objections that can result in financially harmful project changes.

Financing risk: Development projects are financially daring undertakings that require partners with corresponding financial power. Funding might not be achieved or might be stopped due to delays or other problems resulting in the failure of a project and severe financial consequences for the developer.

Building ground risk: The building ground bears high potential for additional cost and delays. This is due to the fact that building grounds bear uncertainty regarding supportable load and contamination that cannot be eliminated completely with preliminary studies.

Cost risk: Cost risks arise mainly from the long time horizon of development projects and the uncertainty regarding the exact specification of the future product. Thus it is often very difficult to predict exactly the production cost of a large-scale development project and additional costs may arise from the other mentioned risk factors.

Market risk: The final test of every project is when it comes to market. Are the potential tenants willing to pay the calculated rents resp. is demand high enough to meet the additional supply at the specified price? How much are investors willing to pay for real estate assets? Real-estate value is driven by the space market that couples demand and supply for space and by capital markets (Arkes & Blumer, 1985). These two markets are already difficult to assess in the present and their behavior is much more difficult to predict for multiple years ahead. Inevitably this leads to large uncertainty when dealing with the market risk of real estate developments. Besides that real estate markets behave in long lasting cycles that are characterized by periods of strong growth in prices and high construction activity followed by phases of stagnation and price decline (Nofsinger, 2001). Developers need to anticipate markets correctly and make the right preparations and decisions based on their estimations.

Why are we telling here about real state risk, its uncertainty, complexity, and financial behavior of market? This is all about the understanding of financial issues whether it is real state or any project or any firm/company and your business success and failure is based on your financial acumen. People developed many models, algorithms, and simulated processes to control the financial uncertainty and risk but it is your financial acumen which decides where you will take your business in future.

There is one famous quote by Alexa Von Tobel that "Good financial management is a road map that shows us exactly how the choices we make today will affect the future."

- How do you drive shareholder value?
- What is capital budgeting?
- How do you read and understand financial statements?
- How do you calculate total shareholder return?
- What exactly is EBITDA?

All simple but crucial tools mean basic financial concepts which are used in analyzing company's financial performance.

Businesses, more than ever, are faced with rapid change, disruption, and intensified expectations. Your managers, emerging leaders, individual contributors, and leaders, many of whom have no business or financial background, need to fully understand the implications of their business decision-making and how they can directly affect your company's performance metrics and drive for shareholder value.

There are lots of customized training programs provide nonfinancial and financial personnel with new skills and tools needed to measure the success of their decision-making on strategy execution.

These comprehensive learning helps participants to understand and apply financial principles, methods, and techniques. These include a series of flexible learning objects including assessment, pre-work, live or virtual learning sessions, computer-based business simulation, application tools, and measurement and impact studies, etc.

These finances focused learning integrate a deep understanding of financial topics by using a business simulation to help develop new skills and interactively illustrate how to understand a company's finances. These business simulations provide a risk-free environment that strengthens skills and improves the effectiveness of learning and are proven and effective tools for driving the application newly learned concepts.

These business simulations present participants the opportunity to develop and execute a business strategy. By making a series of cross-functional operational decisions, learners are able to see and experience the immediate effects of those decisions on the simulated financial results and outcomes. The simulation can be calibrated to reflect any industry or business model by adjusting the market dynamics and elasticity of the base conditions.

Finance-related examination is records that give information of the affiliations fiscal status. It sum delineates the money-related soundness of the association benefit. The appraisal of association's likelihood and possibility with the true objective of settling on decisions is taken for growth of business. The objective of financial analysis is to give information about the cash-related position. Execution and alter in budgetary position of an endeavor that is profitable; to a broad assortment of customers in settling on monetary decisions. Cash-related clarification should be sensible, huge, and trustworthy. They provide a correct photograph of an association's condition and working results in a thick from declared the assets, the liabilities and esteem clearly to an affiliation's budgetary execution examination and interpretation of cash-related explanation helps in choosing the liquidity position whole deal dissolvability, financial achievability, productivity and soundness of a firm. There are four major sorts of money-related clarification: resource report, wage verbalizations, wage announcements, and clarifications of held benefit.

Cash-related clarifications are the formal record of the funds-related exercises of a business. Individual and underground creepy crawly component means what accounting and financial statement shows does not mean it is so simple or straight means. Fiscal enunciations give a graph of a business or individual's budgetary condition of short and whole deal. All the vital financial details of a business try presented in a sorted out route and in a kind of straightforward way to deal with fathom is known as the funds related clarifications.

7.4 UNDERSTANDING OF FINANCIAL STATEMENTS

There are four financial statements those plays vital role in decision-making. They are Balance sheet, Income articulation, Statement of held income, and Cash stream articulation.

7.4.1 BALANCE SHEET

Financial record is a cash-related explanation that packs an affiliation's focal points, liabilities and theorists' a motivating force which is in particular point of time. The three cash-related record pieces give inspectors a thought with respect to what the affiliation affirms and moreover aggregate given by money related authorities. Bookkeeping report is moreover cleared up as a review of affiliations cash-related condition. Of the four guideline financial decrees, the bookkeeping report is the primary clarification which relate to a singular point in time. Affiliation fiscal record has three segments: assets, liabilities, and ownership esteem. The basic class of advantages are generally recorded the first and after that took after by the liabilities. The refinement between the advantages and liabilities is called as esteem or the capital of the affiliation. It is called bookkeeping report since its counterbalance the two sides.

It chips away at the accompanying equation:

Resources = liabilities + Shareholders' Equity

7.4.1.1 CONTENTS OF BALANCE SHEET

The balance sheet discloses reveals real classes and measures of an organization's benefits and also real classes and measures of its monetary structure, including liabilities and value. Real groupings utilized as a part of the announcement include:

Assets—anything claimed by an organization that has money related esteem, including financial assets which are present and plausible future value. They are Current resources (cash, attractive securities, records of offers or commitment owed to an association, stock, and prepaid expenses), Investment, Fixed resources (property, plant, and equipment), Intangible resources (licenses, copyrights, liberality), and Conceded charges or diverse assets.

Liabilities—present and conceivable future commitments owed by an association against its advantages, including the responsibilities of a business to trade assets or offer organizations to various social affairs later on due to past trades. They are Current liabilities (lender liabilities, notes payable, pay payable, and appraisals payable), Long-term liabilities (bonds payable, advantages, and lease duties) and Different liabilities.

Proprietors' value—the advantages place assets into an association by the proprietor. Proprietors' esteem is equal to the focal points resulting to deducting the liabilities. They are Capital stock, other paid-in capital in excess of standard and retained acquiring.

7.4.1.2 PROFIT AND LOSS STATEMENT

An advantage and mishap announcement (Profit & Loss) is a cash or verbalization gathers pay rates, and costs accomplished amidst a particular time designation, frequently a budgetary quarter or year. These records give data around an affiliation's capacity to make benefit by developing pay, diminishing expenses, or both. The Profit and Loss illumination in like way intimated as "verbalization of preferred standpoint and hardship," "pay declaration," "articulation of operations," "elucidation of money related outcomes," and "pay and cost illumination."

7.4.1.2.1 CONTENTS OF PROFIT & LOSS ACCOUNT

Expenses—Cash spent or cost acknowledged in an alliance's endeavors to convey wage, tending to the cost of coordinating. Costs might be as bona fide money divides, (for example, wages and pay rates), an enrolled snuck past part (debasing) of inclination, or a whole ousted from pay, (for example, ghastly responsibilities).

Turnover—the fundamental components of the salary for association is its turnover, which contained deals and its items and administrations to the outsider customers.

Sales—It is a trade between two social affairs where the buyer gets stock (unmistakable or unimportant), organizations or possibly assets as an end-result of money. It can in like manner insinuate a comprehension between a buyer and trader on the cost of a security.

Other Operating Expenses—These costs are not specifically identified with the creation procedure but rather adding to the movement of the association. These contain the costs like; Distribution cost and offering cost, Administration cost and innovative work costs.

Other Operating Income—The other working salary comprises of all other income that have not been incorporated into alternate parts of the benefit and misfortune account. It does not contain offers of merchandise or administrations, announced turnover or revealed inside the net intrigue class.

7.4.1.2.2 FINANCIAL RATIO

Finance-related extent is a quantitative examination of data contained in an affiliation's money-related articulations (Figure 7.1). Degree examination depends upon line things in budgetary verbalizations like the preferred standpoint report, wage elucidation and pay declare; the degrees of one thing—or a mix of things—to something unique or mix are then figured. Degree examination is utilized to review assorted parts of an affiliation's working and money-related execution, for example, its benefit, liquidity, favorable position, and dissolvability.

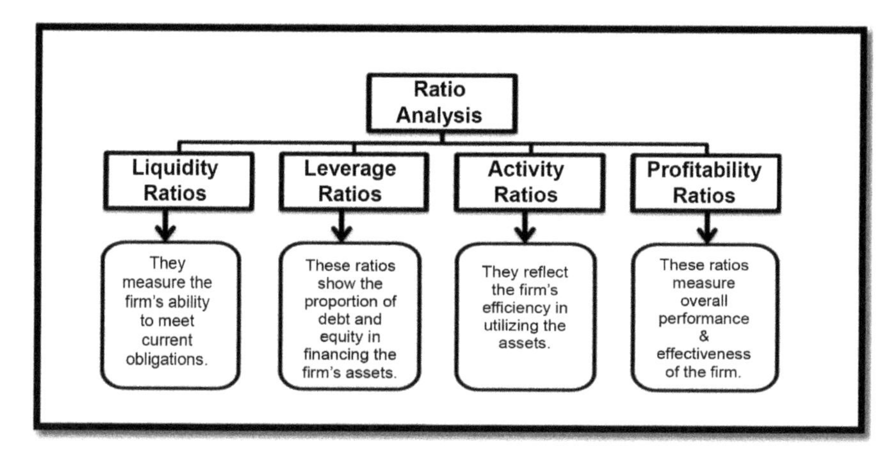

FIGURE 7.1 Ratio analysis.

Examination and elucidation of money-related affirmation is an essential contraption in evaluating affiliation's execution. It uncovers the quality and lacks of a firm. It influences the customers to pick in which firm the hazard is less or in which one they ought to contribute with the target that most exceptional purposes of premium can be earned. It is in addition that setting resources into any affiliation fuses a considerable measure of dangers. So before setting up exchange out any affiliation one must have careful getting some answers concerning its past records and execution. In the context of the information accessible, the case of the affiliation can be predicated in not too far-removed future. This meander of monetary examination and appreciation in the age concern is not just a work of the undertaking however a smaller learning and experience of that what to look like at the finance related execution of the firm. The examination got a handle on has gotten to the light of the running with conclusion. As exhibited by this undertaking I came to comprehend that from the examination of finance-related elucidation unmistakably have been acknowledging advantage amidst the period of study. So the firm should concentrate on getting of more points of interest in coming a long time by taking idea inside and furthermore outside elements. Moreover, with respect to assets, the firm is take usage of inclinations fittingly. Likewise, in addition the firm has a kept up low stock. This undertaking fantastically concentrates on the beginning of various sorts of finance related declaration. Asset report and Profit and Loss declarations have been considered. From allot examination of Balance sheet and P&L Statement, it was mulled over that liquidity position of the affiliation is amazing. Current disperse, responsibility regard degree, brisk degree, net general wage, working net wage, net favorable position apportion, return on resources, level of profitability, and return capital utilized supposedly were unsatisfactory. The degrees that are believed to request are Current circulate, Return on speculation, and Return on working capital, and Debt Equity Ration.

7.5 FINANCIAL ACUMEN

"Improving Financial Acumen Increases Strategic Engagement by Increase your financial understanding and confidence, Analyze data and interpret key performance indicators, Diagnose what is working and what needs to improve, Stay up to speed on your company's financial performance and Engage in more strategic conversations."

For the company to succeed, all employees must understand that the money is returned to the company. When employees are not trained in financial management, poor business decisions are made. Financial understanding is a skill that can be learned and that ensures the integrity of numbers, because everyone uses the same measure

Warren Buffett once said: "Only when the wave rises, you can find out who swims naked.." The patron is a very good cause why we should improve our financial knowledge and your vision of the business. A good financial culture does not mean that you should be a junior accountant, but you can "speak the language." Although you cannot master the word "finance," you can develop a sense of conversation to continue it, the language of business. Financial leaders should not dominate the conversation; the best decisions and coherence are achieved when all departments are balanced throughout the organization.

Cash flow, balance sheet, profit management, organization of growth, and people are the five factors on which successful companies believe that their continued success is based. To improve their performance and achieve long-term sustainability, successful companies find ways to overcome the contradictions between these factors. There is a natural contradiction between profitability and growth. At the same time, how should you balance short-term decisions with long-term decisions?

7.6 THE IMPORTANCE OF FINANCIAL STATEMENTS

Three main financial statements tell the story of commercial transactions. Each statement tells a separate part of the story. For example, a smart businessman knows that there is a difference between cash flow and income. Due to this why it is important to familiarize yourself with the balance sheet, income statement, and cash-flow statement.

First, you have to understand the finances at ground level. Profit and loss statements are a business marker. They provide information on what has just happened on the ground. Although they report past events, they can also be predictors of the future.

There are also many forms of income, and each tells its own story: raw, operational, clear, but if you do not know how to increase your profitability, you probably will not be able to, which will affect the bottom line of your business. This is where the importance of a cash flow statement comes in.

Third is the financial report—Balance sheet. Managing the balance can be more complicated than understanding the income statement, but the basic equation analyzes what the organization has less than it owes to others. As Buffett's candidacy suggests, companies can hide behind their income statements, but in a recession, the balance sheet will indicate whether the organization can seize the opportunity or survive another day.

Too often, the concepts of accounting and finance are considered a black and white science. However, there is really an art of financial understanding. For example, the figures in any financial report reflect estimates and assumptions, and the art of preparing financial statements uses limited data and estimates from the future to provide a clear description of performance. In addition, to fully understand the business, executives need to know what analytics tools will complement the image. Some proportions are crucial in some areas and less important in others. Although a causal analysis can help to make decisions, it can also cause paralysis of the analysis.

Although the increase in financial education seems simple, the numbers have no meaning throughout history. Why management may think that company staff already understands the financial resources considering at the time of decision-making and hence no financial training is required. The budget may not provide enough space for this type of training. Or maybe it is due to the fact that the appropriate financial experience is applied at the auditing and approval from the treasury, accounting, or finance, which ensures that process.

Whatever the reasons, I think:

- Institutional managers, even those with many years of experience cannot function fully without applying a financial analysis to the decision-making process.
- Any company that produces, transports it, and distributes with intensive use of capital can develop a long-term competitive advantage by spreading financial knowledge among its various management positions.
- According to recent studies, investment in human capital management, of which training is an important element, will add value to shareholders.
- Speaking of manufacturing plants in the country, we are talking about solutions in which operating, maintenance, and capital improvement costs are higher than in any other sector. As a result, a small error in decision-making can have a significant negative effect on financial performance.

Another equally punitive action by rating agencies such as Moody's, Fitch, and S&P is to lower the company's credit rating. A deteriorated credit rating could increase utility company interest costs by several hundred thousand dollars a year, says Osborne. If interest expenses increase without compensating for reduced operating and operating costs, profits fall, so profit forecasts are no longer true. This may lead to another downward adjustment of the credit rating, therefore, which will lead to an even greater increase in interest expenses. This cycle is called the spiral of death of the credit rating. After the California energy crisis and the Enron bankruptcy, many utilities and electricity producers experienced this.

Every decision that involves spending money, capital or O&M at company, whether in a regulated or deregulated environment, must be a good decision from a financial standpoint. That happens when all employees have good business and financial acumen knowledge.

A variety of best practices defines this as follows: the ability to understand commercial discourse: why it exists, its underlying principles, its catalysts and threats. The ability to read financial statements and understand strategies for generating revenues, penetrating markets, and increasing profits, and how the organization's structure and processes support these strategies. The acquisition of this set of skills requires improving your intuitive, theoretical, and practical understanding of the strategic perspective, compensation decisions, cash-flow factors, market direction, and the competitive environment of your business. Although statistics and measurements are important, they are not just numbers. From the inside, it knows how the numbers reflect the company's decisions about generating revenue, managing expenses and responding to competition. From the side, we are talking about reading commercial, technological, political, regulatory, and demographic trends and their impact on business models, as well as design, development, and marketing of products and services. For financial professionals, this means that there are many ways to contribute and help implement business strategies. This requires the ability to determine where and how diversity and inclusion can anticipate, shape, and promote business and operational strategies. It is about being able to quantify how diversity and engagement contribute to the top line and help manage the bottom line. At a more tactical level, the commercial and financial sense also depends on the ability to interpret the financial statements in order to determine the strategies and directions of the company. Let us move on to these issues. Learn the language of finance. Understanding business and finance is not an empty space for many professionals, but staff costs and difficult economic times are transforming organizations and increasing the need for employees. At all levels of understanding of business dynamics, some studies show that too many employees are not ready to help their enterprises achieve their business goals. Skills Inconsistency Report: Analysis of the effectiveness and strategy, the strategy consulting firm and the Economist Smart Unit, showed that most companies simply do not need a strategic watch. Follow a business strategy. The survey showed that responsible employees at all levels do not have the necessary knowledge about how their business makes money. And about 67% of respondents said that this lack of understanding of their activities limits their ability to achieve their strategic goals. The lack of commercial criteria, as well as the change many companies have made to define their employees as "assets" (which are also becoming more diversified), may be factors for which there is a growing movement on the part of CFOs.

An article published in 2012 by the Human Resource Management Society discusses the "growing rumor" that senior financial professionals are entering the realm of human resources and diversity. The article says:

"In times of economic hardship and in the conditions of saving services, personnel costs and labor productivity are the focus of management's attention." This can help determine the return on investment (ROI) and increase the chances of HR professionals. Finance professionals take this coveted seat at the table. "Such a business movement presents risk and opportunity, and the risk is that financial leaders can overcome the functions of human resources and diversity with the depth and subtlety of this work and just look at initiatives in favor of diversity as an expense that contributes little to the end results." Diversity Best Practices defines it as: The ability to grasp the discourse of business—why it exists, its underlying principles, its enablers, and threats. The ability to read financial statements and understand strategies to generate revenue, penetrate markets, and increase profit margins, plus how organization structure and processes support these strategies. Acquiring this skill set requires honing your intuitive, theoretical, and practical understanding of your company's strategic perspective, decision tradeoffs, cash flow drivers, market orientation, and competitive landscape.

While the statistics and metrics are important, it is not just about the numbers. From the inside out, it is about how the numbers inform corporate decisions about revenue generation, expense management, and reaction to competitors. From the outside in, it is about reading business, technology, political, regulatory, and demographic trends and their implications on business models, and the design, development, and marketing of products and services. For professionals, this means there are plenty of ways to contribute to and to help execute business strategies. It requires the ability to identify where and how diversity and inclusion can anticipate, shape, and advance business and operational strategies. It is being able to quantify where diversity and inclusion contributes to the top line and helps manage the bottom line. On a more tactical level, business and financial acumen also depend on the capacity for interpreting financial reports in ways that discern business strategies and direction. Let us deeper into these topics of learning the language of finance. The business and financial acumen is not just a gap for many practitioners. As staffing costs and challenging economic times transform organizations and increase the need for employees at all levels to understand business dynamics, several studies indicate that too many employees are woefully unprepared to help their organizations meet business goals.

The 2013 Skills Mismatch: Business Acumen and Strategy Execution report by BTS, a strategy implementation-consulting firm, and the Economist Intelligent Unit, showed that most companies simply do not have the business intelligence needed to execute business strategies. The survey found that critical employees at all levels lack the essential knowledge about how their companies make money. And about 67% of respondents said that this lack of understanding surrounding their business limits their ability to accomplish strategic goals. This lack of business judgment, as well as the shift so many companies have made to identifying their people as "assets" (who also happen to be increasingly diverse) may be factors in why there is a growing move for CFOs to become involved in HR and in diversity. A 2012 article in the Society for Human Resource Management discusses the "growing buzz" around senior-level financial professionals entering the realm of human resources and diversity.

The article states, "Particularly in tough economic times and a service-driven economy, staffing costs, and productivity move to the forefront of executive concerns. The perspective that a CFO might bring to HR issues can help identify ROI and boost HR professionals' [and diversity practitioners'] chances for getting that coveted seat at the table." Such a corporate move represents both risk and opportunity. The risk is that CFOs can be overwhelmed by diversity and human resources functions, which may not understand the depth and subtleties of this work, and simply see diversity initiatives as costs that are not significant to the results. Personnel and diversity specialists have the opportunity to improve their game, get acquainted with the financial and commercial areas of their business and become reliable consultants on strategic business issues. Considering this next change as an opportunity also shows that many diversity professionals do not have formal financial background and often come from other disciplines such as talent management, hiring, regulatory compliance, or training and development. To embrace your arms around an area where you do not feel safe requires courage and a thirst for continuous learning.

However, one caveat just describe your conversations in financial terms such as ROI, return on equity, return on assets, cash flow or net income, and the Net result is insufficient. Fulfilling this without real understanding of the true financial meaning of these terms can mislead anyone who believes that they "understand" the financial strategy or that their efforts to ensure diversity and integration will improve the business, reduce vulnerabilities and reduce business risks threats. For example, during a 2011 session on best practices in diversity, Gudrun Granholm, CEO of Box One, Inc., a financial advisory and training company, former CFO of Hannah Andersson, Smithsonian Institution and financial analyst at *The Washington Post*, explained diversity professionals often misused financial terms such as ROI. Too often, he says, diversity experts use ROI to discuss the value created by inclusion activities. However, in the financial sector, ROI is real and measurable financial results, such as an increase in income and profits. Terms that imply cost avoidance or perceived benefits may be considered fuzzy financial language, but not valuable data. Such language failures and misleading oral exercises can really only demonstrate the lack of financial value that it actually has. Specialists in the field of finance are of particular importance for the ROI, which originates in cash flow. In its simplest form, ROI is used to measure the ROI. The basic ROI subtracts the value of the investment from the ROI and divides it by the value of the investment. Income from investments (in dollars)—Investment costs (expenses). This basic formula does not take into account issues such as time (time value of money or recovery period), general expenses (support staff, administrative expenses, IT, training, and management, development, etc.). Also there are several ways to evaluate the ROI. For example, a diversity professional might consider two different diversity initiatives. After studying the direct costs of each program, a professional learns the benefits. Does it generate income, reduce real costs, or add value in another measurable way? Then the professional simply divides the benefits into direct costs in order to get value for money. A finance specialist can consider the same two programs, taking into account revenue generation or cost reduction, and divide them by the total cost of any resource or expenses, such as IT support, marketing, human resources, operations, etc. If applicable, the recovery period of investments, depreciation, and other factors can be included in more detailed ROI estimates. This process can be even more complex for diversity and integration professionals, because the nature of our business rarely brings direct revenue or reduces real costs, but may include several corporate resources.

Timothy Kenny, Executive Vice President and General Manager of Freddie Mac, explained how he worked at the same session on best practices in diversity in 2011: he talked about some of the financial indicators. A report published in November 2012 in American Progress examined 30 case studies of direct and indirect staff costs from 1992 to 2007. The main finding was that companies spent about one-fifth of employees' annual salary to replace employee replacement costs. Around 20% remain unchanged for workers earning less than $30,000 a year at $75,000. As soon as the annual salary was over $75,000 or the employee had specific skills, the turnover rate was about 30%. These direct costs are valuable indicators that can be used to determine the ROI of the employee—a diversity initiative that takes into account staff turnover. Diversification professionals do not need a formal education in corporate finance to become a strategic contributor to the business. But they need to understand how the financial aspect of the business works, the process of budgeting and financial reporting, and speak the language of finance, even if they focus on D&I. Business Information and financial are also important for the protection of annual D&I budgets. No organization has unlimited resources to refer to high-value initiatives. Each annual budget cycle is a choice between competing needs, priorities, and market rates. Granholm explains that only about one-third of project applications receive funding and approval from the company. Therefore, the overall odds are already against anyone making a budget. In addition, D&I professionals face even greater prejudices because of the often external status of D&I. So how can diversity professionals use their business and financial skills to protect both diversity and organization? Consider three main ways: (1) as the driving force of your business; (2) as a strategic partner; (3) as a reliable consultant

If you are not a numbers specialist, finances are scary. But it is important to understand terms like EBITDA and net present value no matter where you are in the organization chart. How can you increase your financial acumen? How do you decide which concepts are most important to understanding your work and your understanding of the business? And who is best placed to give advice?

Commercial and financial understanding is also important in protecting annual budgets. No organization has unlimited resources to manage all and all of the company's valuable initiatives. Each annual budget cycle represents a choice between the needs, priorities, and rates of a competing market. Granholm explains that only about a third of project applications receive corporate funding and approval. Thus, the overall odds are already against anyone who makes a budget release.

By creating budgets or business plans requiring financial support from the organization or its business units, professionals with financial knowledge rely on these documents with credible tax assumptions, reasonable forecasts and sound financial models. To do this, many skills are needed, ranging from understanding your organization's budget process to forming alliances with the company's finance team for help with the technical aspects of budgeting and templates until strategic decision-making, which identify resources that can support initiatives.

Obstacles to business and financial literacy are often related to problems with poor performance by professionals: poor communication skills, inconsistent work, inefficient leadership, and inability to combine talent management and performance metrics, business, and employees. Whether at the initial, intermediate, or senior career level, diversity professionals need to be able to understand how the financial and business processes as well as the diversity and inclusion process work together on behalf of the organization. At the beginning of his career, unorganized and inaccurate financial documents illustrate the fact that he does not understand the importance of this contest. Unclear, vague, and unapproved communications with your organization's external diversity partners regarding financial commitments will affect your efforts to be more expert in this area. For mid-career professionals, it is essential that you have firm control over the essence of your organization, its business objectives, its mission, its strategies, and the countervailing forces that affect your financial reputation. Your upward mobility as a diversity professional will be hindered without this understanding. Moreover, senior diversity leaders who do not have a deep understanding of their company's activities, their sector and how the profits are generated undermine the value that diversity can bring to an organization because they cannot make any argument. Financial strength is a language that financial leaders understand. It also means that these diversity leaders probably do not have a relationship with the OFC or other financial sector players and may not be able to properly explain the link between D&I and financial issues. While possessing the capacity for financial and business know-how is a skill that can be acquired by diversity professionals at the initial and professional levels, it is absolutely necessary that leaders of higher levels of diversity possess this capability. Conclusion: As you have seen, it is essential that you are well anchored in the financial leaders' worldview in order to be able to present budgets and investment plans so that they speak with the spirit of the CFO and allow it to thwart the inevitable setback. The financial managers of your company are not the "bad guys"; they simply see all the requests for financing as a strict financial year that analyzes whether a specific demand brings a measurable value. Developing your financial and business knowledge will enable you to show convincingly how diversity and inclusion help the company increase revenue and manage the bottom line through cost management. In the current context of extreme cost management and hyper-competitiveness, diversity initiatives can only survive if they are integrated with the growth and profits of the company. Develop and refine your business and financial sense and see how it becomes an effective way to lead to true cultural change.

The ability to apply financial discipline, budget discipline and analytical concepts to make good judgment, make decisions and act quickly on the business and the sector. This capability applies to a wide range of budget and investment activities, including transactions with suppliers, brokers, third parties, customers, customers and distributors.

Essential Competence:

- Develop a budget/investment plan aligned with the company's objectives.
- Develop and operate within budget and budgets. Anticipate budget needs and adjustments in response to changing business priorities.
- Establish and manage budgets that reduce sales as a percentage of net sales value.
- Review plans and budgets frequently, anticipate budget needs and proactively adjust budgets/programs according to corporate priorities.

Manage the Budget Effectively:

- Understand the company's financial position (profitability, margin, sales, cost, cash flow, working capital, etc.) and interpret it.
- Understand the client's financial situation (profitability, margin, sales, cost, cash flow, working capital, etc.).
- Understand how the client earns money.
- Apply an understanding of the company's financial statements and parameters when creating recommendations.
- Know how the elements of the company fit in with the impact of this profitability and cash flow.
- Think like a business owner and use financial measures to evaluate ideas.
- Apply an understanding of financial statements and customer metrics when creating recommendations.

7.6.1 UNDERSTANDING FINANCIAL

- Understand the structure and purpose of the contract; negotiate agreements within the agreed framework.
- Establish agreements and contracts that create mutual value for suppliers, brokers, third parties, customers, customers, distributors, and the company.

- Negotiate new agreements with suppliers, brokers, third parties, customers, customers, and distributors that maintain the integrity of our contract and participate in the appropriate development.

7.6.2 UNDERSTAND AND COMPLY WITH THE TERMS OF CONTRACTS WHOSE COMPETENCE SPEAKS

- Review and analyze best practices, reports, and relevant information in accordance with the appropriate methodology, and its effective implementation.
- Use appropriate methods to analyze financial data and other information to determine the impact on sales, particularly net sales and growth.
- Develop plans that have a positive impact on net sales and growth.
- Income statement, annual customer report, etc., for vendors, brokers, third parties, customers, distributors, and businesses.
- Sales and cash, especially net sales.
- Develop plans that have a positive impact on net sales and cash.
- Compare finances with other consumer goods companies and make recommendations to improve costs for sellers, brokers, third parties, customers, customers, distributors, and the company.

7.6.3 USE FINANCIAL OVERVIEW

- Understand how to leverage the budget for the best investment opportunities that generate the best mutual benefit between suppliers, brokers, third parties, customers, customers, distributors, and the business.
- Conduct assessments before and after all core activities to achieve the objectives (financial and operational).
- Implement successful activities based on the results of the evaluation.
- Provide ROI analysis to management and communicate on efficiencies, results and its business impact (sales, cash flow, and profit sharing).
- Create appropriate business plans/models to improve the mutual profitability of the clients and the business.
- Analyze the business portfolio with financial and strategic criteria and look for ways to optimize the overall investment.

- Communicate and share best practices in financial management of the account.

7.6.4 *DELIVERY OF INVESTMENT PERFORMANCE*

- Demonstrate an ability to consistently practice disciplined and inspired financial thinking to make high quality, timely, and cost-effective decisions.
- Explain to others financial instruments, fiscal discipline, and analytical concepts in order to make informed decisions and make quality decisions. Facilitate the understanding and application of practices.
- Train and encourage other users to use financial information, manage budgets, enforce contracts, and formulate and deliver an effective ROI.
- Serve as a resource for financial information, advice, and guidance.

7.7 BEHAVIORAL FINANCE

We examine how individuals assess the risk and return of a financial asset and the factors that can influence it. The development of a consensus between behavioral researchers and the biopsy process can be used to determine the dynamics of this problem. It is a theory of treating an individual and exacerbating (or sometimes improving) the bias on an individual level. Given this warning, markets can be expected to reflect their behavior as long as individuals behave in a consistent manner.

The term "bias" is used for better classification to take into account systematic deviations from normative models, rather than pointing out that decisions are "wrong" or inaccurate. Theoretically, biased perceptions of financial risk can occur at any stage of information processing before a verdict is made. A typical information processing model includes the levels of perception of existing information, the combination of multiple sources of information, and the types of observable and observable behavior (Benesh & Peterson, 1986). (Noteworthy information that can be used instead of or in combination to make a decision.) Also retrieving existing information and leading to non-normative and heuristic integration of the judgment rules, resulting in biased judgments.

Branch (1977) in his research paper stated similarly and could be conceptualized in consumer-focused investment funds. Distortions due to the way this information is retrieved from memory are called memory distortions.

Prejudices due to the way in which different sources of information are integrated before an assessment is made are referred to as prejudices with regard to information integration. The latter captures fixed action rules, including simply repeating (or not repeating) actions based on habituation. Perceptual and recovery distortions were less taken into account in the financial literature than criteria or criteria distortions. However, following the explanation of perceptual bias, it is the short sighted or excessively short horizons that were raised in the context of stock price divergence relative to fundamentals (Chen et al., 2004).

Cognitive biases are distortions that are due to deficiencies in the process of perception, reflection, and reasoning, and are based on a range of information that needs to be processed. These distortions have already been taken into account in the financial literature.

The perception that the impact is because people with inadequate cognitive abilities (e.g., mental resources) to process the available information results in an incomplete selection of available information or a lack of information, is growing insufficient integration and information submitted for assessment. In addition, since there are many aspects of the data, one of these aspects may be referred to leading to an overestimation of these aspects of the data during the experiment (Brown, 1999). For example, in certain circumstances, trends may attract more attention than noise, and context signals may exacerbate the importance of the relative performance of a financial instrument relative to its absolute performance.

Sample distribution information from decision makers, the most important points perceptual (such as the distribution parameters) that are most likely to be selected as the initial anchorages, being used as inputs in the method of decision-making. Decision makers focus more on the perceptual aspects of stimuli and inappropriately adapt to other information, including general information.

7.8 DIFFERENT CONSIDERATION OF PERCEPTUAL ASPECTS

Policy makers focus differently on the elements of the subset of relevant information. For example, a graphical representation of data may contain nearly infinite information. People with limited information processing capacity would reach their limits and could not use all existing information. As a result, attention can be directed to the centroids of the distribution to simplify the task of information processing. Based on previous literature on the treatment of visual information, it is likely that the open points are the start and end points of a graph (Kristensen & Garling, 1997) and the minimum and maximum points on the graph of space (Jureviciene & Jermakova, 2012). The confirmation of Kahneman and Tversky (1979), developed as a model of the theory of perspective, that the numbers in themselves have no meaning, but reach only a single point, is a good one. As a result, a price does not equal the amount you pay, but the profit earned on the benchmark or a loss based on the benchmark. In this context, the selected specific benchmarks can be critical to the perception of risk and performance.

Shefrin (2007) found that long positions remained open longer to lose positions compared to profit positions. Fama (1998) found the same effect on year-end trading volume: those who gained shares outperformed those who lost shares; and Morone (2008) found that losses on the government bond futures markets remained longer than the winners. He also found encouraging that this was more likely to be the case for failed traders, implying that capacity factors mitigate decision makers' ability to correct the use of benchmarks in the integration of appropriate regulatory decisions. However, it may be that the price at which an action was purchased is not the only reference point. As alternative reference prices the indices could be used, with which a current share is compared. As a result, it should be noted that the A-share outperformed the index while the B-share outperformed the index. If the return is below share B, this can be considered as a stock that performs better on its benchmark. When users use index information as a reference (so to speak), fund managers can use misleading practices to improve consumer perceptions of the Fund's performance in the past. Likewise, the presence of other contextual information can influence risk perception. For example, the risk of a higher risk in one market can cause traders to underestimate the risk in other markets, simply because the risk is less than they would have traded. Understanding and including benchmarks in decision models seems to be a promising approach. In summary, therefore, capacity reduces distortions in the processing of information, with when. Experts are less inclined to deal with non-normative information than beginners. Yes, more regulatory decisions are possible with less time.

7.9 CONCLUSION

Foundation of financial acumen is based on financial knowledge and learning and your understanding with numbers of financial statement. Also the experience you gain during decision-making and results of your decisions.

In behavioral finance, we deal with repeated biases, heuristics, and pricing inefficiencies present in financial market with limited, uncertain, risky information having time constrains, and strategic in nature.

The thing is that psychological biases, emotions, stress, and individual differences influence financial decisions which further impact on financial acumen. So we need to understand how behavioral finance impact on decision-making and can we control these impacts.

However, we will able to develop such a training module which will minimizes the impact of biases and improves the decision-making of an individual, whether investor or advisor.

Technology advances especially data analysis and artificial intelligence can further play an important role for decision-making on financial analytics based on machine learning but final decision would be taken by human being. For that such a training module will help to make rational financial decision.

Behavioral finance impacting not only your decision-making but also impacting your financial acumen as well as business acumen. To minimize the effect of behavioral finance or financial anomalies and biases, we should analyze financial data through predefined algorithms and also implement some applications based on artificial intelligence so that human intervention should be minimized. Due to involvement of emotions and alternative solutions, humans are irrational especially in financial acumen. So we should try to overcome the impact of behavioral finance by using IT solutions based on artificial intelligence for financial acumen.

Further, it is important to analyze impact of financial market anomalies and personality traits on financial decision-making, so that financial acumen will improve and new methods of financial acumen will develop for the betterment of the individuals and the society.

KEYWORDS

- **behavioral finance**
- **financial acumen**
- **pricing inefficiencies**
- **psychological biases**
- **financial decision**
- **heuristics**

REFERENCES

Ariely, D., Huber, J., & Wertenbroch, K. (2005). When do Losses Loom Larger than Gains? Journal of Marketing Research, 42, 134–138.
Arkes, H. R., & Blumer, C. (1985). The Psychology of Sunk Cost. Organizational Behavior and Human Decision Processes, 35, 124–140.
Belsky, G., & Gilovich, T. (1999). introduction.behaviouralfinance.net. Retrieved August 10, 2013, from behaviouralfinance.net.
Benesh, G., & Peterson, P. (1986). On the Relation Between Earnings Changes,Analysts' Forecasts and Stock Price Fluctuations. Financial Analysts Journal, 42(6), 29–39.
Bernerd, B., & Thomas, J. (1989). Post Earnings Announcements Drift: Delayed Price Response or Risk Premium. Journal of Accounting Research, 27, 1–36.
Bernstein, P. L. (1998). Against the Gods: The Remarkable Story of Risk. USA: John Wiley and Sons. Birau, F. R. (2012). The Impact of Behavioural Finance on Stock Markets. Retrieved February 20, 2014, from researchgate.net: http://www.researchgate.net/publication/258698903_THE_IMPACT_OF_BEHAVIOURAL_ FINANCE_ON_STOCK_MARKETS
Bloomfield, R. (2010). Traditonal vs Behavioural Finance. Retrieved February 2014, from psu.edu: http://www.personal.psu.edu/sjh11/ACCTG597 E/Class02/ BloomfieldWP10TraditionalVBehav ioural.pdf
Brabazon, T. (2000). Behavioural Finance: A New Sunrise or a False Dawn? Limerick: University of Limerick.
Branch, B. (1977). A Tax Loss Trading Rule. Journal of Business, 50(2), 198–207.
Branch, B., & Chang, K. (1985). Tax Loss Trading—Is the Game Over or Have the Rules Changed? Financial Review, 20, 55–69.
Brown, G. W. (1999). Volatility, Sentiment, Noise Traders. Financial Analysts Journal, 55(2), 82–90.
Chandra, A. (2008). Decision-making in the Stock Market: Incorporating Psychology with Finance. Retrieved 2015 8, January, from http://ssrn.com: file:///C:/Users/hpcompaq/ Documents/My%20Music/Downloads/ SSRN-id1501721%20(1).pdf
Chen, G. M., Kim, K. A., Nofsinger, J. R., & Rui, O. M. (2004). Behavior and Performance of Emerging Market Investors: Evidence from China. Retrieved December 29, 2014, from www.ccfr.org.cn: www.ccfr.org.cn/cicf2006/cicf2006paper/2006 0104133427.pdf
Dehnad, K. (2011). Behavioural Finance and Technical Analysis. (Shojai, Ed.) The Capco Institute Journal of Financial Transformation, 32, 107–111.
Fama, E. F. (1970). Efficient Capital Markets: A Review of Theory and Empirical Work. Journal of Finance, 25(2), 383–417.
Fama, E. F. (1998). Market Efficiency, Long-Term Returns and Behavioural Finance. Journal of Financial Economics, 49, 283–306.
Gustavo, B. (2010). Herbert A. Simon and the Concept of Rationality: Boundaries and Procedures. Brazilian Journal of Political Economy, 30, 3(119), 455–472.
Herbert, A. S. (1979). Rational Decision-making in Business Organisations. The American Economic Review, 69(4), 493–513. Indian Journal of Commerce & Management Studies EISSN: 2229–5674 ISSN: 2249–0310 Volume VI Issue 2, May 2015 18 www.scholarshub.net
Hertwig, R., & Hoffrage, G. G. (1997). The Reiteration Effect in Hindsight Bias. Psychological Review, 104, 194–202.

Huberman, G. (2001). Familiarity Breeds Investment. The Review of Financial Studies, 14(3), 659–680.

Jureviciene, D., & Jermakova, K. (2012). The Impact of Individuals' Financial Behaviour on Investment Decisions. Electronic International Interdisciplinary Conference, (pp. 242–250).

Kahneman, D. (2003, December). Maps of Bounded Rationality: Psychology of Behavioural Economics. The American Economic Review, 93(5), 1449–1475.

Kahmeman, D., & Tversky, A. (1973). On the Psychology of Prediction. Psychological Review, 80(4), 237–251.

Kahneman, D., & Tversky, A. (1971). Belief in Law of Small Numbers. Psychological Bulletin, 76(2), 105–110.

Kahneman, D., & Tversky, A. (1979). Prospect Theory: An Analysis of Decision Under Risk. Econometrica, 47(2), 263–291.

Kahneman, D., Knetsch, J. L., & Thaler, R. (1990). Experimental Tests for the Endowment Effect and the Coase Theorem. The Journal of Political Economy, 98(6), 1325–1348.

Kanehman, D., & Tversky, A. (1972). Subjective Probability: A Judgment of Representatives. Cognitive Psychology, 3(3), 430–454.

Kaneko, H. (2004). Individual Investor Behavior. Japan: Security Analysts Association of Japan.

Kristensen, H., & Garling, T. (1997). Anchor Points, Reference Points, and Counteroffers in Negotiations. 27(7).

Lin, H. W. (2011). Elucidating Rational Investment Decisions and Behavioural Biases: Evidences from the Taiwanese Stock Market. African Journal of Business Management, 5(5), 1630–1641.

Monti, M., & Legrenzi, P. (2009). Investment Decision-making and Hindsight Bias. Retrieved January 5, 2014, from csjarchive.cogsci.rpi.edu: http://csjarchive.cogsci.rpi.edu/proceedings/200 9/papers/135/paper135.pdf

Morone, A. (2008). Simple Model of Herd Behavior, A Comment. Retrieved January 2014, from uni-muenchen.de: http://mpra.ub.uni-muenchen.de/9586/

Nofsinger, J. R. (2001). Investment Madness: How Psychology Affect Your Investing and What To Do About It. USA: Pearson Education.

Novemsky, N., & Kahneman, D. (2005). The Boundaries of Loss Aversion. Journal of Marketing Research, 42, 119–128.

Odean, T. (1998). Volume, Volatility, Price and Profit When All Traders are Above Average. Journal of Finance, 53(6), 1887–1934.

Opiela, N. (2005). Rational Investing Despite Irrational Behaviors. Retrieved January 8, 2015, from http://catalystwealth.com: http://catalystwealth.com/pdfs/rational_investin g.pdf

Parikh, P. (2011). Value Investing and Behavioural Finance. New Delhi: Tata McGraw-Hill.

Pompian, M. M. (2006). Behavioural Finance and Wealth Management. USA: John Wiley and Sons. Rabin, M., & Thaler, R. H. (2001). Anomalies: Risk Aversion. Journal of Economic Perspectives, 15(1), 219–232.

Ritter, J. R. (2003, September). Behavioural Finance. Pacific-Basin Finance Journal, 11(4), 429–437. Sadeghnia, M., Hooshmand, A. H., & Habibniko. (2013, April). Behavioural Finance Indian Journal of Commerce & Management Studies EISSN: 2229–5674 ISSN: 2249–0310 Volume VI Issue 2, May 2015 19 www.scholarshub.net and Neuro Finance and Research Conducted in this Area. Interdisciplinary Journal of Contemporary Research in Business, 4(12).

Sewell, M. (2007). behaviouralfinance.net/behaviouralfinance.pdf. Retrieved April 2012, from behaviouralfinance.net.

Shane, F. (2005). Cognitive Reflection and Decision-making. Journal of Economic Perspectives, 19(4), 25–42.

Shefrin, H. (2000). Beyond Greed and Fear: Understanding Behavioural Finance and the Psychology of Investors. New York: Oxford University Press.

Shefrin, H. (2001). Behavioural Corporate Finance. Journal of Corporate Finance, 14, 113–124.

Shefrin, H. (2007). Behavioural Finance: Biases, Mean-Variance Returns and Risk Premium. Equity Research and Valuation Techniques Conference (pp. 4–11). Boston: CFA.

Shiller, R. J. (1981). Do Stock Price Move Too Much to be Justified by Subsequent Changes in Dividends? The American Economic Review, 71(3), 421–436.

Shiller, R. J. (2000). Irrational Exuberance. USA: Princeton University Press.

Solt, M., & Statman, M. (1989). Good Companies, Bad Stocks. Journal of Portfolio Management, 15(4), 39–45.

Statman, M. (1995). Behavioural Finance Versus Standard Finance. Retrieved January 25, 2014, from aiinfinance.com: http://www.aiinfinance.com/Statman.pdf

Thaler, R. (1980). Towards a Positive Theory of Consumer Choice. Journal of Economic Behavior and Organisation, 1, 39–60.

Thaler, R. H. (1999). The End of Behavioural Finance. Financial Analysts Journal, 55(6), 12–17.

Thaler, R. H. (2005). Advances in Behavioural Finance (e-Book) (Vol. 11). http://press.princeton.edu/titles/7944.html: Princeton University Press.

Thaler, R. H. (2008). Mental Accounting and Consumer Choice. Marketing Science, 27(1), 15–25.

Tseng, K. C. (2006). Behavioural Finance, Bounded Rationality, Neuro Finance and Traditional Finance. Investment Management and Financial Investments, 3(4), 7–18.

Tversky, A., & Kahneman, D. (1973). Availability: A Heuristic for Judging Frequency and Probability. Cognitive Psychology, 5, 207–232.

Tversky, A., & Kahneman, D. (1981, January 30). The Framing of Decisions and the Psychology of Choice. Science, 211, 453–458.

Tversky, A., & Kahneman, D. (1986). Rational Choice and the Framing of Decisions. The Journal of Business, 59(4), S251–S277.

Tversky, A., & Kahneman, D. (1991, November). Loss Aversion in Riskless Choice: A Reference-Dependent Model. The Quarterly Journal of Economics, 1039–1062.

Uzar, C., & Akkaya, G. C. (2013). The Mental and Behavioural Mistakes Investors Make. International Journal of Business and Management Studies, 5(1), 120–128.

CHAPTER 8

Current Trends in Finance in the Context of Adoption of Principle-Based Accounting Standards in Accounting Education

NEHA PURI* and HARJIT SINGH

Amity University, Noida, Uttar Pradesh 201313, India

**Corresponding author. E-mail: nehabajaj1984@gmail.com*

ABSTRACT

In today's world entering in the global territory has become indispensable for the organizations. These days principle-based accounting standards have been recognized all over the globe. The economies that are industrialized are also maintaining and complying with principle-based accounting standards to maintain records for the general purpose financial statements. The accounting standards provide an overview about recording of the business transactions which can be introduced in accounting education to ensure effective utilization of resources and formulating economic decisions for business prosperity and growth. The overgrowing business conditions in the global and foreign financial markets have encouraged the need for accounting education to be incorporated with the implementation of a uniform set of accounting standards and this will generate the platform for the common language embraced by the entrepreneurs.

8.1 INTRODUCTION

An International Financial Reporting Standard (IFRS) is a typical worldwide language to resolve the business issues such as window dressing, so that organization record is maintained and overcome global limits. IFRSs are dynamically replacing the wide range of bookkeeping gauges. Accounting standards help the administrative organizations in benchmarking the bookkeeping exactness. Bookkeeping principles are being built up both at national and global levels which create the differences in the recording of the financial transactions among the countries of the world which can be covered with the adoption of principle-based accounting standards instead of rule-based accounting. This leads to the introduction of principle-based accounting standards in the accounting curriculum. This will enhance to prepare the financial reports of the organization, which is acceptable in the global business environment. IFRSs reflect standardized accounting for transparency and concept clarity across the globe. IFRS is an accounting framework that sets out requirements for recognition, measurement, presentation, and disclosure of transactions and events reflected in the financial statements. The International Accounting Standards Board (IASB) created IFRS in 2001 in the public interest to provide a single set of high quality, comprehensive, and consistent accounting standards. With the possibility of trading beyond the domestic territories, it has become indispensable for the companies, especially export firms to replace their domestic accounting standards with IFRSs, India is not an exception. With the conversion of rules based on a principle-based set of accounting standards, it has become essential for both the education planners and educators to incorporate IFRS in the curriculum.

With the growth and development in the economy and increasing integration among global economies, Indian businesses are also increasing their assets worldwide owing to diversification, cross-border fusions, investments, or divestments. In these conditions, adopting IFRS for their financial reporting is essential for the Indian corporate globe. The Core Group of "Ministry of Corporate Affairs" of India (MCA) has recommended IFRS convergence in a phased manner from April 1, 2012.

The entities in emerging economies are increasingly accessing the global markets to fulfill their capital needs by getting their securities listed on the stock exchanges outside their country. Capital markets are, thus, becoming integrated, consistent with this worldwide trend. IFRS has become the global standard for the preparation of financial statements by the public companies. The adoption of IFRS creates a means to the investors to compare the results. Around 90 countries have confirmed the usage of IFRS in the financial reports. An investor can access the overall performance of different companies by comparing the financial reports which have been made on the basis of uniform set of accounting standards. This helps the investors to take potential decisions for the investments to be made in the different countries around the world. IFRS helps in fostering trust, quality financial reporting and improved regulation through transparency, accountability and improving investor's ability to analyze the investment opportunities or mitigate risk across a global economy. International Financial Reporting Standards have been fully adopted by 90 countries, whereas the increasing in the number of countries to adopt IFRS will reach more than in 120 nations soon. The adoption of IFRS has brought uniformity in the recording of the business transactions. IFRS has become a global accounting standard for the financial statements preparation by the public entities.

8.2 BACKGROUND OF ACCOUNTING STANDARDS

In early time, the traders were expected to fulfill a small segment of investors including friends, families, and associates about the fiscal modesty of their businesses. In today's era, numerous investors invest their cash into various companies all around the world and it becomes mandatory for the business organizations to prepare financial transactions not for their shareholders, employees, and other stakeholders, but also for government and taxation authorities.

In 1966, International Accounting Standards (IAS) evolution incorporated with a recommendation IASB to establish standards, which was adopted worldwide. The Accountants Group formulated; moreover, it started to publish documents on various topics covering accounting policies, and principles, which helped to form the groundwork for accounting standards, which appeared later. The "International Accounting Standards Committee" (IASC) designed in the year 1973 with the purpose of evolving accounting standards that were monitored worldwide.

The IASC proposed a sequence of standards termed as "International Accounting Standards," titled and came from IAS 1 to IAS 41. IASB in 2001, replaced IASC with IFRS Foundation. The IASB proclaimed that it will monitor the standards already allotted by IASC, but specified that any new standards would be known as part of a series called the IFRS, evolved by the IFRS Foundation. IFRS was made to fulfill the following objectives:

- To frame and develop standards in the interest of the public
- To meet the requirements of high-quality set of comprehensible standards
- To develop standards that are accepted internationally
- To have uniform accounting standards worldwide

The funds to incorporate these standards have been raised by the IFRS Foundation, an independent, nonprofit organization from banks and another desiring organization. The IASB is made up of board members who are accounting experts from around the world who have a good knowledge of standards and academic work.

These standards are composed approach records formulated by accounting bodies, identifying different parts of estimation, taking care of, and exposure of accounting policies and transactions recorded in the accounts. The arrangement of "Generally Accepted Accounting Principles" (GAAP) remains communicated to be standards for bookkeeping approaches and the information about the rules to build up the budget summaries ought to be shed in the records and introduced in the yearly reports. The unpretentious motivation behind standard setting expert is to advance the scattering of suitable and important data to speculators, lenders, loan managers, and other related parties that help in achieving consistency in monetary detailing and to guarantee consistency and equivalence in the financial records distributed by the organizations. In addition, the eminence of bookkeeping principles helps to endeavor budgetary building for the prosperity and growth of the organization. The study focuses on aspects other than accounting standards that revealed to financial engineering, including managers, high-powered incentives, inadequate disclosures, and sources of information outside the financial statements of companies (Dye et al., 2015).

1.3 BENEFITS OF ACCOUNTING STANDARDS

8.3.1 MAINTAINS UNIFORMITY

The standards framed have their own comprehensive set of rules, which are conflicting with each other; accounting standards help reduce such conflicting alternatives and ensures uniformity.

8.3.2 GUIDANCE TO ACCOUNTANTS

Accounting standards are arranged to provide day-to-day guidance to the accountants, which ensures that the operation of the business is conducted smoothly. The role of an accountant is to provide financial information that covers the qualitative aspects, such as relevancy, reliability, neutrality, and comparability—all of which can be attained with the help of accounting standards.

8.3.3 GLOBALIZATION OF BUSINESS

In today's era, globalization has become quite essential for the companies. Many multinational corporations face the problems of maintaining reports due to different GAAP adopted in different countries. Accounting standards bring up the uniformity and avoid confusions and misunderstanding.

8.3.4 IMPROVES RELIABILITY OF FINANCIAL STATEMENTS

Accounting standards help in bringing harmony in divergent accounting practices and protect the interest of the users of that financial information which results in bringing confidence. Financial statement's reliability relies on the accurate financial data obtained with the help of accounting standards.

8.3.5 SUBSTANTIAL FOR MAIN CONTRIBUTORS

Accounting standards help to furnish and understand the financial information in a way, which helps the key contributors like management, the board of directors, investors, and stakeholders to take the important decisions. Well-designed accounting standards improve and boost the confidence of the participants.

8.3.6 REDUCE PERSONAL BIASES

Accounting standards turn as a disciplinarian in the field of accounting. This reduces personal biases and the accountant has no option, but to practice accounting standard as stated. For example, there is a prescribed format for the preparation of Cash Flow Statement.

8.3.6 AVOIDANCE OF MANIPULATION

Accounting standards bring the uniformity in recording the financial transaction of the operations of the business, which helps in identifying the nature and type of transaction. Accounting standards are rigid, which help in avoiding manipulation.

8.3.7 FACILITATE COMPARISON

Accounting standards expedite comparison of the accounting records for different years among the companies belonging to the same industry located in different parts of the world.

8.3.8 RESOLVE CONFLICT OF FINANCIAL INTEREST

Accounting standards aid in resolving the issues related to financial concerns among different groups.

8.3.9 CREATE EFFECTIVE CORPORATE GOVERNANCE PROGRAMS

Accounting standards are a very noteworthy practice, which lead to operational corporate governance programs. It also provides a useful instrument to restructure corporate values.

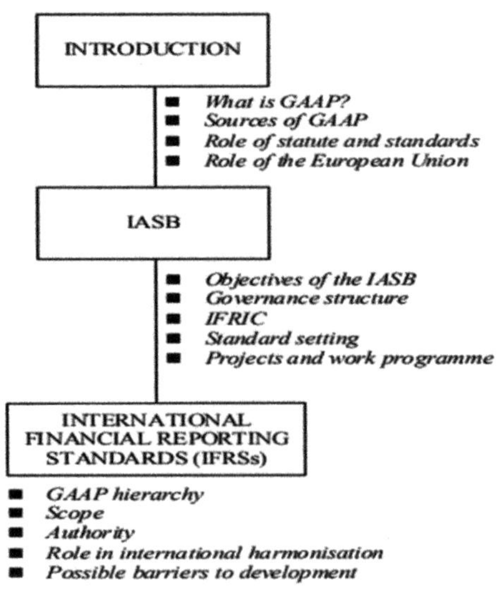

FIGURE 8.1 Overview of GAAP, IASB, and IFRS.
Source: Reprinted from https://www.icpak.com/wp-content/uploads/2015/09/IFRS-Study-Material.pdf

8.4 INTRODUCTION TO GAAP

Generally accepted accounting principles (GAAP) are a set of accounting standards and guidelines that furnish the business records for the organization. GAAP as a specific set of guidelines that have been formulated to help public entities to create and prepare the records for the financial transactions with the help of financial statements.

TABLE 8.1 Insight about Generally Accepted Accounting Principles (GAAP)

Economic entity assumption	Activities of the business must be kept separate from the activities of the owner of the business.
Monetary economic principle	Activities of the business which can easily be converted in terms of money should be recorded in the books of accounts for the accounting period. No nonmonetary transactions will be furnished in the books of accounts.
Time period assumption	Activities conducted in the business can be recorded in distinct intervals of time. For example, month, quarter, or fiscal year. The accounting period can commence from January to December or April to March.
Full disclosure assumption	All the information that is in relation to the business be reported either in the content of the financial statements or in the notes and summary to the financial statements.
Cost principle	The historic cost of the items of the business to be recorded in the financial statements. The amount recorded as per the historic cost of the item will not change.
Going concern principle	Activities or transactions in the business will intent to continue the operations of the business irrespective of any unpredictable future events and not to liquidate the business.
Matching principle	This principle explains the manner in which a business reports furnish income and expenses. The principle adopts accrual basis of accounting and the business transaction matches with the business incomes and expenses. For example, a commission of 5% is entitled to the salesman on the sales for the month of February. The commission was paid in the month of March. The expense related to the commission will be recorded in the month of February.
Revenue recognition principle	The principle explains the manner in which revenue or income is recognized.
Materiality	The concept of materiality explains that the financial transactions of a company to be material from the prospective of the financial statement preparation.
Conservatism	This principle explains that the business records should furnish fair and reasonable presentation of the financial data.

8.5 INDIAN GAAP: AT GLANCE

"Institute of Chartered Accountants of India" (ICAI) issued Indian GAAP primarily comprising 18 accounting standards. To provide help in the interpretation of financial statements, the ICAI has also issued guidance notes and "expert opinions" on specific queries raised by companies and accountants. In addition, the Indian Companies Act 1956 and various other industry-specific statutes prescribe certain minimum disclosures in the financial statements. The listed companies on the stock exchanges also need to comply with a few other accounting rules such as preparing cash flow statements and accounting for stock-based compensation.

- It is precisely considered for the Indian companies.
- Indian GAAP is followed by most of the Indian companies while preparing their accounting records.
- Indian companies only adopt Indian GAAP to show a true and fair view of their financial transactions.
- Indian companies are required to record and prepare their financial transactions in the balance sheet, profit and loss account, and cash flow statement.
- In Indian GAAP, the revenue is considered, when the companies charge for products/services.
- Indian GAAP does not require an exchange rate.

8.6 ROLE OF IASB

IASB stands for International Accounting Standard Board. According to IFRS Foundation, a statement is to develop IFRSs that bring transparency, accountability, and efficiency to financial markets around the world. Our work serves the public interest by fostering trust, growth, and long-term financial stability in the global economy.

8.7 IFRS EDUCATION: AT GLANCE

8.7.1 CONCEPTUAL FRAMEWORK OF IFRS

IFRS frames out the objectives of preparing general purpose financial statements. IFRS assists the accounting setter (IASB) in framing the financial

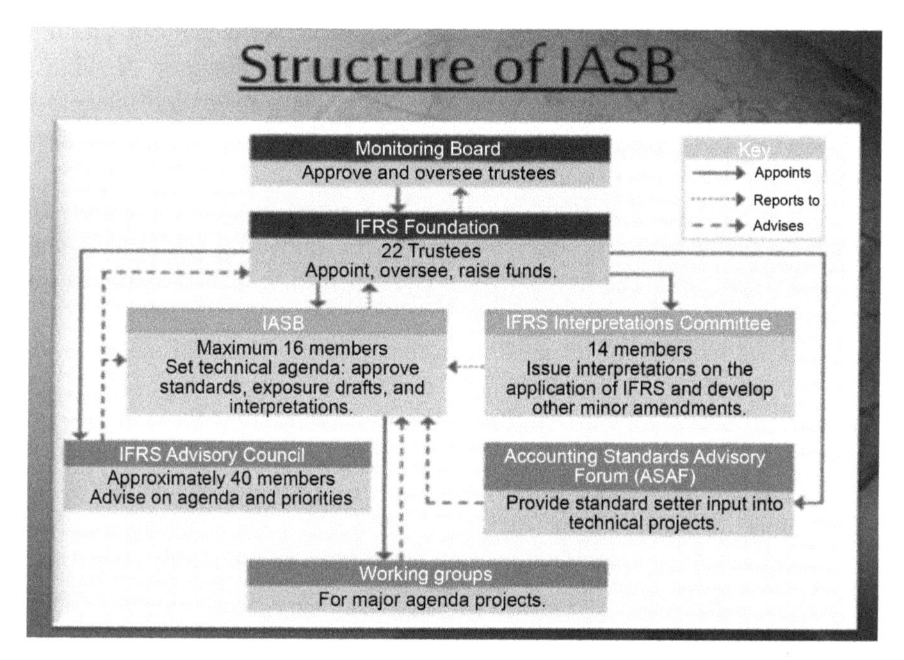

FIGURE 8.2 Structure of IASB.

Source: Reprinted from https://www.iasplus.com/en/resources/ifrsf

statements. These standards provide the outline and financial position about the company to the auditors. The adoption of IFRS by various economies has resulted in progression of the business. The qualitative characteristics pertaining to the adoption of IFRS or principle-based accounting standards result in:

- Usefulness of financial statements
- Comparability
- Understandability
- Timeliness
- Verifiability
- Recognition
- Faithful representation
- Relevant
- Consistent
- Fair value measurement.

The enrichment of the above qualitative characteristics of IFRS brings fair value measurement in the recording of the financial statements. These standards take into consideration the material aspects based on the magnitude or nature or both which can influence in formulating economic decisions. The principle-based accounting standards represent information that must faithfully develop economic phenomena. It purports to represent complete, neutral, and free from error business information.

IASB formulates IFRSs, once known in the past as IAS, for setting up standards, interpretations, and framework for preparing and presenting financial statements. The new IASB took control of the IASC's duty to set IAS on April 1, 2001.

The ICAI has clearly released a summary of its references on the IAS selection agenda, which are largely combined with IFRSs. The ICAI proposals considered, by the "Indian Corporate Ministry" (MCA) in its final decision on conclusion on usage of the accounting standards.

IASB Chair Hans Hoogervorst in his discourse facilitated by ICAI talked about the selection of IFRSs can prompt a solitary arrangement of top-notch worldwide bookkeeping guidelines. In an ongoing blog entry (link to ICAI site), K Raghu, President of the ICAI, noted in the accompanying accordingly:

"Here I must mention that we have very consciously decided to converge with IFRS and not fully adopt it, mainly because a resurgent India is very different from its peers in the West or elsewhere, with its own priorities and preferences. There is no change in that stand as of now. However, in the longer term, we may consider his viewpoint, but only after sorting out some crucial issues and testing the impact of our converged Ind AS on [the] corporate sector."

Tracy (1999), in his book, reveals that cash is called the "lubricant of business." He emphasized that accurate records are important to be maintained to safeguard the effect of financial warning signs, which includes high debt, high, expenses, etc. (Socea, 2012). In his study, he discussed the relationship between managerial decision-making and financial accounting information; it is found that the managers' decisions must be inclusive of subjective and irrational elements. The study reveals that how incomplete business transactions results in the failure of the business within the retail industry (Nyathi and Benedict, 2017). He found that accurate financial records are insights for taking quality business decisions. In the study conducted by Kapellas and Siougle (2017) reveals that investment efficiency, cash flow sensitivity, stock market efficiency, and over and under-investment are important factors for the cash flow statement. The study focused on the disclosure of the financial statements and cost of equity (Daske et al., 2008). In the study conducted by Poongavanam (2017), it was found that the information disclosed in the financial statements helps to understand and judge the profitability and financial soundness. A similar study conducted reflects that the overall financial performance depends on financial reporting practices, financial trend, financial health which constitutes major factors to report the financial position and performance of the company by using various tools of financial analysis such as ratio analysis comparative balance sheet and common size income statement (Ganga et al., 2015). For better and effective reporting quality of the companies, the IFRS adoption can utilize the domestic accounting standards (Kim, 2012) and the adoption will allow uniformity that brings greater comparability and consistency of the financial reports (Ahmed and Neel, 2013; Barth et al., 2008; Brochet et al., 2012). The major reason for changing to a uniform set of norms has been explained by the author such as comparability, creditability, relevancy, and global market access of financial reporting. The paper examines the IFRS selection system in India. In conclusion, the paper finishes up for those routes through which these issues may tend to be addressed (Kaur, 2011). The author discussed that the fundamental interest for selection alternately merging about IFRS will be callous business feeling. Expanding cross fringe contributing need posed a test should organizations as they face numerous norms. Adoptions of IFRS will promise better accounting performance which results in giving benefit to several related parties such as academicians. The author focused on Indian accounting standards which requires harmonization with worldwide bookkeeping standard (Jadhav, 2014; D'Souza, 2016). Revealed Indian bookkeeping guidelines and justification of IFRS adoption in India and describing the arrangements to be incorporated in making relevant financial statements and preparedness of IFRS adoption in India (Muniraju & Ganesh, 2016). The author focused on adoption and selection of IFRS in India. IFRS adoption will lay down investment trading which leads to growth and prosperity of the business. The globalization may be making a merging of investment trading, political, and social procedure. Another study examined the cross-listing role in United States for adoption of IFRS (Alali & Cao, 2010). Moreover, the author focused on the importance of implementation of IFRS. The implementation process results in adoption of uniform set of quality standards by the organizations (Muniraju & Ganesh, 2016). Studied about the financial statements for more than 3600 multinational organizations which create business in the distinctive parts of India. The various elements of research models in the literature on the study strategy (Struyven et al., 2003) examined the impacts of research methods on teaching efficiency at distinct concentrations of ACCA. The previous studies have distinct views of teaching evaluation methods and do not regulate for the potential endogeneity of the accounting approaches.

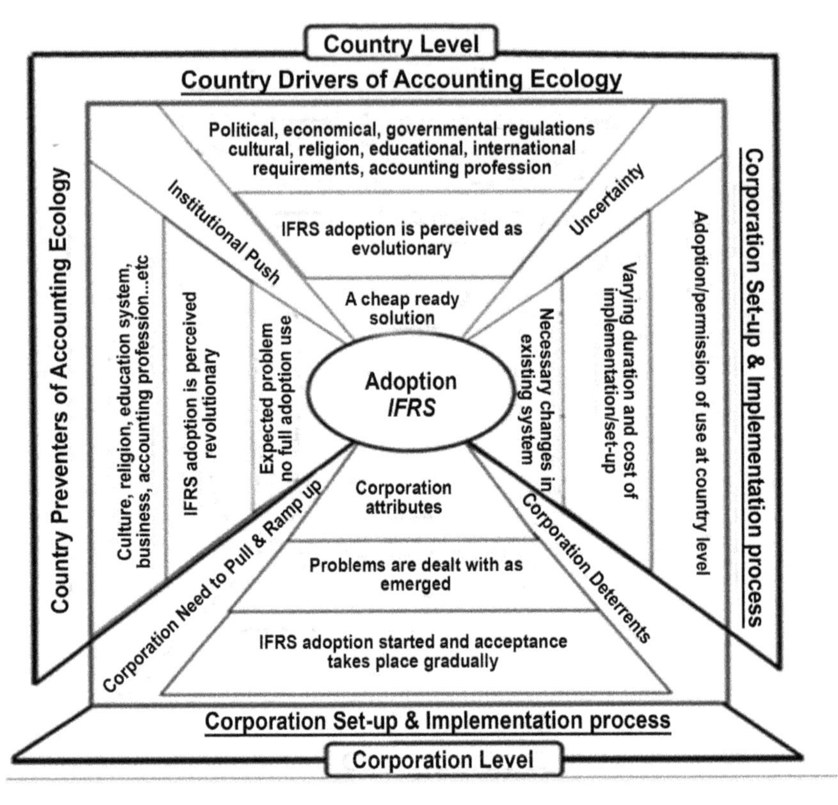

FIGURE 8.3 Adoption of IFRS.
Source: Reprinted with permission from Al□Htaybat (2017). © John Wiley.

In India, many people are still not aware of IFRS, its complexities, and its effects. An adjustment in the revealing setup will involve awareness, preparation, and training of these new norms and frameworks. The reality is that despite gaining balance, the book's rules have not been continuously seen in our country and people are seeking advantage balancing the books. There is a need to investigate some established lacunae of our common balancing of books, such as purchasing per share, information on future money torrents, amalgamation, mergers, acquisitions and perhaps even the balancing of books regulations, which are tough to follow due to the lack of a reasonable test principle such as verifiable expense, input concept, etc. Before IFRS is joined with as and rehearsed in the Indian condition, the instructive framework ought to set itself up to coordinate IFRS into its educational module so Indian graduates would get a handle on it and can make most ideal use of review money-related reports in light of IFRS. Indian GAAP required including the essentials in balancing the books educational module and in our training framework.

In spite of the fact that the endeavors to compose uniform norms of money-related detailing are principally determined by their immediate and quick effect on capital markets, the companies additionally have major instructive results, which have diverse level of extent and that they merit more consideration from scholastics and those accused of the duty to create guidelines.

The need to bring changes in reading material, course content, classroom talk, and examinations system alongside the expert examination for Certified Professional Accountant accreditation is required. Without a legitimate standard for some class room sessions, course readings, class dialog, and examinations, the learning outcomes will not be effective. The classroom talk draws less ability to balancing of the books and eventually the balancing of the book records can be undertaken by adopting live projects. This will enhance to understand present business situations and possibilities to take effective decisions. Uniform set of accounting standards measures contribute a significant role in the field of accounting.

To close, finding harmony between uniform benchmarks and standards, and characterizing the degree of their separate parts in monetary revealing, are difficult errands. Standard-setters think that it is hard to know which principles are unrivaled, and what ought to be the criteria for positioning the option norms. Social orders that rely on standards and convention additionally can stall out in wasteful arrangements and it might take change developments, even outfitted uprising, to discharge them.

Despite the fact that the improvement of a uniform arrangement of balancing the books measures has its own avocation, however considering the statistic components like, training and mindfulness level of various economies of the world, the use of IFRS to every single open firm crosswise over most nations of the world through administrative fiat does not appear to be suitable.

The disclosure and development of better strategies for money-related problems or utilizing better techniques for balancing the books are required to be discussed in the classes through different university programs on balancing the books in college educational module. The expected advantages as the expanded likeness of monetary reports universally. To figure accounting standards with a view to helping the council of ICAI is developing and setting up in India. Identifying the accounting standards, which opens avenues for the organizations, the classes, have to be moved toward concentrating on line-and-verse use of those guidelines, and not on a basic examination of the benefits of giving option on balancing the books remedies in business schools.

8.7.2 IFRS THEORETICAL FRAMEWORK

The IFRS Theoretical Framework is as follows:

- Develop future accounting norms and review current accounting requirements to guarantee that the IASB is consistent across norms.
- To help the IASB in encouraging records, laws, accounting norms, and process synchronization by providing a base for the number of alternative accounting transactions permissible as per the accounting standards.
- To help the creation of domestic standard bodies of domestic accounting norms.
- To support, financial statement preparers in applying "International Financial Reporting Standards" and addressing issues related to financial statements.
- Support related parties of the financial statements in interpretations of the financial information enclosed in the financial statements that remain in accordance with IFRSs.
- Assist auditors in assessing whether economic statements comply with global accounting norms.
- Provide data on its methodology to those interested in the job of the IASB in formulating accounting requirements.

8.7.3 FEATURES OF IFRS

- IFRSs are principle-based accounting standards as compared to rule-based accounting standard. It means that the companies have to report the essence of each small transaction, which is to be finally audited and approved. In this manner, the transaction cannot be manipulated easily.

- IFRSs are drafted in a clear and simple language and are easy to understand and apply.
- Under the IFRSs, the "historical cost concept" of recording the fixed assets has been replaced by "current cost system" for a more accurate and realistic position of the business enterprise.
- Under IFRSs, useful life of the assets has to be reassessed or computed repeatedly until the asset is actually removed from the books of accounts.
- Under IFRSs, the assets, liabilities, revenues, and expenses are not reported in the local currency but in its functional currency. Well-designed currency means the currency of the place or environments where the entity operates, which may be different from the local currency.

8.8 CONVERGENCE OF ACCOUNTING STANDARDS TO IFRS

8.8.1 BENEFITS OF CONVERGENCE TO VARIOUS STAKEHOLDERS

There are numerous benefits of convergence of accounting standards which is stated as below:

1. Investors

Indian Accounting Standards convergence with IFRS makes bookkeeping data more convenient and uniform which results in arranging and encouraging financial specialist to pool resources in the interest of investors. It will likewise create a better understanding of money related articulations, which expand the certainty among the individuals as financial specialists. Foreign investors would get a better chance to recognize the accounting processes required in the recording of the business transactions as IFRS will bring up uniformity in the accounting standards.

2. Industry

IFRS will increase confidence among industry specialist in handling financial reports. The procedure adopted in IFRS is less complex and will reduce the expense of setting up money-related articulations.

3. Accounting professionals

There would be at first numerous issues, however, union with IFRS would unquestionably help the bookkeeping experts, and it will be useful for them to offer their ability and skills over the globe. This adoption will bring a common platform for all the accounting professionals throughout the world to have a healthy discussion.

4. Corporate scenario

IFRS convergence would bring together the Indian corporate world's notoriety and lasting connection with worldwide currency components. In addition, it would profit the corporate substances back in India for a few reasons. The bigger, the quantity of consistency maintained in the external and internal reporting, results in better access to global budget markets, because it will make the corporate world more focused.

5. The economy

IFRS would offer assistance to the industry, which is useful to the corporate world as the comparability, reliability, and uniformity in the accounting standards would help in understanding the profit generation of the companies. In addition, this would lead to knowing about the exact tax to be imposed. Adoption of IFRS will create a common platform for the existing economies.

6. Enlarge comparability

Companies using comparable norms to prepare financial statements can compare more completely with each other. This is very helpful in comparing businesses based in distinct nations; as otherwise, in the preparing of these papers they may have distinct processes and instructions. This higher comparability assisted investors to better recognize where their investments should be profitable.

7. Gainful for prospect and existing investors

By enhancing and simplifying reporting norms, IFRS can assist fresh and small investors by placing investors in a comparable position with professional investors who were not achievable under prior norms. This also includes a decreased danger for these investors when trading, as due to the nature of financial statements, the experts will not be prepared to take benefit of it.

8. Flexibility

Instead of rules based, IFRS uses a philosophy that is principles based, which set the standards in a way that help in achieving a goal with a reasonable assessment of different ways of performing tasks. This would give businesses the freedom to adopt IFRS in their particular circumstances, resulting in more readable and convenient financial statements.

8.9 CHALLENGES IN THE PROCESS OF ADOPTION OF IFRS

Many people say that IFRS is in good sink in a way that it has led to more complexity and also may be for auditors, the compliance would be difficult. The biggest challenge is the adoption of IFRS by the management of the organizations. The challenge in the adoption of IFRS into caters the needs of stakeholders/investors with finite resources. The step toward the movement of national standards to IFRS will bring a change in the country language. This may result in facing more difficulties from the perspectives of the accounting bodies. The magnitude of the change that will brought by the adoption of IFRS depends upon the different entities/enterprise. Tremendous burdens that Indian corporate may confront:

- **Awareness of extensive practices**

The entire procedure of money-related reporting sharpens necessities to encounter a splendid change after the collective event of IFRS education to vanquish the measure of complexities between the accounting standards. It would be a test to perceive thoughtfulness regarding IFRS education and its impact among the customers interested in cash and cash equivalents.

- **Training**

The best obstruction for the teacher in executing IFRS education is the knowledge about the benefits of imposing IFRS educational courses. IFRS has been perceived with impact from 2011, yet it is monitored that there is noteworthy nonappearance of designed IFRS trainers. The ICAI has begun IFRS training programs for its members and other contributed individuals. However, there exists an incomprehensible isolated between trained professionals required and designed experts.

- **Amendments to the present law**

Different happenstances with the present laws are other veritable trouble, which is found in the Companies' Act 1956, Securities and Exchange Board of India heading, saving money laws and controls and the insistence laws and controls. Beginning at now, the reporting necessities are tended to be controlled by comptrollers in India and their methodologies supersede definite laws. IFRS education does not see such repudiating laws.

- **Taxation**

Presently, Indian tax laws do not see accounting standards. In this way, a total stimulates of tax laws are not governed by the Indian lawmakers quickly. In July 2009, a planning get-together was incorporated by the MCA of the government of India, with a view to see the shifting true blue and honest to goodness changes required for joining and to set up a guide for satisfying the same.

- **Fair regard**

IFRS, which uses a sensible assistant as an estimation base for concerning most of the things of budgetary illuminations, can bring a gigantic measure of irregularity and subjectivity to the cash related statements.

- **Management pays form**

This is an immediate aftereffect of the budgetary results under IFRS that are no doubt going to be unmitigated not precisely the same as those under the Indian GAAP. The assertions would be renegotiated by changing terms and conditions relating to the affiliation of IFRS.

- **Reporting structures**

The disclosure and uncovering necessities under IFRS are by no means whatsoever, the same as the Indian declaring essentials. Affiliations would need to ensure that the present business-uncovering model is changed as per necessities of IFRS education. The information structures should depend upon to get new essentials related to settled assets, divide, related gathering trades, the closeness of truthfulness, control, and constraining the risk of business unsettling dominance while modifying or changing the information structures.

- **Complexity in the assignment**

The pros feel that the best danger in the convergence of Indian GAAP with principle-based education is the way that will bring the change in the books. The changing over to IFRS will create the disperse quality with the introduction of insights, for instance, exhibit regard and sensible regard. In IFRS structure, treatment of costs like premium payable on the recovery of debentures, markdown allowed on the issue of debentures, ensuring commission paid on issue of debentures and so forth is not the same as per the present framework used. This would bring a complete adjustment in pay illumination bringing complexities in the present way of recording financial transactions.

- **Time**

It is one of the major constraints that adoption and learning of IFRS are time consuming. The accounting experts are inclined toward the usage of a customary way of maintaining the books of accounts, which may result in more time consuming if IFRS will be applied for maintaining financial transactions.

8.10 NONINTEGRATION OF IFRS EDUCATION

- The most basic obstacle of IFRS relates to the costs of the application by multinational companies which require the change in their internal structures to make it better as per the new critical standards.
- The issue of sorting out IFRS in all countries, as it will not be possible by prudence of various reasons past IASB or IASC control, as they cannot keep up the utilization of IFRS by all countries of the world.
- To incorporate IFRS education it requires a lot of consistency and rigorous knowledge of standards.
- Small- and medium-sized enterprises (SMEs) have an issue in a change to IFRS that has stressed with high importance is the usage of a sensible driving force as the fundamental another inconvenience of IFRS is that it is incredibly perplexing and costly. The collecting of IFRS requires an imperative soft spot for SMEs as they will be hit by the tremendous move costs and the level of the strangeness of IFRS may not be eaten up by SMEs.
- Workshops and seminars are required to be conducted to keep up-to-date knowledge of IFRS standards, which results in an increase in the cost of training programs.
- There is a lack of adequate funding, provision of appropriate materials, modification of significant legal frameworks.
- Lack of use of ICT as a means of achieving effective IFRS education.
- In addition, the examiners comprehend in making the change in their outmoded curriculum.

8.11 POLICY RECOMMENDATIONS FOR EDUCATION INSTITUTES

1. The accounting curriculum has to be revised to the minimum standards so that the integration of IFRS can be possible into accounting education.
2. The study material IFRS should be circulated in the business schools to serve as an orientation document for accounting lecturers.
3. Legal frameworks should be amended to accommodate IFRS.
4. Organizing seminars, workshops, and train-the-trainer programs to be implemented to bring up-to-date accounting knowledge.
5. Government funds should intervene in IFRS education by way of sponsoring to be provided to the accounting lecturers to attend overseas training, seminars, and workshops.
6. All accounting education lecturers should be computer literate in order to easily learn IFRS.

8.12 CONCLUSION

IFRSs are accounting standards based on fairness that provide less guidance on prescription, interpretation, and compliance than some regional standards do (Hodgdon et al., 2011). As far as accounting standards are concerned, such as going concern concept, materiality, and relevant disclosures, and problems experienced in the implementation of the majority of IFRSs, such as presentation and disclosure, classification, recognition/de-recognition, and measurement. The adoption of IFRS in the curriculum will broaden the scope for the students to understand the financial statements prepared by the companies located outside national territory. Students will be benefitted in a way that more job opportunities can be facilitated with IFRS incorporation in education.

KEYWORDS

- **accounting standards**
- **education**
- **financial statements**
- **IFRS**
- **GAAP**

REFERENCES

Ahmed, A. S., Neel, M. N. (2013). Does mandatory adoption of IFRS improve accounting quality? Preliminary evidence. Contemporary Accounting Research, 30(4), 1344–1372.

Al-Htaybat, Khaldoon. IFRS Adoption in Emerging Markets: The Case of Jordan. Australian Accounting Review. 12 July 2017. https://doi.org/10.1111/auar.12186

Alali, F., Cao, L. (2010). International financial reporting standards—Credible and reliable? An overview. Advances in Accounting, 26(1), 79–86.

Barth, M. E., Landsman, W. R., Lang, H. M. (2008). International accounting standards and accounting quality. Journal of Accounting Research, 46(3), 467–498.

Brochet, F., Jagolinzer, A. D., Riedl, E. J. (2012). Mandatory IFRS adoption and financial statement comparability. Contemporary Accounting Research, 30(4).

Daske, H., Hail, L., Leuz, C., Verdi, R. S. (2008). Mandatory IFRS reporting around the world: Early evidence on the economic consequences. Journal of Accounting Researc, 46(5), 1085–1142.

D'Souza, D. A. (2016). Indian Accounting Standards (IND AS). Snow White Publication.

Dye, R. A., Glover, G. C., Sunder, S. (2015). Financial engineering and the arms race between accounting standard setters and preparers. Accounting Horizons, 29(2), 265–295.

Ganga, M., Kalaiselvan, P., Suriya, R.. (2015). Evaluation of financial performance. International Journal of Scientific and Research Publications, 5(4).

Hodgdon, C., Huges, S. B., Street, D. L. (2011). Framework-based teaching of IFRS judgements. Accounting Education, 20(4), 415–439.

Jadhav, S. B., A study of International Accounting Standard & Indian Accounting Standard, Indian Journal of Applied Research, 4(2). ISSN - 2249-
555XRESEARCH.

Kapellas, K., Siougle, G. (2017). Financial reporting practices and investment decisions: A review of the literature. Industrial Engineering & Management, 6(4), 235.

Kaur, J. (2011). IFRS: A conceptual understanding and framework. Indian Journal of Finance, 5(1), 49–56.

Kim, J. L. (2012). The impact of mandatory IFRS adoption on audit fees: Theory and evidence. The Accounting Review, 87(6), 2061–2094.

Muniraju, M, Ganesh, S. R. (2016). A study on the impact of international financial reporting standards convergence on Indian corporate sector. Journal of Business and Management, 4, 34–41.

Nyathi, M., Benedict, O. H. (2017). An analysis of bookkeeping practises of micro-entrepreneurs in the retail clothing industry in Cape Town. Journal of Entrepreneurship & Organization Management, 2017, 6, 2

Poongavanam, S. (2017). A study on comparative financial tatement analysis with reference to das limited. IOSR Journal of Humanities and Social Science (IOSR-JHSS), 22(10), 9–14.

Socea, A.-D. (2012). Managerial decision-making and financial accounting information. Procedia - Social and Behavioral Sciences, 58, 47–55.

Struyven, K., Dochy, F., Janssens, S. (2003). Students' perceptions about new modes of assessment in higher education: A review. Optimising New Modes of Assessment: In Search of Qualities and Standards. Springer, the Netherlands, pp. 171–223.

Tracy J. A. (1999). How to Read a Financial Report Wringing Vital Signs Out of the Numbers, 5th ed, Wiley.

CHAPTER 9

Effect of Corporate Social Responsibility on the Firm's Financial Performance

SHALINI SRIVASTAV[1*] and AHMED. A. ELNGAR[2]

[1]*Amity University, Greater Noida Campus, Uttar Pradesh 201308, India*

[2]*Beni-Suef University, Beni Suef City 62511, Egypt*

**Corresponding author. E-mail: shalini_15@yahoo.com*

ABSTRACT

In today's scenario, companies are working a lot on the concept of corporate social responsibility (CSR) and trying to evaluate the firm's performance on such basis. We all generally talk that the competitive action in terms of CSR should be thought as an important activity on the financial performance of the firm. We all understand that the competitive pressures can be the major reason to affect the CSR activities on the firm's financial performance. Recent researchers have found that there are two types of CSR. Positive CSR increases the firm's performance when the firm's competitive level is high, whereas negative CSR is good for people where there is less competition or no competition. In order to update knowledge and understanding, this study deals with the study of CSR on the performance of the firm.

9.1 INTRODUCTION: CORPORATE SOCIAL RESPONSIBILITY (CSR)

This study has considered many people's attention across the world and has given a new dimension and a new direction to the global economic concept. Companies are taking lot of interest in CSR because in the current scenario due to globalization and increase in the international business, CSR has created new demands due to increased competition worldwide. Every organization needs to be transparent in its approach in order to sustain in the market. Also, the government is taking a lot of effective measures to increase the lifestyle of everyone and also the need of the society and its overall development which consists of living standards of people, improving the basic society needs, and environment in order to enhance the living standards of the people. It is seen that various analysts are giving their major changes in terms of giving all the favorable and good ratings to all companies with good CSR strategies. As a result, the spot light is also turning its impact on the business activities in the society and all the companies are trying to find their business by means of CSR.

The understanding of CSR is more than 30 years old and is an integral part in understanding the history of any business organization. The major objective of this research is to understand the importance of CSR and operating competitiveness of any firm in terms of money, operations, no time foundation, product dispatch, complete package, and other benefits from the gradual improving environment of a developing country (Famiyeh, 2017).

CSR has created a positive impact on all the consumer minds and as a result of which can create a positive environment and will encourage people to improve to performance of any company. Moreover, there is also seen a comparative data by the authors from Korean manufacturing industry (Lee & Jung, 2016).

Therefore, the authors require more activities in the company's financial performance and CSR relationship. The study deals with the following activities.

First: It is advised to follow future policies of environmental protection rather than looking for past trends and feeling bad about it. It is always better to limit and reduce our research to the current and future study as we all are in the position to think about the firms' economic position and its environmental support.

Second: In this case, we understand the environmental investments both ways like thinking about the past and the future trends both short term and long term. Although previous studies, are not considering the environmental effects in the long run, so it can also be considered to have a major role in the future prospects and in other performances (Nakamura, 2014).

Also, in the future there are various scholars who are responsible and who are paying more attention to CSR than corporate reputation.

In the current scenario, there are a large number of financial institutions that are actually accepting the concept about changes in the increase in social and environmental responsibilities. The major commitments to all the practices are having interesting aspects and applications. The coordination between CSR and the banking sector can be interpreted via all the social and political activities, which is trying to find out CSR as a financial tool which is available to almost all companies in order to have their legacy for all the shareholders and in order to improve their investor investment criteria (Antonio Lorena, 2018).

Bank managers should be encouraged to implement and disclose socially responsible practices, to the extent that this creates an opportunity to more effectively fulfill the mandate given by owners. This contrasts with the agency vision of CSR. Second, thanks to the contribution of CSR in terms of differentiation and reliability, a win–win model removes the consequences of CSR that are seen in the trade-off approach, making the expected impact of CSR more consistent with stakeholder theory and the resource-based view. Third, our findings contribute to an evolutionary vision of the interaction between banks and customers (depositors and financed companies), leading from a transactional to a more relational approach. Fourth, the results of the current study should attract the attention of policy makers and authorities (Francesco Gangi, 2018).

It is seen that CSR is majorly about business organization as it is highly competitive and tries to enhance its firm. This has given a path way to researchers to find out the impact of CSR in business. As a result, the work in this area has produced major results in business. The major perspective tells us that CSR is providing us a competition which is finally strengthening any business in itself (Shafat Maqboo, 2018).

All those companies which are as small-sized companies, medium, and private companies are not added in this study. There are chances to understand research in a broad way. Moreover, it is interesting to understand the research activities of all small and medium-sized enterprises in order to understand the competition in the market. There are various scientists who are trying to study on the combination of CSR and company by inculcating various features of CSR in government organizations. It is seen that all the companies change their CSR features as per their requirements and there is a positive relationship in the CSR activities and a company's growth (Abdul Rahim Othuman, 2012).

Due to the presence of direct and connected relationship in the development of various CSR activities, the company performance is calculated on the basis of a multiple shareholders' investment point of view for a long lasting relationship in this context. In order to understand this activity, a sample size of approximately 500 small and medium enterprises was taken and analysis on the basis of partial least square technique is used for the calculation. The major results are telling us that the development of CSR initiatives are contributing to increase the competition in the market by performing both indirectly and directly, to manage the CSR activities by shareholders investment and also the ability of the company to perform. This paper is discussing about the impact of the society on investment alternatives and its intangible support which creates a major effect of CSR on its emerging competition (Jesus Herrera Madueno, 2016).

CSR disclosure majorly deals with all the activities affected, all the targets which are to be achieved, company expenditure in the stakeholder's interest are all the things in relation to CSR. It mainly deals with other reports like citizenship report, corporate citizenship report, or sustainability report. The main idea is to provide lot of information to various stakeholders, so that it becomes easy for them to calculate long- and short-term businesses which include cash flow, risk, also changes in finding societal and environmental issues (Richa Gautam, 2016).

This research tells us that there is a very strong relationship between CSR activities in Saudi Arabia (Zakat) and corporate financial position in the same country. This also tells us that Zakat also helps in contributing to both the major firms' value chain and also profitability to be taken care as a major process to increase profits and upgrade the performance while understanding and thinking about the benefit to the society. The changes to the various methods of calculation are a result (Nizar Al-Malkawi & Javaid, 2018).

For better understanding, the study tells us that internal CSR idea is playing a major role in understanding the implementation of the CSR contents to all the external stakeholders. Majorly, the study reveals that CSR involvement in all the major external efforts can improve the competitiveness of the bank due to the relationship between changes in citizenship and the positive reputation of the bank (Francesco Gangi, 2019).

Various other types of investors have various other types of CSR preferences and tastes for CSR firms. All the advertisers of various business groups whether affiliated or nonaffiliated have different behavioral patterns toward the CSR firms. Diversified behavior of various institutional investors is clearly depicted through this type of study. Various banks and other foreign institutions are very helpful and supportive of various CSR investments in any organization. Various other promoters in family and affiliations in the group are also interested in the CSR activities (Panicker, 2017).

It is very important to establish and create a very positive culture in order to reduce the corporate fraud which is very threatening and is impacting our societal norms. It is also very important for the investors' corporate board of directors and also regulators to consider CSR as a major concern for the top management, and all the moral values which is related on a negative and also affecting the occurrence and seriousness of corporate fraud. We are trying to find the bond between the regulators, corporate board of directors, and also business regulators in order to find out CSR as a major part of top management which is affecting the moral values and is all negatively related in the occurrence and impact of the corporate fraud. In order to strengthen the moral values among top executives and all the corporate employees by encouraging all the CSR activities in our society in order to excel the major outbreak of frauds done in corporate (Harjoto, 2017).

As we all are aware that nowadays CSR is becoming a part of our business; thus, it is becoming difficult to analyze its performance on separate basis. Initially, it is important to keep all the constant features and calculate the impact of financial performance of the company, after the implication of CSR practices and also before the same. In case, if it is not the case various empirical methods are adopted to identify the company's conduct and financial performance. This paper majorly deals with a study of selected research gaps in literature and it can finally contribute in the understanding of any business concern (Krishnan, 2012).

The major issues understanding the major terminologies are used in order to explain CSR since the beginning. It also tells us the meaning which is related to understanding the concept of CSR will merge in sync with all the business, political, and social issues in the situation of globalization and major changes in the areas of mass communication. The effect of globalization and mass communication majorly means that such kind of impact will consider all the situations like global trends and various changes in the International law (Taneja et al., 2011).

We actually cannot define CSR in one definition, but each word that majorly exists has a major impact on the society and there are various social implications for the same. It is seen that the CSR effect exists in every business like various activities as donations, working for relief of people globally and also in corporations, charity, donations boxes, etc. Besides this other activities like, philanthropy, strategic philanthropy, sustainability in corporate, and all the business and corporate activities. CSR activities are finding all the activities as beneficial to the society. It also finds the relationship and difference between organizations, very large organizations and societies in order to interact with various groups that maintain an interest in all the organizational activities as a whole. CSR is a kind of group which talks about the society and the company as a whole in order to balance the environmental, economical, and societal aspects of its operational activities which is taking care of the organization interest. One writer has described CSR as "the firms activities and internal issues which are related to the major issues rather than economic, environmental, and all the other issues of the company."

Against this backdrop of increased corporate engagement in social activities worldwide and given the substantial prior evidence in the context of developed countries, the special issue of emerging markets review aims to inculcate a highly advanced research level which can create a learning for the determinants and consequences of CSR practices of firms (financial and nonfinancial) and the role CSR plays in emerging financial markets. In particular, we are interested in innovative papers that exploit unique sources of CSR data and apply rigorous empirical methods.

According to Michael Hopkins, "CSR is concerned with treating the stakeholders of the firm ethically or in a responsible manner, meaning treating stakeholders in a manner deemed acceptable in civilized societies. The wider aim of social responsibility is to create higher and higher standards of living, while preserving the profitability of the corporation, for people both within and outside the corporation."

According to CSR Asia, a social enterprise, "CSR is a company's commitment to operate in an economically, socially and environmentally sustainable manner while balancing the interests of diverse stakeholders."

According to World Business Council for Sustainable Development, "Corporate Social Responsibility is the continuing commitment by business to behave ethically and contribute to economic development while improving the quality of life of the workforce and their families as well as the local community and society at large."

The above viewpoints clearly state us that:

- The CSR is an integrated idea of improving any business financially and socially and also creating a positive impact on its business ideas.
- CSR also requires checking the benefits of its shareholders and also all those people who are the stakeholders of the company.
- All these activities are not only benefitting CSR but also being good to any business activity creating various business benefits by leaps and bounds.

9.2 OBJECTIVES AND IMPORTANCE

The main aim of this article is to analyze the concept of CSR which has gained huge popularity in recent times, and its impact on the financial performance of the firm. Companies have become quite conscious of the negative impact that they might have to face if they do not contribute to the needs and well being of the community and society at large. They have realized that an adequate transfer of profits to its CSR segment can not only help it to satisfy all its stakeholders, but also help it to gain huge advantages over its competitors. In the light of the above statement, an analysis of India's largest FMCG company—ITC Ltd. has been done toward the end of the chapter in order to get an idea of the major effect that what CSR is working for a company, which is awarded for the best CSR activities in the past 2–3 years which may not have only profits, but also capitalization of markets and variation in stock prices on its brand value, reputation, and word of mouth.

9.3 CSR AND SUSTAINABILITY

Sustainable development of resources is majorly another name for corporate sustainability. It means that it is the main role which companies are playing in the development of new markets and creating a balanced approach in the economic growth of the country, social growth and also environmental sustainability is the main agenda for companies.

Indian CSR deals with the major focus on the profits of the company. Also, it is seen that sustainability of the business is of the top most importance and it consists of various factors like, the social and environmental aspects of any business and this is also another reason to be profitable. It is seen that in India CSR is considered to be an important factor and also it is making its say internationally and across the globe. Although, there are various ways of putting CSR across the global organizations, the idea of this convergence is observed frequently from the recent rules which is existing and also it is intermingled in the Companies Act, 2013 which also tells us about all the stakeholders and mingling it with social, environmental, and other objects related to the environment. It consists of the triple bottom approach and has also acknowledged on the guidelines of CSR for various public sector enterprises in year 2013.

9.4 INNOVATIVE IMPACT OF CSR IN INDIA

It is seen that all the in line compliance of CSR is increasing year after year and it is suggested to increase up to 97%–99% by financial year 2019–2020. It is said that education is considered to be the most preferred area of all the organizations and they all are trying to grow with a minimum contribution of CSR and it is considered to be the most preferred concept in the coming years. Every year the CSR level of involvement in companies is increasing and also the detailed documents in the company's annual reports can be considered as a natural practice for the betterment of the society.

The CSR projects are above average and they all are showing losses and less improvement from 2014–2015 to 2016–2017 onwards, but it can also change. It is seen that almost top 10 companies in India are covering the CSR circle and top 20 companies are working around 45% of CSR activities and it can also be constant for the next 3–4 years. All the government associations are going for majorly large partnerships related to CSR and are also planning to align their CSR programs with all the government organizations, allotted various schemes and other business activities with special emphasis on various mission like Skill India mission, National Nutrition Mission, Ayushman Bharat, and Aspirational District program. Also, some major work is done on collaborations like Business to Business, various joint designing and all the project implementation is majorly getting idea in other rooms and we are going to have more such type of collaborative projects in the times to come. The regular practice of transformation of CSR activities in the organizations only working for CSR foundations will finally come to an end as nowadays CSR is effective everywhere.

9.5 ADVANTAGES OF STRONG AND HEALTHY CSR PROGRAM

In order to check the complex business scenario, all the shareholders will become happy about their needs and desires, their outstanding performances, efficient CSR practices can be considered good and can also contain benefits given below.

9.6 OPERATIONAL LICENSE PROVIDED BY COMMUNITIES

As far as the internal sources of ethics and values are taking place, many stakeholders also influence CSR activities and their major areas are government organizations which are following the law and order, people who want to invest and also layman as customers. It is seen that the fastest growing community in India is the stakeholder community, and there are many companies who are seriously involved in it and their operating license is given by the government organizations and also various other communities who are supporting a profitable company or business. Therefore, it can be seen that a fast speed CSR program is able to meet the desires of various communities, which in itself provides them with the operating license and also a license of trust.

(1) *Employee retention and attraction:* Various human resources have tried to experiment and link a company's capacity to retain efficient employees with the help of CSR and its activities. All those activities which are required to encourage and boost the employees' morale are also participative for the growth of the organization.
(2) *Communities as suppliers:* There are majorly certain issues which are emerging through the CSR initiatives and are majorly investing in the activities for the betterment of the community by adding them in the supply chain. This has majorly benefitted many people and had increased the income level of various companies by providing them an additional benefit to secure the supply chain in the organization.
(3) *Enhancing corporate reputation:* It majorly deals with the traditional aspects of generating a positive image and all the branding benefits which are continuing to work for those organizations that are majorly focusing on an efficient and effective CSR program. This is also benefitting to various organizations that have to start a CSR committee which will consist of new board members which will have one independent director also.

This process will encourage all the companies to spend some minutes in various companies in order to look after them as a very effective corporate people.

9.7 UPDATED CLAUSE 135 OF COMPANIES ACT, 2013

In India, the concept of CSR is governed by Clause 135 of Companies Act, 2013. The main provisions of CSR within this Act are majorly applicable to various companies having an annual turnover of INR 1000 crore or it can be a net worth of INR 500 crore or more. There are various new rules which are formed and minimum amount of their net profit on CSR activities, and that amount can be up to 2%. All organizations can implement these activities after the approval. Below is a list of some activities which are specified under CSR category:

- Promoting the educational facilities.
- Remove poverty level and increase the living standards.
- Opportunities for women.
- Looking after the pregnant women and health of the baby.
- Proper vaccinations for HIV–AIDS, malaria, and other diseases.
- Helping the relief funds of state and central levels.
- Various business projects for the upliftment of society.
- Environmental sustainability.

9.8 BOARD INITIATIVES FOR CSR

- Formation of carious CSR committees.
- Approval of CSR policies for implementation.
- Confirm the implementation of CSR activities.
- Minimum 2% of the amount should be spend for CSR activities.
- Communication of various reasons for not confirming the minimum amount spend.

9.9 MAIN OBJECTIVES OF CSR COMMITTEE

- The committee has mostly three directors and one head to monitor the team.
- Make and guide a new CSR policy to the board members.
- Suggest various processes and the finances used to be covered in a limited time frame.
- Checking the CSR policies on regular basis.

9.10 CSR MODELS

The history of CSR is divided into four different models, which are as follows:

- *Ethical model*: It deals with the philanthropic contributions by various companies. The father of our nation Mahatma Gandhi is able to force various companies by his charismatic approach in building the nation and also to help in promoting the social-economic development of the country. The history has told us that corporate has various community investments, and provision of essential activities like development of schools, infirmaries, etc.
- *Statistical model:* This model was adopted in India after independence where the changes in the economy are adapting the social framework and after independence also India became a mixed economy with large public sector and state-owned organizations. Also, the legal aspects of various Indian laws and ownership majorly reflect the CSR of any company.
- *Liberal model:* It says that companies have shareholders responsibility, and this model tells us that it is important for any new business to work on the lawful conditions which by means of taxation and private choices can be through social needs.
- *Stakeholder model:* It majorly deals with the changes in investors, customers, suppliers, and also their focus on other aspects which deals with the commercial accessibility, long-term value and successful business. It aims to create loyal relationship with investors, customers, suppliers, employees, and their commitment which leads to long-term value of the company, commercial viability, and business success. The growing awareness among public compels the organizations to take serious look over their consequences or else face public campaigns or actions against irresponsible behavior. Therefore, the companies are answerable to all the people associated directly or indirectly with the company.

9.11 MEASUREMENT OF CSR

Various previous studies on CSR are under criticism for conducting CSR inappropriately. All the research analysts are using various corrective measures to assess CSR as a unistructural measure like structured ranking of companies on control measures of pollution, social responsibility factors of Markowitz and Fortune 500 companies' index. All these standards have been criticized for not able to increase the number of shareholder's investment.

In order to rectify such issues, various companies only evaluate CSR by the number of shareholders.

Major standards like ISO 26000, UNGC guidelines cover a very wide range of effective business practices which are related to the environment, health, hygiene, safety of people, corruption control, human effective rights,, etc. Also, in the past few years many companies are adding various CSR annexure in their annual reports. Moreover, all the Indian companies are leaving behind their global subsidiaries in terms of their CSR. In India, there is no organized database on CSR. Two majorly used databases on various Indian companies are mainly two secondary databases in CSR.

The former is responsible to give information to various CSR reports and the latter is majorly containing articles published in newspaper on corporate sessions like CSR activities. This study consists of a comprehensive study of primary stakeholders consisting of social, ethical, legal, and global issues as per the basic norms.

9.12 MEASUREMENT OF PERFORMANCE OF THE FIRM

Various researches have shown that in order to measure the firms performance, it is important to assess the size of the firm, the return on the assets of the firm, return on equity of the firm, sales return of the firm, and age of the assets are most frequently used measures of the financial performance of the firm. Earlier, the most authentic measure of firm's performance is ROA. It is different from other systems of accounting like return on equity or sales. ROA is not directly affected by the changes in the level of leverage which is present in the financial performance of the firm. ROA is mostly related with the changes in the prices of the stock which is mostly related with the increase in shareholder value.

All the measures of financial activity are mostly slow indicators and they consider the historical variations coming from majorly all assets except intangible, as they are not able to record their performance in terms of customer behavior and perception, innovation, investments related to growth and research, and measurement of employee satisfaction in current scenario.

All the investments in assets that you can explicitly calculate, for instance, things such as growth books are used at present and there is immediate use instead of documenting in historical account books.

Such type of treatment has a less effect on the profits of the company in the present year, but the long-term benefits of such investments are very large.

As per the accounting standards, nonfinancial planning (NFP) measures are providing all the indirect ways of the performance of the firm. It is because of the major focus on the performances than on consequences, al, the NFP measures are considered as the "Lead Indicators." Financial planning measures are objective in nature and NFP measures are subjected to nature's viewpoint which consists of the managers' perception on the company's performance in terms of market share, health of the employee, investment in areas of research and infrastructural development, etc. Therefore, all the measures of financial planning are going hand in hand with the NFP, which are a part of firms' performance.

Based on the CSR analytics through CSRBOX and our engagement with key stakeholders in the CSR ecosystem, we have analyzed the various program-types in each of the broad theme of the Schedule VII of the Companies Act.

9.13 EFFECT OF FIRM'S PERFORMANCE ON THE CSR

In order to understand the relationship between CSR and financial performance of any firm, the study comprises of mainly two types. The first type is using the method of conduction of an event in order to find the abnormal returns while conducting of an event. The second type of event consists of the study which tells us about the impact of corporate social performance and its impact on the long-term performance by considering financial tools in order to measure and support its profitability.

The relationship between the financial performance and corporate responsibility are majorly important and various positive activities have been reported about various CSR activities and firm financial performance. It is, therefore, suggested that the social connection with the business where CSR is majorly an important tool to increase the economic activities of the firm. All the managers in the firm see this concept as a value addition to the organization and it also helps in increasing the stakeholders' numbers and contacts. All the management people are agreeing that in order to improve the financial performance of the firm, it is important to inculcate CSR activities in it. The major impact of stakeholders on the firm through CSR can be explained by three theories, which are as follows:

(1) *Consumer inference making theory*: This theory tells us that if the consumer is aware that the product manufacturer of any product which they use is a good firm, then in that case it becomes easy for

the customer to rely on the quality of the product. Such activities increase consumer retention in the organization and impact their intention to purchase.

(2) *Signaling theory*: Suggests that in places where there is asymmetric information between the buyers and sellers, all the consumers are looking for various information which impacts the company's performance and is providing information which actually differentiates one company from other companies which are performing poorly.
All the warranties indicating the benefit, reliability, and higher products indicating better quality which tells consumers to differentiate between the companies.
Please break the above sentences for clear understanding.

(3) *Social identity theory*: All consumers associate themselves with a much higher quality of the product and some people who are looking for job also consider the CSR records of companies for better opportunities. This theory states that every company has a right to be involved with various social groups besides their own organizational activities which enhances the CSR activities of the company. Employee's self-image is also taken care by their employers, and also all the investors are looking forward for their association with socially responsible firms. Such types of bonding encourages the positive evaluation of a firm to judge its market value among other firms and also their value increases in terms of loyalty in business, positive brand management, and reduces negative information. In other words, the negative behavior disturbs the shareholders of the company. Negativity results in boycotting the organization, no purchase of company's products, reduction of goodwill of the company, and irresponsible businesses.

Boycotting of Nike products in relation of human rights under the bad working conditions also creates problem for consumers.

9.14 HYPOTHESES

H1: CSR has a major effect on the financial position of the firm.
H2: CSR has a tremendous impact on the net profit of the firm.
H3: CSR is largely affected by the total assets of the firm.

9.15 THE FINANCIAL IMPACT ON CSR PERFORMANCE: AN ANALYSIS OF ITC LTD.

ITC is considered as one of the best private sector companies with a market capital of US $45 billion and also a turnover of US $7 billion approximately. It is considered as the best big companies of the world and the most reputed companies by Forbes magazine and is also one of the most valuable companies. As per the current data, ITC is considered as the India's most precious organizations as per Business Today magazine. ITC is considered to be one of the most active companies in terms of performance as per the Economic times, Business Week magazine, etc.

ITC's constant efforts to be the best for its country and its shareholders to find in its portfolio all the traditional and greenfield business activities consisting of fast moving consumer goods, hotels, special documentation, agriculture, packaging of products, and information technology. ITC's strong business set up lies on their strong foundation and a very connected network among various institutions which is derived from the cutting edge technologies and strong infrastructure and excellent human resource system.

ITC's main objectives is big and they want to ensure value for the country and its shareholders in order to start a portfolio of traditional and Greenfield business consisting of innovative and new strategies for business which is majorly synergizing the livelihood of people and trying to increase the living standards of the people by creating sustainable livelihoods and the consideration of natural business with increase in shareholders' value. There is a triple line strategy of being economically strong, environmental healthy, and socially competent to survive in the market.

9.16 CSR ACTIVITIES AT ITC

ITC is the only organization in the world which can be considered as filled with carbon, water, and all the solid waste products which can be recycled, as its business is supporting million and millions of households.

The new programs of the company involve Wealth out of Waste, etc.

Because of all these contributions, ITC has won many awards and has gained constant improvement in its rankings. Business world FICCI CSR award in large enterprise category, AIM Asian CSR Award by the Asian Forum on CSR, and Best Overall CSR Performance by Institute of Public Enterprise are some of the awards that have been won by ITC for its exemplary contribution to the triple bottom line.

ITC was considered as the most efficient company in CSR for consecutively 3 years.

ITC has, along with Reliance Industries Ltd., received top rating in Asia for their CSR initiatives, according to report by the Hong Kong based brokerage and investment firm CLSA, which has given a rating of five points from 1 to 5 to both of these companies.

Over the period of 1 year from 2013 to 2014, paid up capital of ITC has increased from 790.18 crore INR to 795.32 crore INR, total turnover from 41,809.82 crore INR to 46,712.62 crore INR, total profit after taxes (PAT) from 7418.39 crore INR to 8785.21 crore INR.

Due to a major change in the profits and turnover, its contribution and the total spending on CSR has also increased from 82.34 crores INR in 2013 to 106.63 crores INR in 2014. The list of activities on which the expenditure on CSR has been incurred is shown in Table 9.1.

TABLE 9.1 ITC Developments as per the Companies Act

Schedule VII of the Companies Act, 2013	ITC Current Developments
(i) Reducing poverty, hunger, and malnutrition	Giving good healthcare facilities by the help of good and efficient infrastructure, services, and clean sanitation.
(ii) Increasing educational facilities	Increasing quality education, infrastructure of schools, development of skills, and adding animal husbandry services.
(iii) Increasing activities on women empowerment and gender equality.	Increasing the promotion of small organizations for women and looking for better living.
(iv) Providing sustainable environmental facilities.	Implementing various forest activities for social development, growth of various resources of water, and also sustainable activities by companies like CII-ITC Centre of Excellence for Sustainable Development.
(v) Development of art, culture, and national heritage.	Enhancing Indian classical music.
(vi) Protecting projects on rural development.	Providing agriculture facilities and productive farm enhancement activities.

In lieu of the above details, it is seen that the financial status of ITC can be studied. *H1* is telling us the impact of CSR on the firms' activity which is reflecting in the market price of the company's shares which are shown on stock exchanges.

H2 tells us the effect of CSR on net profits of the company which can be checked from the profit after tax of ITC.

H3 is giving the information of CSR on the total assets of the company. These figures of ITC over the last 5 years have been shown in Table 9.2.

TABLE 9.2 Share Price of ITC

Year	Share Price as on 31 March of Respective Year (in INR)	PAT (in crores)	Total Assets (in crores)
2019	292.70	12,464.32	57,957.68
2018	274.38	11,223.25	51,411.20
2017	227.44	10,200.90	45,358.96
2016	235.45	9844.71	30,405.19
2015	269.80	9607.73	30,721.99

It is clearly seen from the performance calculation of ITC for 5 years that this company is growing very fast over a certain period of time. As a result, we can see that there is constant growth over a certain time period. It is seen that the financial performance of the companies are improving, it is also having more and more activities in its FMCG section and has a good ranking and rating which has made TCS a more accepted company. It is seen that the profits and the share prices of a company is more likely affected by its equity return which is reduced from 25.2% in 2015 to 21.7% in 2019.

The ROI has not reduced alarmingly because the increase in liabilities is majorly due to expansion and diversion of the business, which is also a positive change. Due to which, we observe a minor fluctuation in the ROE of the last 5 years.

9.17 CONCLUSION

By doing the detailed analysis of the sustainability reports and also the final accounts of ITC Ltd., it is clearly seen that the CSR has a goodwill about the financial performance of various companies. All the companies are involved in various CSR activities and are having an added advantage over other organizations in financial terms, also in easy tax policies and also an added advantage in other financial issues. Also, there is lot of benefit in terms of brand image and trust of the customers and their stakeholders.

Therefore, it is clear that the CSR initiatives followed by a company should be majorly followed by all those organizations that are looking forward for a financial benefit in comparison to other organizations in the long run.

KEYWORDS

- **economy**
- **business**
- **corporate**
- **financial statement**
- **positive and negative CSR**
- **competitive action**
- **financial performance**

REFERENCES

Abdul Rahim Othuman, S. P. (2012). Corporate social responsibility and company performance in the Malaysian context. *Procedia—Social and Behavioural Sciences*, *65*, 897–905.

Antonio Lorena, P. C. (2018). The relation between corporate social responsibility and bank reputation: A review and roadmap. *European Journal of Economics*, *4*, 7–19.

Famiyeh, S. (2017). Corporate social responsibility and firm's performance: Empirical evidence. *Social Responsibility Journal*, *13*, 390–406.

Francesco Gangi, M. M. (2018). Corporate social responsibility and banks' financial erformance. *International Business Research*, *11*, 570–575.

Francesco Gangi, M. M. (2019). The impact of corporate social responsibility (CSR) knowledge on corporate financial performance: Evidence from the European banking industry. *Journal of Knowledge Management*, *23*, 277–285.

Harjoto, M. A. (2017). Corporate social responsibility and corporate fraud. *Social Responsibility Journal*, *13*, 762–772.

Jesús Herrera Madueño, M. L.-M. (2016). Relationship between corporate social responsibility and competitive performance in Spanish SMEs: Empirical evidence from a stakeholders' perspective. *Business Research Quaterly*, *19*, 55–72.

Krishnan, N. Impact of Corporate Social Responsibility on the Financial And Non Financial. PhD Thesis, D.Y. Patil University, Oct, 2012.

Lee, S. & Jung, H. (2016). The effects of corporate social responsibility on profitability: The moderating roles of differentiation and outside investment. *Management Decision*, *54*, 1383–1406.

Nakamura, E. (2014). Does Environmental Investment Really Contribute to Firm Performance? An Empirical Analysis Using Japanese Firms. *Eurasian Business Review*, *1*, 91–111.

Nizar Al-Malkawi, H.-A. & Saima, J. (2018). Corporate social responsibility and financial performance in Saudi Arabia: Evidence from Zakat contribution. *Managerial Finance*, *44*, 648–664.

Panicker, V. S. (2017). Ownership and corporate social responsibility in Indian firms. *Social Responsibility Journal*, *13*, 714–727.

Richa Gautam, D. A. (2016). Demystifying relationship between corporate social responsibility (CSR) and financial performance: An Indian business oerspective. *Independent Journal of Management & Production*, *7*, 1034–1057.

Shafat Maqboo, L. N. (2018). Corporate social responsibility and financial performance: An empirical analysis of Indian banks. *Future Business Journal*, *4*, 84–93.

Taneja, S. S., Taneja, P. K., & Gupta, R. K. (2011). Researches in corporate social responsibility: A review of shifting focus, paradigms, and methodologies. *Journal of Business Ethics*, *101*, 343–364.

CHAPTER 10

Utilizing Corporate Social Responsibility as Financial Investment Opportunity in Conserving Local Environment Toward Sustainable Business

DIVYA AGARWAL[1*] and ANIL K. GUPTA[2]

[1]*Jesus and Mary College, University of Delhi, New Delhi, Delhi 110021, India*

[2]*Division of Environment and Disaster Risk Management, National Institute of Disaster Management, New Delhi, India 110001, India*

**Corresponding author. E-mail: divya.sustainable@gmail.com*

ABSTRACT

This chapter is an attempt toward appreciating and fostering the broad understanding of financial management principles not only to seek just success or profit but also to maintain the same for a longer passage of time for future generations. This chapter describes the relationship of investments in terms of environmental conservation, business continuity, and seeking long-term profits. The corporate social responsibility (CSR) mechanism needs to be appreciated as an opportunity to sustain the business. Many corporates do not find its correct pathways and merely do shortcuts to fulfill the mandatory step toward CSR, namely, providing scholarship to a group of selected youngsters for attending an international summit, awareness campaign, and plantation drives. These steps do count for environmental conservation but may not directly benefit corporate production. However, investments in conserving natural resources, which are the raw materials in production unit, relate concrete ideas of a sustainable profit management approach, which emphasizes importance of ecosystem services in maintaining healthy and integrated industrial management. Financial investments in CSR, for conserving the locale-specific environmental conditions, serve as a tool to maintain organizational profit, management, and sustainability. Importance of the same is cited by studying the impacts of Kerala floods on business, agriculture, and tourism industries, and importance of mangrove vegetation in maintaining ecological balance.

10.1 INTRODUCTION

A business organization or enterprising unit works into commercial, industrial, or professional activities, namely, profit entities or nonprofit organizations to benefit social, economic, and environmental development. It has a wider perspective from an individual, small-scale industry to an international corporation. Simply stating, business is a venture of strategic and organized efforts related to buying, selling, and production (Klewitz and Hansen, 2014). Idea of this chapter is to develop and follow a concrete business strategy goals and objectives to include locale-specific environmental conservation efforts. Maintaining profitability ensures higher revenues as compared to the costs incurred. Costs in production and operations need to be controlled, and the profit margin is applied on products sold. Employee training, equipment maintenance, and new equipment purchases do affect the company productivity. Productivity has a direct relationship with the availability of resources to the employees. The holistic idea to provide resources lies in strengthening the ecosystem services, which is in closest proximity. Ecosystem is a pool for providing air, water, productive soil, minerals, medicines, and goods and forms the baseline toward integral social, economic, and environmental health. Customer interactions, responsibility toward community, and employee satisfaction are related to environmental conservation. This chapter is an attempt to study the importance of environmental conservation as a part of business proliferation opportunity, instead of a rice variety through advanced molecular techniques (Adams et al., 2016).

10.2 SUSTAINABILITY

As per the definition by the United Nations (UN) World Commission on Environment and Development, development carried out in such a way to fulfill the needs of all the individuals of present generation without compromising the needs of all the individuals of future generations is known as sustainable development. Thus, we recommend strong correlation between economic and ecological development. Equity and social justice are also prerequisite to it. Intergenerational equity development is carried out in such a way that needs of all the individuals of present and future generation are to be fulfilled. *Ecological justice:* All the individuals have a right to live in harmony with nature, that is, none of the living organisms should be disturbed (Figure 10.1).

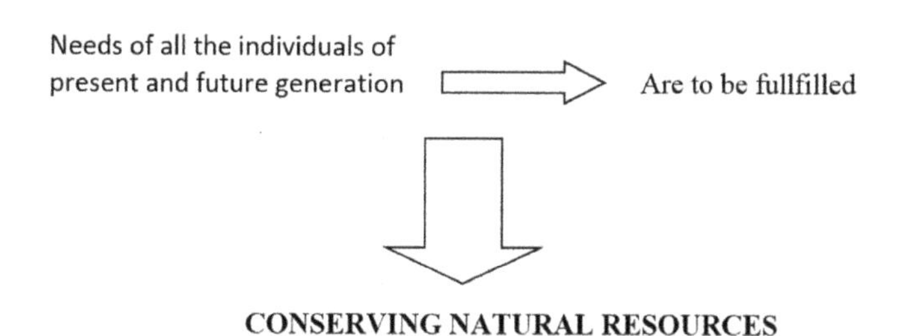

FIGURE 10.1 Objectives of sustainable development.

10.2.1 SUSTAINABILITY IN INTERNATIONAL CONFERENCES

Stockholm Declaration on the Human Environment, 1972, which is an important international conference for indicating global environmental relations, states that "Man has the fundamental right to freedom, equality, and adequate conditions of life, in an environment of quality that permits a life of dignity and well-being." Environmental conservational is, thus, linked with human rights. It also linked policy and law with environmental perspective. Agenda 21 is the voluntarily agreed action plan of UN for socioeconomic and environmental development in the 21st century, in UN Conference in Environment and Development, popularly known as earth summit held in Rio de Janeiro, Brazil, in 1992. Principle 1 of the declaration indicates toward our fundamental rights of life, viz., freedom, equality, and adequate life conditions of dignity and wellbeing. At the same time, it also emphasizes the responsibility of protection and improvement in environment, as a step toward sustainability. Principle 6 relates prohibition on discharge of toxic substances in ecosystems damaging the ecological balance. Principle 15 states importance of urban planning for settlement and industrialization. Principle 18 incorporates the "precautionary principle" and states that urbanization should avoid environmental risks. Global concern for environment is implemented in various treaties, protocols, and conventions time to time. Principles of the Stockholm Declaration on Human Environment provide risk avoidance, risk reduction mechanisms, and integration of environment as a part of the disaster risk reduction and sustainable development. An international agreement held in Kyoto, Japan, in December 1997 by different countries to take required measures to reduce their overall greenhouse emission to a level at least 5% below the 1990 level by 2008–2012 is Kyoto Protocol. Montreal Protocol is an international agreement signed by 27 industrialized countries in 1987 to ensure protection of ozone layer present in stratosphere. More than 175 countries have signed the agreement with the mandate on sustainable lifestyle, finding out a substitute of ozone-depleting substances. World Summit on Sustainable Development or Johannesburg Earth Summit held in Johannesburg, South Africa, from August 26 to September 4, 2002. It reinforced the call to action locally and thinking globally for poverty alleviation and sustainability. Rio +20 Earth Summit at Rio De Janerio in June 2012 was held to discuss issues of sustainable development at a global level, but there occurred lack of concrete global settlement of sustainable development. Cop 14, a global UN meeting, was held in Greater Noida, India, emphasizing on the need to stop land degradation, desertification, and drought by developing an insight into real-time basis problems such as persistent dry-land, resulting poverty, soil erosion, recurrent droughts, and sudden drop in market economic values for smallholder farmers and herders. It is a need of hour, bringing various stakeholders from academia, businesses, scientists, nongovernmental organizations, publishers, trainers, etc., to bring them at a single platform, sharing innovative recent research and ideas to initiate their effective implementation (Cop 14, 2019). The UN Convention to Combat Desertification (UNCCD) and *National Institute of Disaster Management* (*NIDM*) played a significant role to the conference. The first global UN Conference on Desertification was held in Nairobi in 1977. It acted as a catalyst toward such international events, as an important step to secure life on earth, when a severe drought occurred in Africa during the late 1960s/1970s. It is a step to combat drought, desertification, and land degradation. Land degradation, desertification, and resulting droughts are major threats to the existence of life on the planet earth. Productive lands are getting turned into wastelands due to overgrazing or excessive tillage and use of chemical fertilizers and pesticides. A sustainable framework for land management involving eco-restoration processes with improvised agriculture, including modern agricultural techniques such as plant-biotechnological-based transgenic solutions and traditional agricultural techniques such as mixed cropping, role of meditation in agriculture, terrace farming, agroforestry, drip irrigation, dairy farming, use of biofertilizers and biopesticides, crop rotation, scarecrow, water-shed management, and the optimal use of manure and chemical that conserves physicochemical and biological properties of soil, needs to be adopted as a universal model. There exists a need to aware farmers that such a sustainable agricultural framework is a most reliable friend to boost their economy because it will be reducing small agricultural farmer's risk with respect to land degradation and drought.

The international conferences indicated toward integrated approach toward quality of life, natural resource conservation, and the management of waste and sustainable economic growth. The sustainable framework toward agricultural production and natural resource management (water and energy and biodiversity conservation) is integrated as environmental protection and development. Environmental protection and development are integral factors.

10.2.2 ROADMAP OF ACHIEVING SUSTAINABILITY

The principles of sustainability, that is, necessary steps to achieve sustainable development, include population control, conservation, recycling, and wasteland reclamation. Population increasing at exponential rate puts pressure on natural resources to fulfill the growing demands of population (air, water, soil, crops, forest, biodiversity, and minerals) for fresh air, water, food, shelter, etc. Efficient use of natural resources, minimizing their wastage, is the key to conservation. Recycling is an important step, which reduces pressure on raw material from which the product is generated. Planting trees on wasteland, that is, revegetation, is the most viable environmental option. The trees should be tolerant with respect to the pollution, fast growing, and should contain high biomass (Figure 10.2).

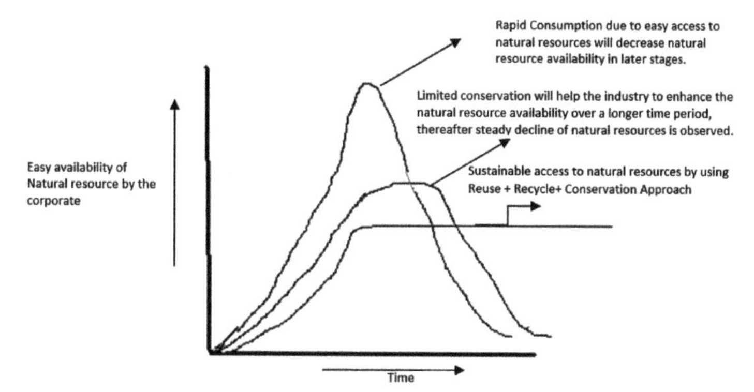

FIGURE 10.2 Principles of sustainable development understood in terms of corporate production. The concept of natural resource consumption versus sustainability is interpreted from Miller's environmental science.

10.3 ECOSYSTEM DYNAMICS

An ecosystem may be defined as a structural and functional unit of biosphere constituting living beings and the physical environment both interacting mutually to form a "self-sufficient unit," for example, pond, desert, forest, and river ecosystem. There are various growth-promoting factors that maintain a state of balance in an ecosystem known as homeostasis: reproduction, adaptation, suitable climatic conditions, and growth-reducing factors, namely, competition, disease, unsuitable climate, etc. The term biosphere indicates a global ecological system or a global ecosystem.

Ecosystem is a unit of steady-state equilibrium in the biosphere. The property of living organisms being in a steady state is known as homeostasis. In other words, homeostasis is the property of self-sufficiency of an ecosystem measured in terms of continuous energy flow and nutrient flow. The ecosystem is a smallest unit with variety of interactions in living and nonliving components taking place in such a way to maintain stability. Similarly, complex series of interactions among the various components governs biogeochemical cycling and maintains the ecological balance in the biosphere. Organisms survive in a range of environmental conditions by undergoing series of adaptations known as a range of tolerance. Organisms survive to their best in the optimum range. At the outer margins of optimal range, zones of physiological stress exist where survival and reproduction are possible with difficulty, that is, hibernation, etc. Outside these zones, there are zones of intolerance, where the degree of hardness of abiotic factors leads to death of living organisms. Variety of factors influence the degree of tolerance, that is, age, species, and behavior, and natural events such as floods, tropical storms, and volcanic eruptions all change conditions for varying periods. Living organisms are the biotic components of ecosystems. Bacteria, algae, plants, fungi, and animals are various organisms present in various ecosystems. A group of organisms of same species within an ecosystem is known as population, and population of different organisms coexists in an interdependent biological community. Interactions among the living organisms are studied in ecology.

10.4 ROLE OF ECOSYSTEM SERVICES IN STRENGTHENING BUSINESS DEEDS

Consumerism and nature's regenerating capacity need to be balanced in such a pace that there is steady-state equilibrium. Expansion of economies must be within the carrying and regenerative capacities of ecosystems. In place of overexploitation of natural resources and attaining faster economic growth, carrying capacity-based industrial planning and innovative technologies with efficient consumption tends to maintain business more profitable and sustainable (Halme and Korpela, 2014)

10.5 MAINSTREAMING LOCAL ENVIRONMENTAL CONSERVATION INTO CORPORATE SOCIAL RESPONSIBILITY (CSR)

CSR is a buzz word with respect to corporate's environmental credibility. Most of them follow it as a mandate. Albeit, a smarter corporate develops and leads long-term sustainable frameworks to implement the sustainable development goals (SDGs). The UN declared 17 SDGs in 2015. Target is to be achieved by 2030, which includes various socioeconomic and environmental benefits toward a sustainable future. SDGs also ensure integrated and sustainable growth of businesses, governments, and communities leading to the shared prosperity of people, nature's reservoir, and the biosphere. Without natural resources, none can flourish. Corporation's development with sustainability in terms of "environment," "social," and "governance" have been accepted together and understood globally (Franceschini et al., 2016). Society and corporate are strongly interrelated; both are actually interdependent. Securing human rights of all stakeholders and supporting local communities across entire supply chain is another important initiative toward a long-term business strategy (Hall, 2002). Incorporating sustainability in industrial processes and operations is the key for the successful global management strategy. A deeper understanding of business risks coming up due to social and environmental deficits emphasize the need of sincere efforts by corporate for their own sustainable growth (Bahena-Álvarez et al., 2019).

Albeit, the sustainable corporate also gets into exploring the new business opportunities by achieving sustainable society goals. For example, in Germany, green labeling business is initiated toward recycling one-time plastic of the corporate. Many such sustainable ventures we come across are palatable notepads, pencils, eatable cutlery, "Karo Sambhav" electronic recycling Producer Responsibility Organization venture, "Sustainable Initiatives," "Chintan." Hitachi is pushing to reduce carbon emissions to combat climate change through innovation and business solutions. They are committed to reduce 80% of CO_2 emissions by 2050 with respect to 2010 levels. 50% reduction is targeted by 2030. Sustainability-based business initiatives need to mention how they shall be contributing toward fulfillment of the SDGs, in an internal application form designated for a business startup (Adams et al., 2012) (Figure 10.3).

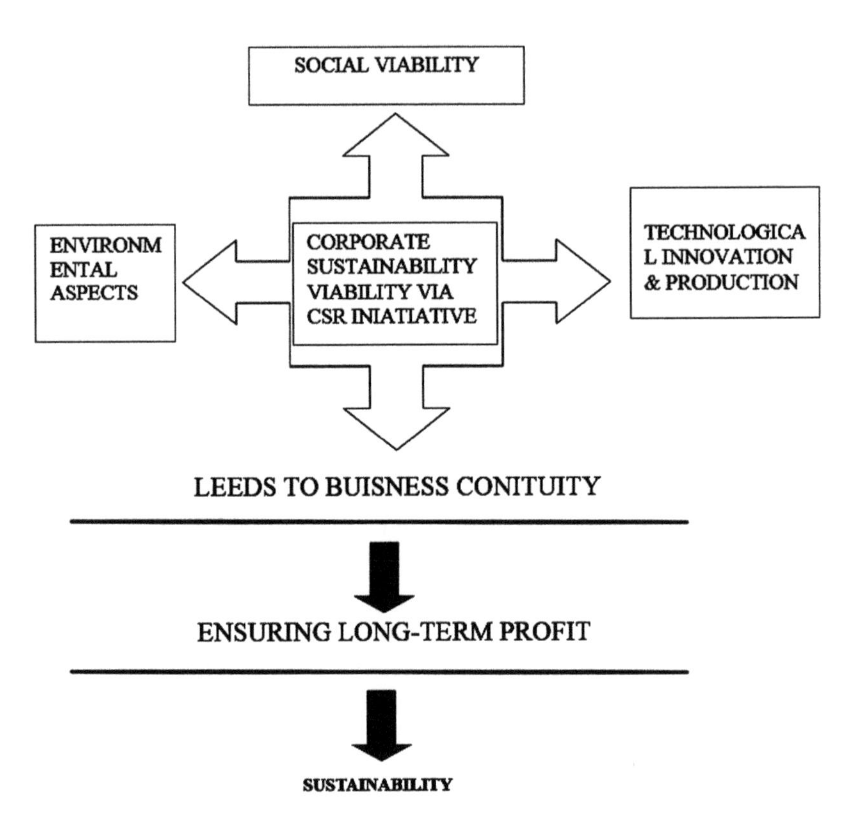

FIGURE 10.3 Factors for executing environmental conservation as a part of CSR plan.

By definition, Environmental Impact Assessment (EIA) is a generic process, wherein pros and cons of a developmental project are analyzed during the whole lifecycle of the project and negative impacts are minimized. EIA is a major instrument for the assessment of developmental activities with respect to the regional carrying capacity. Innovative projects to have solutions and products from CSR help to expand economy, social, and environmental integrity. The CSR projects reduce negative social and environmental impacts to help the corporate in EIA clearance.

10.6 EFFICIENT FINANCIAL INVESTMENTS FOR ORGANIZATIONAL GROWTH TOWARD SUSTAINABLE MANAGEMENT VERSUS SUCCESSFUL MANAGEMENT

Sustainable Business Management revolves around identifying and solving the key social challenges by adopting innovative strategies, which ultimately help fulfill Sustainable Development Goals. Successful management relates to maximizing profit to present stakeholders. It is observed that most of the big industrial projects take land, water, and minerals from nature, adversely affecting the growth of farmers. Thus, if a corporate is achieving SDGs, it is having an insurance to grow successfully as well as sustainably. Insurance of successful and sustainable management lies in adopting sustainable business strategies motivated by local environmental conservation (Figure 10.4).

The cost of the product depends on the expenditure incurred in terms of energy and monetary consumption, known as economic cost, also includes the potential of the labor in manufacturing of a product as well as environmental cost. It depends upon the amount of natural raw material consumed and adverse environmental impacts of manufacturing and disposal of waste products. The extent of depletion of natural resources and compensation to reduce the impact of environmental pollution decides the environmental costs (Figure 10.5).

The setting of industry adversely affects local environmental setting and agriculture. Opposite is also true. Big industrial management also maximizes their profit for present as well as future stakeholders (Hockerts, 2003). Some big industrialists, including Reliance Indutries, Essar Group, and Sanghi Industries, which are among the largest single-location refineries and cement plants in Gujarat, carried out eco-restoration project in acres of wasteland and successfully used the land to proliferate their business. They converted the wasteland into mango orchards. Therefore, they utilized the environmental conservation as an opportunity and generated jobs for local farmers as well.

Many multinational telecommunication industries have also developed environmental conservation strategy, in which they are practicing recycling, to reduce greenhouse gas emissions, manufacturing products with minimal environmental impact. The smart sustainable corporate world now understands that their step toward circular economy actually strengthens the business economic viability. Various successful sustainable industries focus on reducing their carbon footprint, measuring the entire carbon calculated from corporate facilities, product manufacturing, product use, product transportation, and product end-of-life processing, so as to understand where to focus reduction processing (Wüstenhagen et al., 2008). Responsible innovation forms the basis of sustainable entrepreneurship, which links innovation and social values indicating the overall impacts of the innovation over the societal relationships (Baregheh et al., 2009) (Figure 10.3).

The Bill & Melinda Gates Foundation owned by Microsoft founder Bill Gates is serving toward a societal sustainable future. Huge amount of donations and contributions in various disaster relief and rehabilitation attempts, poverty eradication programs, global health care and development systems, global growth, and opportunity divisions is appreciated by people. That is how they gain a respectable image in market, which pays the company in terms of maintaining excellent business relations. Microsoft CSR aims "to serve globally the needs of communities and fulfill our responsibilities to public" (Lubberink et al., 2017) (Figure 10.6).

FIGURE 10.4 SDGs adopted by the UN.

Source: https://www.un.org/development/desa/dspd/2030agenda-sdgs.html

10.7 CASE STUDY OF KERALA FLOODS ADVERSELY IMPACTING ECONOMIC GROWTH

We define flood as a natural calamity, which causes submergence of a land area by overflowing water. Floods are one of the adverse disasters leading to tremendous socioeconomic and environmental losses (Table 10.1). Excessive rainfall is a common reason due to which water bodies such as rivers, lakes, and oceans may overflow. The case study of devastating floods in the state of Kerala in 2018 is cited to understand the ecosystem management approach with respect to the business perspective. It was the worst flood in Kerala after 1924. The continuous and heavy rainfall (2346.6 mm) in the Kerala state from over a short span of time (from June 1, 2018 to August 19, 2018) led to severe flooding in most of the districts (Kerala Floods Study Report by Central Water Commission, August 2018).

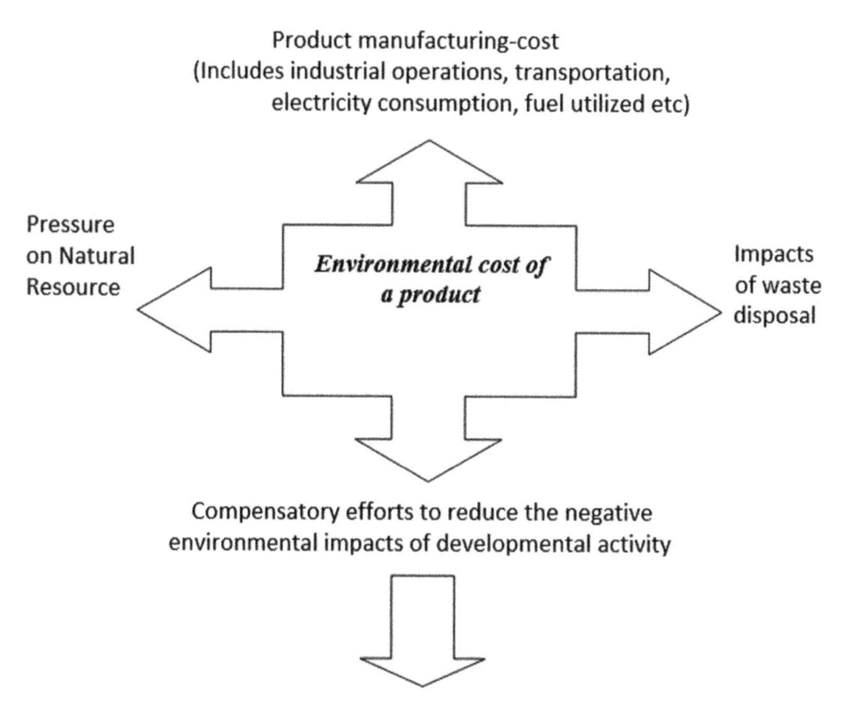

FIGURE 10.5 Costs involved in manufacturing a product with its emphasis on environmental costs. Environmental costs need to be addressed by the manufacturer and should be utilized to reduce the negative environmental impacts of the products.

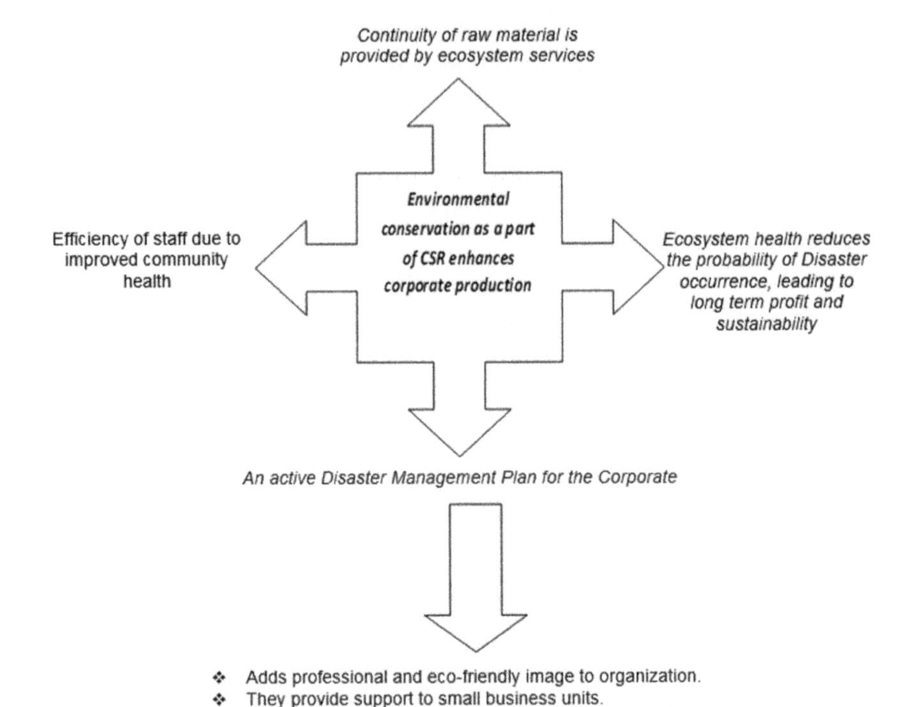

FIGURE 10.6 Benefits of CSR for a corporate.

TABLE 10.1 Tremendous Socioeconomic and Environmental Losses Due to Floods

(i) Infrastructure
• Weakening of building foundations and piles
• Burying and sliding of infrastructure and constructions
(ii) Displacement
• Many people get displaced
• Loss of livelihoods
• Loss of culture and custom
• Adverse effects to society
(iii) Agriculture, Forestry, and Ecology
• Destruction of crops
• Damage located in lands, planting, and forest areas
• Soil erosion and landslides
(iv) Tourism
• Scenic beauty of the place is destroyed
(v) Education and Health Care
• Loss of basic amenities due to collapse in infrastructure

Global climate change is tending to aggravate disasters including extreme precipitation events. The same statement is valid if we find the root cause of exacerbating floods in Kerala. The processes of forecasting, predicting and assessment, analyzing, prevention and mitigation, and relief and rehabilitation measures are the important steps of disaster management. Important disasters of concern are drought, floods, volcanoes, cyclones, and earthquakes (Figure 10.7).

Prof. Madhav Gadgil of the Indian Institute of Sciences, Bangalore, headed an initiative toward preventive measure of such disasters in 2011 in a government-commissioned study by Western Ghats Ecology Expert Panel. The "Gadgil Report" clearly describes Western Ghats (Maharashtra, Kerala, Karnataka, and Tamil Nadu) as ecologically most fragile regions. Goa and Gujarat regions were also kept in the same category. Western Ghats are categorized as one of the hotspots of biodiversity, with 20 rivers and tributaries acting as water source and tending to absorb the excessive rainfall. The report has classified Western Ghats in three important zones to proceed toward developmental activities in the region within the ecosystem carrying capacity. It is a sustainable developmental criterion in the region (Gupta and Nair, 2012).

10.9 CASE STUDY OF MANGROVES AVERTING DISASTERS IN ODISHA

The case study by Prof. Sudamini Das, Institute of Economic Growth, University of Delhi, has established findings that mangroves can provide protection against socioeconomic losses from storms in terms of providing protection to even far away buildings, averting deaths (Das and Crépin, 2013). Mangroves role in wind protection was studied with respect to a theoretical model. Data studied in the model were of 1999 cyclone in Odisha calculated the wind protection economic value of mangroves to approximately US$177 per hectare (Das & Vincent., 2009). Studies need to be discussed in international business forums as well, rather than discussing the same in environmental events. Many national and international organizations are organizing workshops and conferences keeping in view the importance of ecosystem services (Figure 10.7). The National Institute of Disaster Management has recently conducted a workshop on the importance of mangroves as a nature's prudential mechanism toward protection against disasters. This indicates the importance of indigenous vegetation toward strengthening socioeconomic and ecological framework. Idea is appreciating the importance of ecosystem services for business continuity to relate investments in ecosystem health maintenance. Mangroves are providing a concrete example by strengthening coastal ecosystems.

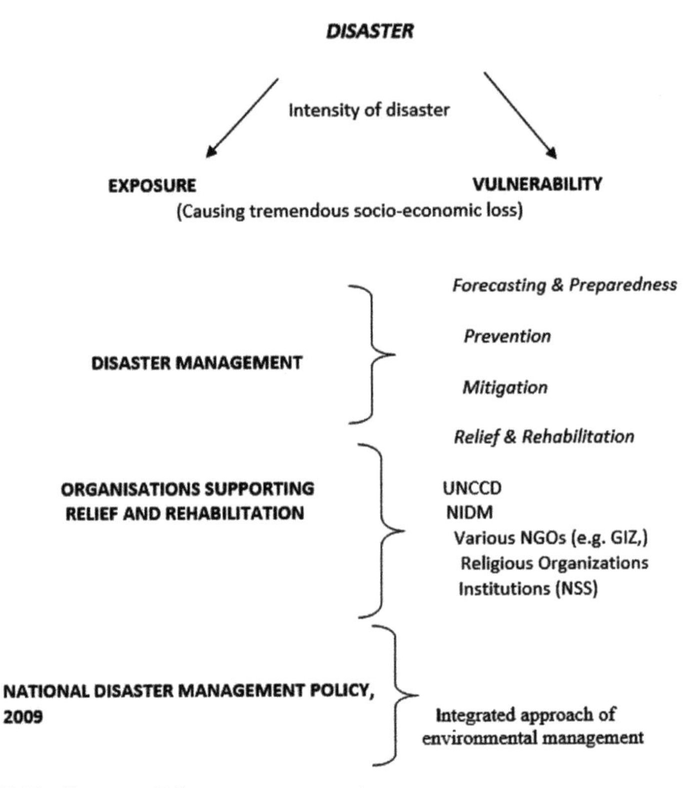

FIGURE 10.7 Concept of disaster management.

KEYWORDS

- **corporate social responsibility**
- **sustainable business**
- **environmental conservation**
- **business continuity**
- **environmental costs**
- **disaster management**

REFERENCES

Adams, R.; Bessant, J.; Jeanrenaud, S.; Overy, P.; Denyer, D. Innovating for Sustainability: A systematic review of the body of knowledge; *Netw. Bus. Sustain.*, London/Montreal, QC, Canada, **2012**. 66.

Adams, R.; Jeanrenaud, S.; Bessant, J.; Denyer, D.; Overy, P. Sustainability-oriented innovation: A systematic review. *Int. J. Manag. Rev.* **2016**, 18, 180–205.

Bahena-Álvarez, I. L.; Cordón-Pozo, E.; Delgado-Cruz, A. Social entrepreneurship in the conduct of responsible innovation: Analysis cluster in Mexican SMEs; *Sustainability*, **2019**, *11*(13), 3714.

Baregheh, A.; Rowley, J.; Sambrook, S. Towards a multidisciplinary definition of innovation. *Manag. Decis*. **2009**, *47*, 1323–1339.

Kerala Floods Study Report by Central Water Commission, Central Water Commission Hydrological Studies Organisation Hydrology Directorate, **August 2018**. [Online]. Available: https://reliefweb.int/sites/reliefweb.int/files/resources/Rev-0.pdf

Cop 14, [Online]. Available: https://www.unccd.int/conventionconference-parties-cop/cop14-2-13-september-new-delhi-india. **2019**.

Das, S. and Crépin, A. S.; Mangroves can provide protection against wind damage during storms. *Estuarine, Coastal Shelf Sci.* **2013** *134*, 98–107.

Das, S., Vincent, J. R., Mangroves protected villages and reduced death toll during Indian super cyclone. *Proc. Nat. Acad. Sci.* **2009**. *106*, 7357–7360.

Franceschini, S.; Faria, L. G. D.; Jurowetzki, R. Unveiling scientific communities about sustainability and innovation. A bibliometric journey around sustainable terms. *J. Clean. Prod.* **2016**, *127*, 72–83.

Gupta, A. K., Nair, S. S.; Ecosystem approach to disaster risk reduction. A report. National Institute of Disaster Management, New Delhi, India. **2012.**

Hall, J. Sustainable development innovation: A research agenda for the next 10 years. Editorial for the10th Anniversary of the Journal of Cleaner Production. *J. Clean. Prod.* **2002**, *3*, 195–196.

Halme, M.; Korpela, M. Responsible innovation towards sustainable development in small and medium-sized enterprises: A resource perspective. *Bus. Strat. Environ*. **2014**, 23, 547–566.

Hockerts, K. Sustainability, innovations ecological and social entrepreneurship and management of antagonistic assets. PhD dissertation, University of St. Gallen, Bamberg, Germany, **2003**, 64.

Klewitz, J.; Hansen, E. Sustainability-oriented innovation of SMEs: A systematic review. *J. Clean. Prod.* **2014**, 65, 57–75.

Lubberink, R.; Blok, V.; Van Ophem, J.; Omta, O. Lessons for responsible innovation in the business context: A systematic literature review of responsible, social and sustainable innovation practices. *Sustainability.* **2017**, 9, 721.

Odum, O. P. *Fundamentals to Ecology*, Cengage Learning: Boston, MA, USA. **2004**.

Wüstenhagen, R.; Sharma, S.; Starik, M.; Wuebker, R. *Sustainability, Innovation and Entrepreneurship: Introduction to the Volume*; Edward Elgar Publishing: Cheltenham, UK, **2008**, 65.

CHAPTER 11

Crowdfunding in Financial Acumen

SHALINI AGGARWAL[1*], NITIN KULSHRESTHA[1], and SAURABH MITTAL[2]

[1]*Chandigarh University, Chandigarh 160036, India*

[2]*GL. Bajaj Institute of Management and Research, Greater Noida 201306, Uttar Pradesh, India*

[*]*Corresponding author. E-mail: shaliniaggar@gmail.com*

ABSTRACT

Crowdfunding is small-time patronage supported by little amounts of contribution by a large number of people through a web-based platform or social networking site for a specific project, business venture, or social cause. The chapter is preceded by the history of crowdfunding and then its types that include peer-to-peer (P2P), equity, donation, and credit-based crowdfunding. After that various types of risk associated with crowdfunding platform is studied that includes substitution of institutional risk by retail risk, risk of default, central role of the Internet, risk of fraud, information asymmetry, systematic risk, and many others followed by the applications of crowdfunding. Crowdfunding for personal use, real estate, and crowdfunding for startups, crowdfunding for business, loans, and college debt, SMEs, etc. are the applications which are studied in the chapter.

Cost-benefit analysis of crowdfunding is an important aspect that found a place in the chapter proceeded by the startups of the crowdfunding platform worldwide. The next important aspect is P2P lending rules in different countries of the world followed by crowdfunding websites such as Indiegogo and Kickstarter attract thousands of people that hope to invest their funds in the next big thing. In the last famous P2P lending crowdfunding ventures of the India that includes Rang De, Faircent, Ketto and Wishberry are studied.

11.1 INTRODUCTION

Crowdfunding is the process of supplication of meager amounts from conjunct investors. It is done through a social networking site depending upon the project, its nature, and its purpose (Gadre et al., 2018). This has been made possible by the enhancement of technologies that enable online funding through a website or a social site. The main idea behind crowdfunding is to work upon a humane and beneficent project of public interest. Think of a community-based social work or a cooperative society to benefit people. This funding project is also evolved to raise funds for a creative aspiration which could be anything like a music album, a short film, or publication of a book. To ensure the authenticity of this kind of venture, crowdfunding in India remains under the lens of Securities and Exchange Board of India (SEBI) and Reserve Bank of India (RBI), who not only govern the financial aspect but also regulate the same (Arumugam, 2018).

11.2 HISTORY OF CROWDFUNDING

The two names Kickstarter and Indiegogo, who enjoy the credit of being leaders of the crowdfunding revolution in the West, were the fathers of zillion cogitations which sprung up abruptly to dazzle the crowd's interest in advocating ingenious, outlandish and out-of-the-way products and projects. The crowdfunding platforms in the USA stand now at 191 in number, which number has grown from only two a decade back? This growth signifies for sure the expansion of SME industry and mushrooming of startups who are indulged into selling everything from a smartwatch (*Pebble*) to the world's first nine-dollar computer (*CHIP*) to the first solar sailing spacecraft! (*LightSail*). The crowdfunding platforms have changed into Shark Tanks for the small-time investors where each one finds something of interest for investment.

The concept of crowdfunding was brought to India by those Indians who were either educated in the West or a few of them had also been the cofounders of some of the top crowdfunding websites for the European countries and America. They thought it viable to implement this concept in India, which is considered one of the most rapidly growing economies of the Third World. They were well convinced of its power to financially fuel the budding businesses as well as solve their problems in medical and social fund building.

As a result about 15 such platforms have come into formation since 2010 and almost all of these are doing well.

11.3 EXAMPLE OF COMPANIES THAT MADE USE OF CROWDFUNDING PLATFORMS AND MADE MONEY

(a) **Oculus VR**: It is an American-based company that manufactures hardware and software products. It was founded by Palmer Luckey and others in 2012. The company used the platform Kickstarter to raise funds to design virtual reality headsets for video gaming for developers. They received $2.4 million for their venture. In 2014 Facebook took over the Company Oculus VR for $2.3 billion.

(b) **M3D**: The company was founded by two friends David Jones and Michael Armani. They manufacture 3D printers. In 2014, the company raised $5 million for the manufacturing of Micro 3D printer through the platform of crowdfunding in three campaigns (Feldman, 2017). At present the company is creating its own crowdfunding platform, that is, FitForLaunch.

(c) In April 2019, Critical Role, a game that features a group of prominent voice actors has raised $4.7 million in only 24 hour for its latest animated special "The Legend of Vox Machina." This is the first time that through the Kickstarter platform campaign, someone has raised a huge amount of money within 30–60 days.

11.4 HOW CROWDFUNDING WORKS

Crowdfunding is an assemblage of dinky amounts from numerous people for a distinct program including a business or social cause, the modus operandi being a web-based platform or social networking sites which make it easier for the fundraiser to canvass the desired sums by sharing the venture/project to the potential resources, who could be either a Maecenas or an investor. As per the rules, equity-based crowdfunding is illegal in India. At the same time, P2P lending is regulated by the RBI. The projects/ventures could be listed on the crowdfunding websites which would help in joining donors/investors to the fundraiser. You simply need to design a juggernaut with well-defined targets and kick the same to get desired results. There are websites which permit a cost-free set up of campaigns and you may have to pay 5% share of the funds raised for using their platform. In case you want their services for driving the campaign for you, they would ask for an additional fee, for which standards are not defined. However, these services help you to attain your targets faster than envisaged for obvious reasons (Gadre et al., 2018).

11.5 TYPES OF CROWDFUNDING

Donation, reward, P2P lending, and equity are the four major categories in which crowdfunding is divided (Figure 11.1).

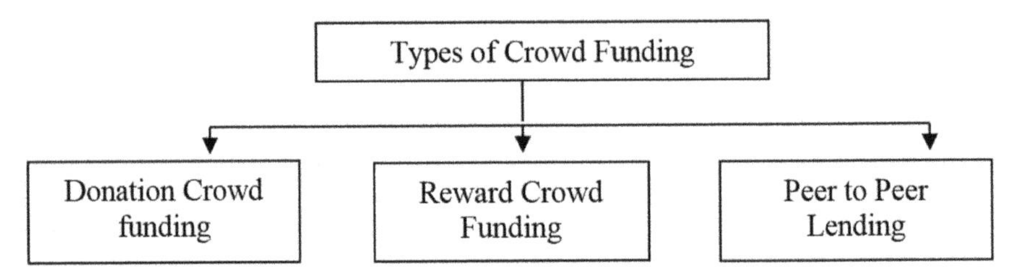

FIGURE 11.1 Types of crowdfunding.
Source: OICU-IOSCO (2014).

(a) Donation crowdfunding: When the adjuration of funds is done for social, artistic, philanthropic, or other purposes which do not have in exchange any discernible value, it is called donation crowdfunding. As donation crowdfunding works on grants and a donation, there is no return of any kind and therefore is exempted from income tax under Section 80G.

For example: In USA there is Kickstarter, Indiegogo, etc. (Kalra, 2019).

(b) Reward crowdfunding: When conjuration of funds is done in consideration of some reward to the investor, it is called reward crowdfunding. The reward can be in the shape of a consumer product or some membership reward. As no monetary benefit is given to the investor therefore it does not comply with SEBI guidelines. Although it is legal in India but if the investors do not receive any rewards as promised it becomes a major cause of concern for the investor as he has no recourse, for example, Wishberry, Rockethub, etc.
(c) P2P Crowdfunding: It is the result of alignment of lenders with borrowers on a single online platform. The lenders provide for some

unsecured loans for which the rate of interest too is decided by the site operators. Some of such loans are arranged between individuals while other funds are pooled by the platforms companies with a view to lend the same to medium-sized businesses. Notwithstanding the fact that while no securities are involved in this category of crowdfunding, still loan/contracts can be bought and sold on a secondary market or P2P platform. P2P platform works under the close inspection of RBI along with SEBI. RBI has given rules for online P2P lending platform from October, 2017. The platform will be categorized as Non-Banking Financial Company (NBFC)-P2Ps (Jalan, 2018). Directions given by RBI will be known as NBFC-P2P Lending Platform (Reserve Bank) Directions, 2017 (RBI, 2018).

P2P market is very successful in the USA, UK, and China. USA and UK have small number of players in P2P market as compared to China which has thousand of players plying in the market (Nirupama and Shankar, 2018). For example: Faircent, Prosper, Zopa, Funding Circle, etc.

(d) Equity-Based Crowdfunding: When equity shares of the company are allotted to investors in deliberation of the financial amount contributed by them; it is called equity-based crowdfunding. This is the process adopted in the infant stages of a business whereby equity interest is offered online to lure investors to be a part of the company. Crowdfunding platforms are the means for raising capital for businesses. As such it is an intermediary between the seeker and the investor which help efficiently the coming up companies who are yet to establish their base in the business world. As per the inherited practices, the budding companies generally got funds from angel investors, private equity, or through financial institution because it is mandatory for a business or product to be commercially doable before notifying an issuance of public equity. The equity-based crowdfunding is an early stage methodology to provide impetus to a business so that it can mature well in due course. For example: Syndicate Room, Crowd Cube, and Seders.

11.6 SEBI GUIDELINES FOR EQUITY CROWDFUNDING IN INDIA

(a) "Accredited investors" are allowed to invest.
(b) Qualified institutional buyers to hold at least 5% of issued securities.

(c) Any retail investor can contribute Rs 20,000 as the minimum amount and maximum he can contribute is Rs 60,000.
(d) The number of retail investor applying in any company can be maximum up to 200.
(e) It is mandatory for the companies to disclose their proposed business plan, usage of fund, audited financial statements, details of the management, etc.
(f) Listed crowdfunding platform need to perform regulatory check, basic due diligence of startups and investors. Each platform need to constitute screening committee consisting of 10 persons with experience in capital markets, mentoring startups, etc (Monika, 2019).

11.7 RISKS OF CROWDFUNDING

(a) Substitution of Institutional Risk by Retail Risk

In the current scenario venture capital funds (VCFs) and private equity (PE) are the two pillars that support the newly born businesses by bearing the financial risk of investors. These two entities put in their best to lure numerous investors through crowdfunding to come forward with pigmy sums. So VCF and PE are in turn substituted with the very low-risk level retail investors who are actually not able to understand the risk involved in these ventures.

The investments in such startup business are a high-risk level game and hence treated aggressively. VCF/PE investments are long term and hence stand a better-negotiating pricing during this duration (SEBI, 2014).

In comparison to VCF/PE investors, banks or other financial institutions, investment in crowdfunding is more prone to an intensified risk of failure in view of the inexperienced players who have neither artfulness nor proficiency required to make judgment prior to investing/lending in this unconventional business model.

(b) Risk of Default

In case of delinquency or scam, there are not enough opportunities available to investors to deal with the problem. Since funds are organized through website platforms, the issuer does not come in the picture and also does not come with an offer document. In such a situation the websites may not be able to conduct the intentions and seriousness of the issuer. If any of these online platforms cease to operate or are under breakdown temporarily, the investors feel duped and at bay as there is no direct relation between the two. Investment risk is high due to failure and loss of equity.

Let us take the example of Bubble and Balm, a fair trade soap company of UK. They were the pioneer company to adopt this model and used a crowdfunding platform named Crowdcube way back in 2011. This startup equity brought them £75,000 in return for 15% of the company's equity from 82 investors, their individual contribution ranging between £10 and £7,500 each. In this venture, the investors lost their 100% money because the company went missing in July 2013 leaving no clues and contacts behind (SEBI, 2014).

(c) Risk of Fraud

Another kind of risk involved is posed by the hackers who could use these websites as bait for their own benefit by alluring the investors to provide details of plastic money and can thus cheat them by misusing their cards (SEBI, 2014).

(d) Systemic Risk

- It may not be possible for investors to have diversification principles due to the "individual" nature of crowdfunding.
- In absence of a secondary market no exit feasible, hence the risk of illiquidity.
- Money laundering could be another possibility.
- As was experienced during the subprime mortgage crisis, these platforms could render other financial sectors to the danger of default. However, these risks could systemize, in case of a continuous growth in P2P lending.
- Some cross-border implications could still affect the deal due to disparities in applicable laws if the funds are solicited through Internet.

(e) Information Asymmetry

- The asymmetrical nature of information is a unique feature of this business model because people invest based on the information that is not known to other investors. The risk intensifies further due to excessive faith in the information presented on the social networking platforms.
- Monitoring of accounts on these platforms is not feasible.

- The issuers are not obliged to share information with respect to the usage of funds raised that leads to lack of transparency, which is not a healthy practice.
- The information is either incomplete or distorted hence the issuer is ignorant about the true amount of investment, resulting in the overestimation of the actual return. This motivated the investors to spend their money in a product unaligned to their risk tolerance.

11.8 APPLICATIONS OF CROWDFUNDING

(a) Crowdfunding for Personal Use: Crowdfunding is beneficial for personal usage. It is another avenue open for many individuals who have been successful in gaining funds to meet their own requirements. It could be for a celebration, travel, medical emergency, education, a social cause, or more.

(b) Real-Estate Crowdfunding: There are ways to invest funds that are secured by real estate to support housing projects (Choudhury, 2018).

(c) Crowdfunding for Startups: There has been a gigantic growth in the startup industry in the last few years as the financial needs are fulfilled by the crowdfunding platform. The platform can assist you to get funds easier than any bank or any other means. Further, the government is also promoting startups. They are providing tax benefits for the new startups for the first three years. Simplification of work, financial assistance, network prospect is the other benefit given to the startups in India (Cleartax, 2019).

(d) Crowdfunding Loans and College Debt: Different types of loans are available in the market nowadays. For instance, you can take an education loan, car loan, health loan, and many more. Crowdfunding is another source that facilitates loan to you. If you are a college student and unable to fund your education or you need financial assistance for your treatment then you can raise funds through crowdfunding source as several generous people contribute to such causes.

(e) Crowdfunding for SMEs: Crowdfunding is a new mode of providing financial fuel to startup companies and SME sector because it helps to intensify the credit flow in the economy of the country. In the event of bank failures that happened way back in 2008, which was the direct result of the financial crisis resulted in the Basel III Capital adequacy norms applicative on banks to arrest the problem for the future. This change of rules have in a way restricted the loaning capacity and freedom which has come in the way of risk new ventures which are considered to be more prone to risk. crowdfunding has come into being as a new option to fulfill the needs of SMEs which does not involve much of paperwork and other formalities because the funds are directly collected through an online website.

11.9 COST-BENEFIT ANALYSIS OF CROWDFUNDING

(a) The architecture of crowdfunding is quite empowering with its ease of operation. It is a panacea for the budding entrepreneurs who have outworn their capital resource or do not have avenues open to procuring funds from conventional sources. This also helps the businesses to concentrate more on work because of the comfort of capital rising.

(b) In crowdfunding some of the expenses like underwriter fees, registrar and transfer agent fees, and legal and accounting fees, merchant banker fees, marketing and advertising fees or distribution commissions, and other fees are automatically eliminated because money lending is online and hence does not involve any middle man or documentation which is an integral part of the existing systems. Hence the target of reducing cost in crowdfunding is easily managed.

(c) Crowdfunding provides a platform that helped the companies to raise funds at low cost. The companies are no longer required to deal with the book-building process. Appointment of merchant bankers, listing required and filing of the prospectus and taking permission can be waived off. Only the fees to the platforms are to be paid and they can raise money within no time.

(d) Crowdfunding helps companies to advertise their products in the market thus increasing their visibility. The platform helped the small entrepreneur to grow and thus benefitting society as a whole.

(e) Crowdfunding platform helps investors to put their funds in a number of opportunities available in the market with relatively little operational costs and attain equity positions in companies that may ultimately prove to be winning and profitable, which is not possible with the present regulations. In return, these platforms charge a nominal fee to all its registered accredited investors for carrying out their due diligence.

(f) It is required for the platforms to pay some fees for recognition. The consultation process may help in crystallizing that money for the new framework.
(g) Some of the crowdfunding platforms promote the participation of investors by offering some rewards at the time of joining the stream. These rewards are in the form of some gift, so many times a product of the company. For example, a food company can send a hamper of food products to the investor. In the new scenario, where youth are deeply attached to technology, one could find more investors with the video games making company because the investor would be happy and proud to get an advance copy of the games.
(h) Central Role of the Internet: The role of Internet cannot be under-played in the process of crowdfunding because it helps in a much wider reach out of the potential investors because a major part of the new India generation uses digital technology and hence are better informed than other means of yesteryears. It is also helpful in connecting with the international lenders because no formalities are required to be fulfilled due to jurisdictional issues as money transfers are online.

11.10 CROWDFUNDING WORLDWIDE

Table 11.1 shows the transaction value in the crowdfunding segment amounts to 6923.6 million US$ in 2019 in the world. It is expected to grow every year. In 2023, it is expected that it will reach 11,985.6 million US$.

TABLE 11.1 Transaction Value in the Crowdfunding Segment

Year	Total Transaction Amount in million US$
2017	3979.4
2018	5319.2
2019	6923.6
2020	8537.3
2021	9963.2
2022	11,113.7
2023	11,985.6

Source: Statista, 2019

The total transaction value in this division amounts to 6923.6 million $ in 2019. It is likely to have an annual growth rate (CAGR 2019–2022) of 0.7% resulting in the total amount of $1.8 million by 2022. China has been top in the list with altitudinous deal valuation @ $5572 million in 2019 over the globe (Table 11.2). USA comes at second position with the transaction value amounts to $718 million followed by UK at $88 million. France is on the fourth position with $79 million and Canada stood at $43 million (Statista, 2019). Crowdfunding is most popular in countries like UK, USA, and China. These three countries are collectively contributing 96% market of crowdfunding in the world.

TABLE 11.2 Transaction Amount of Crowdfunding in the World (Country Wise)

Country	Crowdfunding Amount
China	$5,576 million
USA	$718 million
UK	$88 million
France	$79 million
Canada	$43 million

Source: Statista, 2019.

Table 11.3 shows the average funding volume of the selected market per campaign. The volume is defined as the total funding in the particular segment proportionally to the number successfully processed campaigns in one year time period. The average funding per campaign in the crowdfunding segment amounts to 794US$ in 2019.

TABLE 11.3 Average Funding Per Campaign

Year	Average Funding Per Campaign in US$
2017	765
2018	824
2019	794
2020	780
2021	843
2022	924
2023	994

Source: Statista, 2019.

11.11 P2P CROWDFUNDING STATUS IN WORLD COUNTRIES

(a) **Egypt, Brazil, South Korea, and China:** In these countries, P2P lending is an exempt market or there is a lack of definition in the legislation.
(b) **Australia, Brazil, Argentina, and New Zealand:** The platforms that support the P2P crowdfunding ventures are called as intermediary. In these countries, these platforms are to be registered to start operating in crowdfunding business.
(c) **USA:** In USA, the company going for crowdfunding P2P lending has to get them registered with Securities and Exchange Commission. Further applying for license is mandatory to carry out business in a state.
(d) **Canada:** In Canada, the crowdfunding ventures are categorized as equity and nonequity based. Registration is required for equity-based crowdfunding platforms that include truthfulness, expertise, and solvency needs for the persons dealing in it.
(e) **United Kingdom:** In 2014 Britain came out with regulation for conduct of loan-based and investment-related crowdfunding ventures. The norms for the conduct of crowdfunding are dictated by Financial Conduct Authority.
(f) **Japan:** It has permitted the use of equity-based crowdfunding ventures in the country. In 2014, Financial Services Agency has amended its Financial Instruments and Exchange Act to assist and support equity crowdfunding in Japan.
(g) **Israel**: Equity and P2P crowdfunding is banned in Israel.

11.12 POPULAR CROWDFUNDING WEBSITES

Kickstarter and Indiegogo are the two popular crowdfunding websites. They have helped more than 200,000 companies to get funds to start their business all over the world.

(a) *Kickstarter*: It is the largest crowdfunding platform in 2019. Started in the year 2009, it has funded more than 160,000 projects with a total amount of $4.2 billion undertaken for all Kickstarter projects. Kickstarter provides funds once the company reaches the stage of needing financial assistance (Smith, 2019).

(b) *Indiegogo*: It was launched in the year 2007. It is headquartered in San Francisco, USA. Initially, the Indiegogo provides a platform to raise money for independent film manufacturers but later on it diversified and helped any company to raise money through its platform. It can be for a new idea, donation, or a startup company. The Indiegogo is flexible in taking decisions. It allows the companies to receive money on a pro-rata basis (Smith, 2019) and charges 5% fee for its contribution.

(c) *Milaap*: Milaap was founded in 2010 and is one of the biggest crowdfunding platforms in India. Initially, it raises funds for rural development and small entrepreneurs but now anyone can raise funds for pursuing education, health-related issues, personal use, and disaster relief. It also helps the patients to raise funds for the treatment of cancer and other major ailments. Till now it has raised money over US$12.7 million (Milaap, 2019).

11.13 INDIA'S TOP P2P LENDING PLATFORMS

There are around 15 P2P crowdfunding platforms in India that have come up within last 18 months. These platforms facilitate the rising of unsecured funds. It provides a platform where borrowers and lenders interact among themselves. The lender chooses to whom to lend. The interest rate is decided mutually between the borrower and lender. The lending rate normally goes till 20%–24% (Nirupama and Shankar, 2018).

Rang De

It was founded by Smita Ram and N. K. Ramakrishna headquartered in Bengaluru in 2008. RangDe.org, a P2P microlending platform which helps the rural entrepreneurs across India by providing them low-cost loans out of the funds collected from numerous investors across the country. Around 93% of the participants are women. RangDe.org is a nonprofit organization and this crowdfunding platform has been instrumental in luring 9699 social investors whose contributions formed a disbursement of 50,008 loans for the oppressed and weaker section of the Indian population who are not entertained by the banks or other financial institutions. The amount of money collected is equivalent to approximately 7 million US$ while repaying very close to 5 million US$. For these collateral-free loans and interest between 4.5% and 10% per annum is payable by the borrowers whereas Rang De deducts 2% as interest charges of the total amount on all loan payments by borrowers.

World Bank has facilitated Rang De by providing funds through the Development Marketplace. They have received various other awards for helping society for a cause that includes South Asian International Fund Raising Group's Campaign of the Year Award and 2013 Millennium Alliance Award.

The organizations have 25 field partners and deals in 16 states of India who physically take the money to the borrowers and can contact them if they fall into arrears.

Faircent

It was founded by Rajat Gandhi, Nitin Gupta, and Vinay Matthews and is headquartered at Gurgaon. It provides a platform where investors and borrowers deal directly. They deal with respect to the terms of the loan that includes the time period and rate of interest. There is no involvement of banks.

Every client is required to pay 23 US$ as one time fees along with the fee of administration depending on the size of the loan rose.

The website has attracted 26,000 borrowers and has around 6000 lenders. Total loan amount worth 973,000 US$ has been collected during the last 2 years (Nekaj, 2016).

Ketto

Ketto was founded by Kunal Kapoor, Varun Sheth, and Zaheer Adenwala in 2012. It is headquartered in Mumbai. It helped to raise funds mainly in three categories: nongovernmental organizations (NGOs) for a social cause, charities and not for profit motive; arts and cinematography; and for personal use that includes health-related aspect, higher studies, and traveling abroad.

Companies are also encouraged to look for CSR projects through the platform. NGOs are motivated to use the platform for digital marketing of their products. In return, Ketto charges 30 US$ or 5%–8% of the amount rise as their commission whichever is more. Till now Ketto platform earns revenue of 4 million US$. It has 40 employees. Till now, it has raised 822.7K US$ (Ketto, 2019).

Wishberry

It was founded by Priyanka Agarwal and Anshulika Dubey in 2012 in Mumbai. The platform provides services exclusively for artistic projects that include the production of films, music, arts, dance, cinematography, theatre, and many more. The company provides an adviser, that is, campaign coach for handling each assignment. The coach helps the client to make their campaign more effective. He guides the client in making videos, writing material to be published on the website and sometimes even help in the handling of logistics, and allocation of rewards. This leads to the making of the campaign even more successful. The platform has achieved 67% success rate (Wishberry, 2019).

Wishberry crowdfunding has raised 1.3 million US $ from 600 projects contributed by more than 7750 people worldwide. It has a network in 60 countries. A period of 60 days is allowed to the client to raise the funds and achieve their targets. In return the contributors get nonmonetary rewards. They are invited for film premier, sample of new products, project making experience, and many more.

Wishberry charges a fee amounting to 52.37 US$ and a commission of 10% of funds raised through this platform. The company also provided digital marketing services to the clients for a monthly fee (Nekaj, 2016).

KEYWORDS

- **crowdfunding**
- **Kickstarter**
- **Indiegogo**
- **cost benefit analysis**

REFERENCES

Arumugam, S. (2018), All About Crowdfunding accessed on 31st August, 2019 from https://www.fundstiger.com/crowdfunding/

Cleartax (2019), Benefits of the Startup India Program accessed on 23rd August from https://cleartax.in/s/benefits-startup-india-program

Choudhury, M. (2018), How Crowdfunding Works in India All about Crowdfunding accessed on 23rd August from https://finance.cioreviewindia.com/cxoinsight/how-crowdfunding-works-in-india-all-about-crowdfunding-nid-3503-cid-63.html

Diwan, F. (2018), Crowdfunding Regulations in India: All You Need to Know accessed on 23rd August from https://www.impactguru.com/blog/crowdfunding-regulations-in-india-all-you-need-to-know-

Feldman, A. (2017), M3D Raised Millions on Kickstarter. Now its Founder is Launching His Own Crowdfunding Site accessed on 23rd August from https://www.forbes.com/sites/amyfeldman/2017/06/15/m3ds-michael-armani-built-his-business-on-kickstarter-now-hes-launching-his-own-crowdfunding-site/#5b5f930c57f9

Gadre, G; Bhargava, A. and Mehta, L. (2018), How Crowdfunding Works accessed on 23rd August from //economictimes.indiatimes.com/articleshow/66891983.cms?from=mdr&utm_source=contentofinterest&utm_medium=text&utm_campaign=cppst

Jalan, S. (2018), How the RBI's Recognition of P2P Lending has Simplified Access to Finance accessed on 2nd August, 2019 from https://yourstory.com/2018/09/p2p-lending-rbi-recognition-finance-access-india

Kalra, A. (2019), Crowdfunding in India accessed on 2nd August, 2019 from https://www.investindia.gov.in/team-india-blogs/crowdfunding-india-0

Kaushik, T. (2019), From Building Toilets to Saving Trees, Bengaluru does it Through Crowdfunding accessed on 15th August, 2019 from https://economictimes.indiatimes.com/news/politics-and-nation/from-building-toilets-to-saving-trees-bluru-does-it-through-crowdfunding/articleshow/69645907.cms

Ketto (2019), Ketto accessed on 2nd August, 2019 from https://www.ketto.org/new/browse

Kirby, E. and Worner, S. (2019), Crowd-funding: An Infant Industry Growing Fast accessed on 24th August, 2019 from https://www.iosco.org/research/pdf/swp/Crowd-funding-An-Infant-Industry-Growing-Fast.pdf

Milaap (2019), Overview accessed on 24th August, 2019 from https://milaap.org/about-us/overview.

Monika (2019), What is Crowdfunding? How Can You Do It? Accessed on 1st August from https://blog.ipleaders.in/crowd-funding/

Nekaj, E.L. (2016), India's Top 10 Crowdfunding Platforms accessed on 1st August from https://crowdsourcingweek.com/blog/indias-top-ten-crowdfunding-platforms/

Nirupama, V. & Shankar, S. (2018), How P2P Lending Platforms Like Faircent, LendenClub are Helping People Make Money accessed on 5th August from https://economictimes.indiatimes.com/industry/banking/finance/banking/how-p2p-lending-platforms-like-faircent-lendenclub-are-helping-people-make-money/articleshow/50004155.cms

OICU-IOSCO (2014). Crowd Funding: An Infant Industry Growing Fast accessed on 8th August, 2019 from https://www.iosco.org/research/pdf/swp/Crowd-funding-An-Infant-Industry-Growing-Fast.pdf

RBI (2018), Master Directions - Non-Banking Financial Company—Peer to Peer Lending Platform (Reserve Bank) Directions, 2017 accessed on 1st August from https://www.rbi.org.in/Scripts/NotificationUser.aspx?Id=11137

SEBI (2014), Consultation Paper on Crowdfunding in India, Securities and Exchange Board of India accessed on 5th August from http://www.sebi.gov.in/cms/sebi_data/attachdocs/1403005615257.pdf .

Smith, T. (2019), Crowdfunding accessed on 6th August from https://www.investopedia.com/terms/c/crowdfunding.asp

Statista (2019), Crowdfunding, accessed on 7th August from https://www.statista.com/outlook/335/119/crowdfunding/india

Wishberry (2019), Crowdfunding basic accessed on 30th August from https://www.wishberry.in/how-it-Works

CHAPTER 12

Linkage between Business Strategy of the Value Chain and Industry 4.0 in Indian Context

POOJA TIWARI

[1]ABES Engineering College Ghaziabad, Uttar Pradesh 201009, India

[*]Corresponding author. E-mail: pooja2017@gmail.com

ABSTRACT

In the present context of digitalization across all industries, most of the organizations are investing in that equipment, tools, and solutions which will help their procedures, employees, apparatuses, and also the goods to integrate into the solitary system. This single network will be helpful for data collection, analyzing data, and also evaluating the development of the company and its performance as well. In this chapter, authors have tried to understand the impact of 4.0 on the company using the Porter's value chain model, which is quite crucial for the area which is primarily concerned with the value creation with the customers. It has been analyzed that 4.0 primarily influence the process of value creation, and so far had a larger impact in this area. The main aim of the learning is to explore how corporations functioning in India understand the occurrence of 4.0, what equipment and tools they are using for Internet of Things for supporting the process and challenges faced by them during the adaptation. At the end of this chapter, it has been identified that due to real-time data spreading across companies—probably all the necessary tools and methods are available, it can surely have a larger impact on the company. Additionally, it has been concluded during the study that the majority of the companies have moved toward the evolution in a digital way, and they have already started investing in this area.

12.1 INTRODUCTION

In the present scenario, we can say that 4.0 revolution in the industry is coming in the form of pioneering and qualitative nature. On contrary, it can be analyzed that production process is gradually transforming and an integrated way can be adopted to manage and supervise it, but still flexible in nature. To compete and survive in the competitive globalized environment, companies engaged in manufacturing sector are required to continuously develop their production system and adjusting according to the dynamic customer demands of the arcades (Pedersen et al., 2016). These transformations have a huge effect on the different industry and markets, which in turn influences the entire product life cycle, and provides the different production means of conduction of business and further improve the process as a whole and increases the enterprises competitiveness.

This will not only have an impact on the entire life cycle of the product rather it can have a huge impact on the different industries and markets as well. It will also provide different forms of conducting the business and improving production, which will in turn influence the upsurge in the process of production and providing a competitive edge to the organization (Slusarczyk, 2018). Even in the previous decades, computers, automation, and robots were existing, but due to revolution in Internet many opportunities were available to utilize them due to opportunities provided by them (Deloitte, 2015; Geissbauer et al., 2016; Monostori, 2014). The activities of machine, material and workers and even products can be monitored and data can be collected, utilized, and analyzed in the real-time decision-making. Data is one of the factors which leads to industrial revolution. How the data is collected and analyzed will help to take the right decision and develop to get a competitive edge over the others. The companies will be getting the competitive edge not only on the basis of production produced on the basis of coordination (e.g., additive production), but it also gives the edge as it integrates the production in an integral way, that is, understanding the way information is filtered from the data generated so as to support the process of decision-making (Deloitte, 2015; Geissbauer et al., 2016).

Analyzing recent decades, information technology (IT) is supplemented everywhere in both manufacturing and production systems, so as to control and handle the complex technologies in the field of IT. It is becoming one of the complex tasks to deal with the requirements of multiside production and supportive logistic process.

It is seen that IT has become one of the important aspects in transforming the efficiency and working conditions, and obviously the importance cannot be questioned (Nick and Pongrácz, 2016).

In the context of Industry 4.0 readiness, Berger (2014) has observed that although India is emerging as one of the preferred country by many nations but still they need to focus on the sophistication in the production process, level of automation, readiness of workforce, and innovation intensity.

The appropriate answer to the fourth industrialized insurgency is "Industry 4.0." Improvement in automation and operational efficiency can be achieved by the automation and efficiency in an operational manner and effectiveness (Slusarczyk, 2018). The concept of Industry 4.0 is a new concept and brings the paradigm shift, is an umbrella term which encompasses many developments in the industry (Pereira and Romero, 2017). If we have to improve the manufacturing process than certain these types of technologies have to be adopted which comprises of machines, devices, module of production and products which can exchange the information independently and controlling and triggering an environment which is considered as the intelligent in terms of manufacturing (Weyer et al., 2015).

The main focus of this study is to understand Indian companies and understanding the Industry 4.0 and what tools are used by them to support the production process and during adaptation what challenges they are facing. This chapter has contributed in two major areas in the existing body of knowledge. Primarily, this chapter can provide indications regarding the adoption of Internet of Things (IoT) and its influence on value chain with specific to particular company which can be taken as a reference for other companies as well. In line with this introduction, the chapter goes like this. In the initial stage, the importance of industrialization in a digital way is discussed so as to define it in a uniform way. Additionally, how Industry 4.0 will be discussed and its linkage with the value chain. Finally in this chapter, we will be discussing the concept of Porter's model tools used and the solutions accordingly.

12.2 LITERATURE REVIEW

12.2.1 INDUSTRY 4.0 AND ITS RELEVANCE

According to Wang et al. (2016a), Continental Europe particularly countries like Italy have been facing a problem of an increasingly aging workforce.

It is on this account that automation has been rapidly adopted by European countries. The Internet-based emerging technologies have revolutionized the production process by interconnecting various devices. Information and communications technology (ICT)-based technologies have created a seamless network of devices, people, and organizations, thereby creating a value addition process through the sharing of processes. As a result, the product can be manufactured which can be fully customized. Industry 4.0 involves the networking of devices and equipment used in production processes (Bauer et al., 2014; Rüßmann et al., 2015). Rüßmann et al. were able to recognize nine different technologies based on which the industrial revolution can be described in big organizations. In Industry 4.0, there is more amount of connectivity between people and machines, this leads to demand for advanced techniques of data analytics and prediction tools. (Bildstein and Seidelmann, 2016). Hence, Industry 4.0 may be described as ICT-based changes in production technology and processes. These changes may have widespread implications. (Lasi et al., 2014). According to Hermann et al. (2016), however, Industry 4.0 represents authentic digitization of the industrial sector which shields an extensive zone including the development of technologies related to the value chain. Industry 4.0 enables the creation of a unique structure, which integrates, the physical processes in an organization with the virtual world that monitors and controls the physical processes thereby making the functioning of machines autonomous. Research indicates that investments in Industry 4.0 have as a result lead to an increase in productivity. (Bughin, 2016). However, most of the research in this area was conducted during the 2010s when investments in Industry 4.0 were made by large corporations and hence the results cannot be generalized to the entire economy. Research has also been carried out at a macroeconomic level across several countries, which indicates that the effect of industrial automation and robotics has led to an improvement in productivity, profits, as well as product quality, thereby benefitting all stakeholders (Popp et al., 2018).

Increasing adoption of digital technology by the industrial sector is in response to the changing tastes and preferences of customers who are increasingly demanding new products and services. This increasing demand for new products has led to shorter product life cycles. Hence, innovation and development of technology should be kept up to date. The product as well as the production process needs to modify on a regular basis to keep pace with the changing consumer preferences (Hermann et al., 2014). As a result of increasing digitalization, there may be a noteworthy influence on the supply chain and logistics network which may result in lower inventory, logistics, and transportation costs. Users of Industry 4.0 may be able to improve their capacity utilization and make their products available in the market faster.

Research has shown that in India, the critical success factors in improving productivity and profitability were the development in the use of IT (Oláh et al., 2018). According to Geissbauer et al. (2016), as a result of 5% spending on IT-based tools and skills the annual cost of manufacturing processes can decrease by 3.6% on account of a reduction in supply chain and product quality.

According to Wang et al. (2016a), Industry 4.0 can be implemented which involves the straight and straight-up amalgamation of the supply chain and adoption of digital technology in the supply chain network. Digital technology would include IoT, big data analysis, cloud-based systems, embedded systems, etc.

Hermann et al. (2016) and their co-researchers, were able to identify the different fundamental tools which are essential for the implementation of Industry 4.0.

These tools are Internet, IoT, cyber physical system (CPS), and the smart factory. Each of these tools indicates comprehensive categories and not just specific tools. It can be concluded that Industry 4.0 would include the complete value chain of the organization. However, maximum of the components of the value chain would be based on the production and logistics operations.

Industry 4.0 can expand the corporation's network with the help of networked linked technology, software, and cloud computing and data analysis. Industry 4.0 would require new business models and a new level of commitment on behalf of the companies, like data sharing, continuous innovation, etc. The objective is to create a well-integrated seamless supply chain network which leads to a new level of value creation.

In the current research, it has been reflected how companies have adopted digital tools and processes. With the help of a questionnaire, insights have been obtained into how digital technologies are being used by companies in Industry 4.0. Interviews with experts have given insights into the advantages and limitations of the adoption and use of these technologies.

12.2.2 INDUSTRY 4.0 AND ITS LINKAGE TO PORTER'S VALUE CHAIN THEORY

The entire organization will be influenced by the fourth revolution in the industry, so it is very essential to comprehend how the different functions are able to take advantage of the prospects offered by emerging digital technologies.

Ideally, the core process of a company should be customer value creation and Industry 4.0 affects various elements of value creation, particularly production.

Corporate functions that create value can gain immensely from Industry 4.0. Supply chain and logistics represent such a function and its efficiency has increased as a result. Digitalization is creating value in every stage of the supply chain, thereby creating a competitive advantage for the firm. The company's internal structure also helps in creating value for the company by increasing the benefit or by reducing the price.

One way of creating value in a supply chain is to streamline functions inside a corporation and look for a competitive advantage. The value chains of companies will differ across various sectors and industries, as well as within the same industry. The structure of the value chain depends upon its strategy formulation, implementation, as well as its culture. The value created by a supply chain can be defined in terms of the benefit for the buyer, beyond the cost of acquiring it. This forms a basis for any company to survive.

The main base of the corporate strategy is to understand the need of the customer and serving them using the approach of value based. Different activities of intra-corporate should be systematized, which will help to find a competitive advantage. It has been analyzed that the value chain of companies differs from each other but the different companies operating in the same industry create a different value chain. This structure is created based on the strategy of the company, how it is implemented, and based on the tradition of the corporate. The product that is generated for the buyer is according to its worth, which is created by the value chain. This is the main foundation for the corporation to sustain and persist and the price regarding this must go distant yonder the cost. The main basis of the corporate strategy is properly catering to the needs of the customer and understanding their needs.

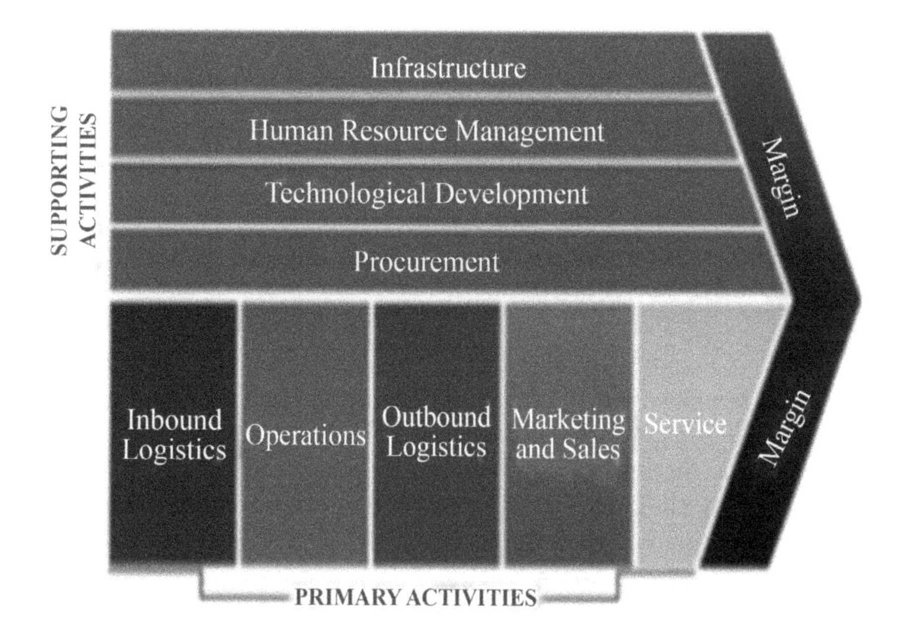

FIGURE 12.1 Porter's value chain.
Source: Porter (1985), Chikán (2017), authors' own editing.

Rayport and Sviokla (1995) have recognized the significance of configuring the activities of the company into a value chain, but it has been recommended that there should be a difference between the chains of physical and virtual. Porter classification of primary function in value creation in customer has been included in the physical value chain (Figure 21.), on the other hand the data of the whole corporation and the evidence taken in the course of physical value creation.

Porter has classified two major functions such as primary function which includes creating the value for the customers which involves the physical value chain on the other hand whole company comes under the value chain and which discusses to capture the material during the physical stage value creation. This is one way how corporations can supervise their entire procedure of creating the value which will help them to perform certain activities which adds value more effectively and efficiently. Even certain activities such as giving product information, about production process, etc. are also well thought-out in the value adding services. The main method is aligning the application of IoT tools and Industry 4.0 technology in the different process of the organization and as well as production and integrating material created services to goods. Since Rayport and Sviokla (1995) have used an almost similar type of activities in their value chain, which were used by the Porter in their process which was primary, and also the virtual chain was not divided into a different set of activities, Porter's value chain has been used to comprehend the IoT tools and Industry 4.0 application in diverse companies.

12.3 FOURTH INDUSTRIAL REVOLUTION AND ITS IMPACT ON COMPANY'S RELATIONSHIP

The fourth industrial revolution has not only influenced the internal business of the company, but it has also influenced the different business relations. In case of supply chain, the most important relationship that needs the mapping is the relationship between suppliers and customers. As per the report by KPMG, they indicated that in future there will be a trend that former competitors will be integrated for working and sectoral alliances to emerge (KPMG, 2016). The strategies of the companies will determine the value-added network at global level, which in turn is driven by the vibes which are derived by the capitalist environment and which tries to minimize the risk through efforts which were present in the environment externally (Shrouf and Miragliotta, 2015). One digital ecosystem is created by the help of Internet which includes all the supplier, manufacturer, and the customer which helps to accumulate all the relevant data which can be retrieved instantly in the cloud so as to efficiently synchronize all the happenings. According to the experts, this cannot be considered as a goal which can be attainable and realistic in the future. This is feasible only when all parties such as a supplier, the factory, and the quite possible customer if they fit into a cluster of corporations, this will help to generate transparency among the subsidiaries of the organization that is central in nature and provides them a platform with opportunities and different standards for performance (Costanza et al., 2014).

In the near future even if it cannot be merged in the single network, but the relationship is changing with the different suppliers and customers. For the supplier, the most important is to meet the need and expectations of the customers. In case of the supplier, the most important thing is the expectation of the customer. In case of development of the product and order fulfillment, the main demand of the customer is speed and flexibility. Additionally, accessibility should be at one place. Due to the technology of cloud computing, the process of production which was earlier isolated is now integrated and automated which has influenced the change in the relationship between different channel partners (Ehret and Wirtz, 2017). According to a survey by PwC (2016), to improve the relationship with the customer and also analyzing the data of the customer response of 72% of the customer is recorded. The relationship with the customers can be improved by the way responses are given to the customers according to their needs which can be achieved by the proper planning of products and services and introducing innovation and customization of the product, and one-piece production volumes. When we try to understand the need of the customer based on the data available, this data can further be used for the development of process of the production, but it will also help to build the supply chain which is centric in nature. Industry 4.0 helps to organize the different sets of people in the value chain horizontally, vertically, and virtually. So, it is recommendable for the suppliers engaged in this to fully collaborate and integrate in the value chain network (Stock and Seliger, 2016). This study basically discusses the offline dimension of service and e-customer satisfaction which can be incidentally linked through the quality of the website and other related issues (Laureti et al., 2018). In total, it is possible to generate the sufficient amount of the data which can be generated through digitization affects in the different areas of business of the company, which helps to improve the transparency, integration, and design ability and other information required such as needs of the customer and individual task to fulfill them. A new business can be created through Industry 4.0, for example, product design and development, and security of the data will be of prime concern in future (Porter and Heppelmann, 2014). Industry 4.0 tool which is used by the corporate in the value chain is presented in Figure 12.2. It can be analyzed that most of the functional boundaries are influenced by the different set of technologies and are influencing the entire process of the value chain or maybe the company itself.

In general, the strategy of the company, process design, and finance cane influenced by changing the real-time data at different intervals of time. According to the generation of data and provided information, it is possible to generate transparency in all the different areas of business. This type of technology will surely be beneficial for both system and process thinking and will surely help the organization for the integration at both the levels, that is, inter and intra level (Ilie-Zudor et al., 2011). Different types of technologies are really beneficial for the company if they want to attain the competitive edge and will also contribute to the upsurge of quickness, compliance, and also how different firms are arranged which have contributed to the network of the value chain (Ketchen and Hult, 2007).

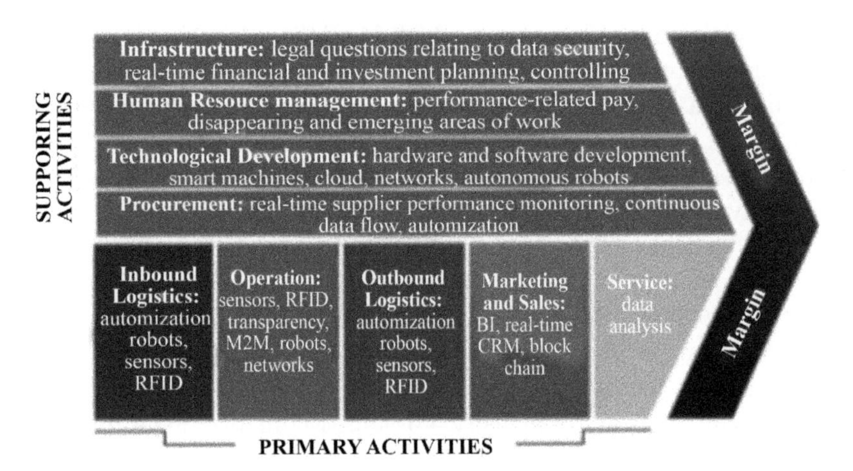

FIGURE 12.2 The tools of Industry 4.0 in the corporate value chain.
Source: Author's own editing based on Porter (1985).

12.4 CHALLENGES TO IMPLEMENTATION OF INDUSTRY 4.0

All the different set of corporations are engaged in the revolution at the fourth level and this still remains the decision-making question. Although this activity will be quite expensive and risky for the companies but still the companies' response toward the revolution will determine that how much cost they will be able to save (Tupa et al., 2017). Additionally, it is not possible for the companies to keep a distance from such progress and dynamic development and implementation of application. Nevertheless, still there are few industries that will surely board on these developments if other companies have paved the track and the required technology is reasonable for different segments functioning with a little margin. At the industry level, there are two variables that will determine the impact of Industry 4.0, that is, its creation and consistency in the implementation of a digital strategy that is corporate in nature (Oesterreich and Teuteberg, 2016).

There are certain factors that can hinder the scope and spread of Industry 4.0. In references PwC (2016) and Porter and Heppelmann (2014) dealing of this question has been done. PwC (2016) has raised the issue of certain factor which are critical and risky in nature, while on the other hand Porter and Heppelmann (2014) and his co-author has emphasized on the loopholes that have to be considered and evaded by the firms.

In 2016, PwC produced a Global Industry 4.0 survey, have considered 2000 respondents for the interview from 26 different countries that how their companies use digitalization for attaining a competitive edge. Based on the result, it indicated that the major hindrance to the execution of Industry 4.0 is processes the absence of a clear digital tactic in value delivery (production and logistics). Most of the companies surveyed (52%), such as corporate executives and support about the introduction of digital technology (PwC, 2016). In digital services, you can join the product to what the consumer is enthusiastic to pay. The truth is that the information is accessible, it is not for sale. Therefore, it is quite essential to consider the factors which are creating value for the buyers.

Security is a prime concern and this development is quite crucial in nature. However, you must have access to security and control of network devices, sensors, etc. for the purpose of your own information that can be encrypted (Hossain and Muhammad, 2016). A model can be implemented which is business centric and for which new entrants will also emerge as with new smart products and services which are related to market and new type of customer. A major concern is where to take the step. If any organization has to wait for a long time, the new operators can adapt to their competitors in the market to achieve a competitive edge in the process of learning. Technology can be very beneficial in producing the new types of smart products, process competencies in the value chain. Table 12.1 lists the factors that prevent the spread of Industry 4.0.

TABLE 12.1 Factors Hindering the Spread of Industry 4.0

Cultural Market Obstructing Factors	Labor Market Obstructing Factors	Organizational Obstructing Factors	Technological Obstructing Factors
Distrust	Inadequate quality workforce	Lack of digital strategy	Expensive technologies
Uncertainty	Shortages among workforce	Risky investment	Lack of standards
Realistic judgment of the abilities of the organization	Old-fashioned training	Fear of loss of control over intellectual property	Security of data, uncertainty regarding the level of encryption,
Lack of demand for continuous learning		Partners do not have the technology	Underdeveloped data analysis
		Failure to develop data-based services	
		Lack of senior management support	

Source: Authors' own editing based on Porter (2014) and PwC (2016).

In order to allow companies to create Industry 4.0 to develop jobs, governments must devote in training to generate a greatly capable, digitally qualified workforce.

12.5 INTERNET PROBLEMS AND SOLUTIONS

Networking of different devices and equipment are the fundamental requirement for the digitalization on an industrial basis. This is generally referred to as the IoT concept of a profession that does not define a unanimous vote. According to the set and enforce Pentek Ball (Hermann et al., 2016), this is the term "portable devices" that were equipped with the chip, RFID, sensor networks, or any additional device talented of, and can exchange and distribute the data and further communicate. The key logic after the IoT is about the last decade, informatics, and telecommunication (Arnold et al., 2016).

For the purpose of information collection and sharing different technological components which are considered as IoT tools helps to enable the connection of the production machine to the network of the corporate. These may include the aforementioned sensors, RFID, 3D scanners, cameras, etc. and those who begin to collect data with the IO can be very useful for the corporate and for most of us to have good use of them. Huge volumes of data that are difficult to handle in predictable inquiry systems are large and continuous data (Wang et al., 2016c; Laney, 2001). Gathering and allocation with authentic human beings can be done by entrepreneurs or data stores of the company that can be stored in the cloud by using cloud computing (such as Amazon, Microsoft) (Rüßmann et al., 2018). They are only valuable if data is really the only tool we have to analyze it and put it in a user-friendly form. It is very beneficial for the attainment of the competitive edge with the help of big data analytics. For the purpose of the correct data and information, companies are increasingly investing in the development of data mining software, algorithms, and enterprise resource planning (ERP) interfaces and which is difficult for the company on the side of the investment, but also appropriate adjustment of the personnel.

CPS is a physical device that is done through cyberspace. A vision for the CPS in the sense that the consumer must be incorporated into the machines, systems, and data storage production facilities can be modified, with independent intelligence (Reinhart et al., 2013). The CPS monitors physical processes and decentralized activation plans and activities, and attention is communicating with real people in it. This fundamental bring improvements involved in processing, manufacturing, marketing, use of materials, and supply chain management in the life cycle (Toro et al., 2015). These things are accomplished by machines with lower significant financial costs (Gubbi et al., 2013) and with greater precision. Consumer devices such as auto-propelled vehicles robot and weapons and also seemed in company exercise carried out in the supply process of production and logistics that the body needs. Connection with the network can be attained by smart devices, interaction with the promoter, and the capability to respond to the transformation you plan to do well (Kortuem et al., 2010).

Industry 4.0 has helped the developer through tools for generating data and essential information for large delivery (Zhong et al., 2015). The interfaces of people's devices are used to connect so often in real-time. Additional tools, such as clouds, a warehouse, but for your knowledge of ERP systems, and those that do not collect, store, and distribute data. If a common base is prepared for so many machines and devices, most of the standards are often updated in the proprietary development software to extract relevant data generate data in what is and it is convenient for users that is, on a phone or tablet such a mobile device (Bologa et al., 2017). The device is the preferred interface, it is not necessary that you cannot specialize in mobile worker. All this is particularly interesting since it is inexpensive and it is sitting on the Smartphone's recess in the upper control and connection, which is like a tool (Woo et al., 2015). The presence of tools, methods, and procedures above the crowd, since it has to start working on the fourth development path created the industrial revolution. Of the latter, which is mandatory for exchange?

1. The application of network tools and technologies for the entire business process to ensure transparency.
2. Manufacturing integration and close cooperation in connectivity define a moment of action.
3. Vertical integration, which primarily involves cooperation with partners in the supply chain, then with allies of the power network, the connection between the digital.
4. Rethink the business model to focus on the spirit of competition, and an efficient organizational structure.

12.6 DISCUSSION AND CONCLUSIONS

It is becoming essential to store and generate the large amount of data which have found on the system. These technologies are difficult to apply and implement, but it surely increases the saving and able to generate the significant revenues for the company. While implementing these applications, it is essential to see the feasibility and how it helps the companies to maintain the competitive advantage to the organization (the automotive and electronics industry),

There are various examples of how the companies are using these practices and different tools in their production and other process and the way it is influencing the business. If we are using a value chain Porter device, it can be observed that the certain IO can not only influence the process of production and Industry 4.0, but it will also influence other major functions of the organization as well. According to the previous discussion, we can conclude that these latest technologies will be beneficial and will have an impact on the business and rather provide more control. This data can further be useful for producing and developing new and innovative products as well. These different processes will also help in maintaining the data that has been shared by the organization with customers and different development partners. It is quite pertinent that 4.0 value chain approach and the appropriate technology in the area of IoT industry can be further extended to the concept of the value chain, and a digital ecosystem in the value chain that is a virtual network. The major hindrance to the execution of Industry 4.0 is the absence of clear value through the delivery of the digital-warfare process and the lack of support from the leader. Many are still afraid of unknown companies and the economic foot level of digital money, and the high cost of these things. Firms that, in effect, are safer for many genres and not willing to, and soon, and investors. Goalkeeper Heppelmann (Porter and Heppelmann, 2014) argue that it requires rethinking industrial digitalization technology, skills, and processes throughout the entire value chain. But ask him out as the pain of a partner familiar with the development and development of the cause of the lakes of reality from what the various types Nullamvelita tends to another. It is too large, and that the methods used in the systems that are built as we can see in the trucks. Companies do not have to develop a number of employees at the front of the work (the great abundance of data storage, analysis: it belongs to the software of the programs), it will remain the basic one, but what do they live in the competitive advantage in the future a. Outside the conversation, since it is the international literature, 4.0 Industry exercises its highest influence on the manufacture and also examines several corporations' current methods and procedures.

KEYWORDS

- **Internet of Things (IoT)**
- **Industry 4.0**
- **business intelligence**
- **cyber physical system**
- **value chain**
- **sustainable development**

REFERENCES

Arnold, C.; Kiel, D.; Voigt, K.-I. How the industrial internet of things changes business models in different manufacturing industries. Int. J. Innov. Manag. 2016, 20, 1640015.

Bauer, W.; Schlund, S.; Marrenbach, D.; Ganschar, O. Industrie 4.0—Volkswirtschaftliches Potenzial für Deutschland; BITKOM und Fraunhofer–Institut für Arbeitswirtschaft und Organisation: Berlin, Germany, 2014; pp. 1–46. Available online: https://www.bitkom.org/noindex/Publikationen/2014/Studien/ Studie-Industrie-4-0-Volkswirtschaftliches-Potenzial-fuer-Deutschland/Studie-Industrie-40.pdf (accessed on 12 March 2018).

Berger, R. Industry 4.0: The New Industrial Revolution—How Europe Will Succeed; Roland Berger Strategy Consultants: Munich, Germany, 2014; pp. 1–24. Available online: https://www.rolandberger.com/ publications/publication_pdf/ro (accessed on 12 March 2018).

Bildstein, A.; Seidelmann, J. Migration zur Industrie-4.0-Fertigung; Springer Vieweg: Berlin/Heidelberg, Germany, 2016; pp. 1–16.

Bologa, R.; Lupu, A.-R.; Boja, C.; Georgescu, T.M. Sustaining employability: A process for introducing cloud computing, big data, social networks, mobile programming and cybersecurity into academic curricula. Sustainability 2017, 9, 2235.

Bughin, J. Big data, Big bang? J. Big Data 2016, 3, 1–14.

Costanza, R.; deGroot, R.; Sutton, P.; VanderPloeg, S.; Anderson, S.J.; Kubiszewski, I.; Farber, S.; Turner, R.K. Changes in the global value of ecosystem services. Glob. Environ. Chang. 2014, 26, 152–158.

Deloitte. Industry 4.0, Challenges and Solutions for the Digital Transformation and Use of Exponential Technologies; Deloitte: Swiss, Zurich, 2015.

Ehret, M.; Wirtz, J. Unlocking value from machines: Business models and the industrial internet of things. J. Mark. Manag. 2017, 33, 111–130.

Geissbauer, R.; Vedso, J.; Schrauf, S. Industry 4.0: Building the Digital Enterprise. 2016 Global Industry 4.0 Survey. What We Mean by Industry 4.0/Survey Key Findings/Blueprint for Digital Success. Retrieved from PwC. 2016. Available online: https://www.pwc.com/gx/en/industries/industries-4.0/landing-page/ industry-4.0-building-your-digital-enterprise-april-2016.pdf (accessed on 12 March 2018).

Gubbi, J.; Buyya, R.; Marusic, S.; Palaniswami, M. Internet of Things (IoT): A vision, architectural elements, and future directions. Future Gener. Comput. Syst. 2013, 29, 1645–1660.

Hermann, M.; Pentek, T.; Otto, B. Design principles for industrie 4.0 scenarios. In Proceedings of the 49th Hawaii International Conference on IEEE System Sciences (HICSS), Koloa, HI, USA, 5–8 January 2016.
Herrmann, C.; Schmidt, C.; Kurle, D.; Blume, S.; Thiede, S. Sustainability in manufacturing and factories of the future. Int. J. Precis. Eng. Manuf.-Green Technol. 2014, 1, 283–292.
Hossain, M.S.; Muhammad, G. Cloud-assisted industrial internet of things (IoT)—Enabled framework for health monitoring. Comput. Netw. 2016, 101, 192–202.
Ilie-Zudor, E.; Kemény, Z.; Van Blommestein, F.; Monostori, L.; Van Der Meulen, A. A survey of applications and requirements of unique identification systems and RFID techniques. Comput. Ind. 2011, 62, 227–252.
Ketchen, D.J., Jr.; Hult, G.T.M. Bridging organization theory and supply chain management: The case of best value supply chains. J. Oper. Manag. 2007, 25, 573–580.
Kortuem, G.; Kawsar, F.; Sundramoorthy, V.; Fitton, D. Smart objects as building blocks for the internet of things. IEEE Internet Comput. 2010, 14, 44–51.
KPMG. The Factory of the Future; Germany, 2016; KPMG AG: Amstelveen, The Netherlands, 2016; Available online: https://home.kpmg.com/xx/en/home/insights/2017/05/industry-4-0-its-all-about-the-people. html (accessed on 12 March 2018).
Laney, D. 3D Data Management: Controlling Data Volume, Velocity and Variety, META Group Research Note; Meta Group, Gartner: Stamford, CT, USA, 2001; Volume 6, pp. 1–3.
Lasi, H.; Fettke, P.; Kemper, H.-G.; Feld, T.; Hoffmann, M. Industrie 4.0. Wirtschaftsinformatik 2014, 56, 261–264.
Laureti, T.; Piccarozzi, M.; Aquilani, B. The effects of historical satisfaction, provided services characteristics and website dimensions on encounter overall satisfaction: A travel industry case study. TQM J. 2018, 30, 197–216.
Monostori, L. Cyber-physical production systems: Roots, expectations and R&D challenges. Procedia CIRP 2014, 17, 9–13.
Nick, G.; Pongrácz, F. How to measure Industry 4.0 readiness of cities. Int. Sci. J. Ind. 4.0 2016, 2, 64–68. Available online: http://industry-4.eu/winter/sbornik/2016/2/16.HOW%20 TO%20MEASURE% 20INDUSTRY%204.0%20READINESS%20OF%20CITIES.pdf (accessed on 12 March 2018).
Oesterreich, T.D.; Teuteberg, F. Understanding the implications of digitisation and automation in the context of Industry 4.0: A triangulation approach and elements of a research agenda for the construction industry. Comput. Ind. 2016, 83, 121–139.
Oláh, J.; Karmazin, G.; Peto˝, K.; Popp, J. Information technology developments of logistics service providers in Hungary. Int. J. Logist. Res. Appl. 2018, 21, 332–344.
Pedersen, M.R.; Nalpantidis, L.; Andersen, R.S.; Schou, C.; Bøgh, S.; Krüger, V.; Madsen, O. Robot skills for manufacturing: From concept to industrial deployment. Robot. Comput. Integr. Manuf. 2016, 37, 282–291.
Pereira, A.; Romero, F. A review of the meanings and the implications of the Industry 4.0 concept. Procedia Manuf. 2017, 13, 1206–1214.
Popp, J.; Erdei, E.; Oláh, J. A precíziós gazdálkodás kilátásai Magyarországon. Int. J. Eng. Manag. Sci. 2018, 3, 133–147. Available online: http://ijems.lib.unideb.hu/cikk/cikk/5af01cf23a77a (accessed on12 March 2018).
Porter, M. E.; Heppelmann, J. E. How smart, connected products are transforming competition. Harv. Bus. Rev. 2014, 92, 64–88.
Porter, M.A. Competitive Advantage: Creating and Sustaining Superior Performance; Free Press: New York, NY, USA, 1985.

PwC. Industry 4.0—Building the Digital Enterprise; PricewaterhouseCoopers LLP: Berlin, Germany, 2016; Available online: https://www.google.com/search?q=PwC+%282016%29%3A+Industry+4.0++Buildingthe+digital+enterprise.+PricewaterhouseCoopers+LLP+Hermann%2C+M.%2C+Pentek%2C+T.%2C+Otto%2C+B.+%282016%29%3A+Design+principles+for+industrie+4.0+scenarios.+In+System+Sciences+%28HICSS%29%2C+2016+49th+Hawaii+International+Conference+on+%28p.+392&ie=utf-8&oe=utf-8&client=firefox-b (accessed on 12 March 2018).

Rayport, J.F.; Sviokla, J.J. Exploiting the virtual value chain. Harv. Bus. Rev. 1995, 73, 75–85.

Reinhart, G.; Engelhardt, P.; Geiger, F.; Philipp, T.R.; Wahlster, W.; Zühlke, D.; Schlick, J.; Becker, T.; Löckelt, M.; Pirvu, B.; et al. CYBER physical Production-Systeme: Enhancement of Productivity and Flexibility by Networking of Intelligent Systems in the Factor; Springer: Frankfurt, Germany, 2013; pp. 84–89.

Rüßmann, M.; Lorenz, M.; Gerbert, P.; Waldner, M.; Justus, J.; Engel, P.; Harnisch, M. Industry 4.0: The Future of Productivity and Growth in Manufacturing Industries; Boston Consulting Group: Boston, MA, USA, 2015; pp. 1–14. Available online: http://www.inovasyon.org/pdf/bcg.perspectives_Industry.4.0_2015.pdf (accessed on 12 March 2018).

Shrouf, F.; Miragliotta, G. Energy management based on Internet of Things: Practices and framework for adoption in production management. J. Clean. Prod. 2015, 100, 235–246.

Slusarczyk, B. Shared services centres in Central and Eastern Europe: The examples of Poland and Slovakia. Econ. Sociol. 2017, 10, 46–58. [PubMed]

Stock, T.; Seliger, G. Opportunities of sustainable manufacturing in industry 4.0. Procedia CIRP 2016, 40, 536–541.

Toro, C.; Barandiaran, I.; Posada, J. A perspective on Knowledge Based and Intelligent systems implementation in Industrie 4.0. Procedia Comput. Sci. 2015, 60, 362–370.

Tupa, J.; Simota, J.; Steiner, F. Aspects of risk management implementation for Industry 4.0. Procedia Manuf. 2017, 11, 1223–1230.

Wang, G.; Gunasekaran, A.; Ngai, E.W.; Papadopoulos, T. Big data analytics in logistics and supply chain management: Certain investigations for research and applications. Int. J. Prod. Econ. 2016a, 176, 98–110.

Wang, S.; Wan, J.; Li, D.; Zhang, C. Implementing smart factory of industrie 4.0: An outlook. Int. J. Distrib. Sens. Netw. 2016b, 12, 3159805.

Wang, S.; Wan, J.; Zhang, D.; Li, D.; Zhang, C. Towards smart factory for industry 4.0: A self-organized multi-agent system with big data based feedback and coordination. Comput. Netw. 2016c, 101, 158–168.

Weyer, S.; Schmitt, M.; Ohmer, M.; Gorecky, D. TowardsIndustry4.0Standardizationasthecrucialchallenge for highly modular, multi-vendor production systems. IFAC-PapersOnline 2015, 48, 579–584.

Woo, S.; Jo, H.J.; Lee, D.H. A practical wireless attack on the connected car and security protocol for in-vehicle CAN. IEEE Trans. Intell. Transp. Syst. 2015, 16, 993–1006.

Zhong, R.Y.; Huang, G.Q.; Lan, S.; Dai, Q.; Chen, X.; Zhang, T. A big data approach for logistics trajectory discovery from RFID-enabled production data. Int. J. Prod. Econ. 2015, 165, 260–272.

CHAPTER 13

Women Empowerment in HR Industry

VARTIKA CHATURVEDI* and K. P. KANCHANA

Jaipuria School of Business, Ghaziabad, Uttar Pradesh 201012, India

**Corresponding author. E-mail: vartika.chaturvedi@jaipuria.edu.in*

ABSTRACT

The term empowerment implies the transformation of authority and power by which a person has some control over the matters related to life and work, which can impact them directly. The contribution of female fraternity in the corporate is very closely related to the objective of consolidated social and economic development. Through empowerment, they are challenging the ideologies of patriarch society that is reinforcing and perpetuating gender discrimination. The redistribution of power and authority is enabling women for their livelihoods by contributing more in their homes and at the workplace. This chapter emphasizes the crucial position of females in decision-making in family lives autonomy from a financial perspective, their access to education, and contribution to the growth of the nation. In the past decade, there is a feeble improvement in the status and well-being of the female employees; still, there is a huge gap between what is said and what is being done for the actual cause. In the recent scenario where the emphasis is laid on inclusive development, its mandate to heighten the awareness of what actually empowering a female employee means for a change of the economy, behavior, and values in the corporate.

13.1 INTRODUCTION

In the current business world scenario, more women have started contributing with men in all domains, and hence, "women empowerment" has become a hot buzz in the business world. They are exhibiting a self-dependent attitude in both personal and professional life, in the era where they had to change their last name to the time where they are proud to be single mothers and get a family dependent on them. Here, we are discussing about women empowerment; Arumina Sinha had set a benchmark, had debacle all the odds of her life, and became the most powerful name in the decade. There is a significant increment in complete freedom and optimizing it in relevant decision-making for fulfilling their dreams without any bars.

Measurable increment of female work force has led to their economic upliftment. It has contributed in generating a unique sense of confidence to lead their lives with honor and dignity. They are growing in their respective fields either at the level of small-scale industry or to big corporate houses, to name a few, Arundhati Bhattacharya, Leena Nair, and Chavi Rajawat, who are the epitomes of women empowerment where they are benefitting different aspects of the society.

The Iron Lady Indira Gandhi or the Great Politician Smt. Sushma Swaraj had a great command over the human resource (HR) topic, work-life balance, where they not only made noticeable contribution in the development of the country, but also were great homemakers. With equality to their professional life, they did well in that by achieving what all they dreamt of.

It is very important to educate women about the privileges and errands in rural parts where change is still not so significant and visible. However, their contribution along with their male counterparts in agriculture and other household activities has ignited the urge to educate themselves about their rights.

Women empowerment can be achieved only when behavioral change takes place in the social order and not being judgmental to women, providing them respect when there is no discrimination in fairness (black or white), dignity, proper education, and introducing them to the society. The modern era understands the power of women and adapt the change from past to present, and women are able to live their life accordingly.

The first step toward women empowerment is educating a girl child. With good education, a girl can get a high paying job. The Indian Government has also implemented many beneficial schemes for promoting girl education, such as Beti Bachao Beti Padhao, Balika Samriddhi Yojna, etc.

A woman is responsible for everything in her family; therefore, empowering her will pave way to growth and development of all related to her with little effort. Women are better than men in understanding certain problems in the society, for example, economic condition of the family, family nourishment, etc. With the need of increasing women empowerment, their safety is equally important. Indian women are subjected to various crimes such as domestic violence, dowry, etc. Many campaigns are done for women safety where they can speak freely without any pressure from the family and society.

For generations, women have been criticized by men in the field of status, equality, respect, and social status, and as an overall human being, it maybe because women were considered to be "small, less intelligent, physically incapable part of biological species." Civilization has even seen dark times, where women were just a means of reproduction to keep generation continuity. It is difficult to imagine the horrors women faced and to contemplate them. The only silver lining in all of this madness was that, "they were allowed to live." The era of swords and shields only told us how brave, how courageous, and how fearless the king was. The tale of his victory, the fear he imposed, "but no one talks about his mother, his wife, or sisters or daughters." Our country even practiced Sati Pratha, child marriage, and female feticide for a long time. History does not lie neither does present.

We are now beyond the age of kings or women being forced to walk on fire. We are now an advanced civilization, with breathtaking skyscrapers, a currency to depend on, mobile phones, and artificial intelligence. We now have human footprints on moon and our own artificial satellite in space. No doubt we are very advanced civilization, but my question is: "Is the woman's voice heard or it continues to be neglected despite all the civilization advancement?"

Every time question related to women empowerment is asked by intelligent leaders of the world, most of the time, they ignore or walk away by saying some intelligent words, because they are good with playing words. Our generation has also witnessed good times, when people of all gender, religion, caste, color, or creed came together, demanding justice for a teenage girl, whose fate was sealed by a group of men, but here is the harsh truth—"Good times like these do not last long enough to stamp a meaningful impact." It just fades away like smoke, as if nothing happened (Shettar, 2015).

Why do men think that empowering women is mediocre subject, "it is just not worth it?" Maybe because they have an illusion of the so-called greater importance in this world or maybe they forgot that who raised them as a child, who was there to differentiate between good or bad, and who was there when no one else was there. Their cheap mentality despite having all the advancement is questionable.

Campaign such as "Metoo" against sexual assault has gained momentum. We need stricter laws against such incidents of violence. Many unethical practices already have been removed by great leaders such as Raja Ram

Mohan Roy. A society where women feel secure and confident is sustainable in the long term (Barbara, 2015).

As India is growing at rapid rate, it must also focus on increasing women empowerment. The government should increase their focus on the working of their schemes they have already implemented.

The concept of women empowerment is not new; it was sensed much earlier even before independence of our country. Not only in India, women have faced atrocities across the world. Women have the right to make strategic choices of their life like men. Women empowerment is all about equipping women to make choices across every issue in the world. This needs arose due to the discrimination, gender biasness, and male domination in the world.

In order to achieve sustainable development, it is highly imperative to amend the political, economic, social, health, and sanitation status of women. There is a need of equivalent involvement of both males and females pertaining to productive and reproductive life. To maintain pace with all spheres, work as well as home, sometimes with no power and recognition, women face big time peril to their living.

To cope with the mechanisms of life, their abilities and knowledge are often unrecognized, and at the same time, they receive less formal education than men. Vesting entire female fraternity with adequate learning's, to boost self-confidence requires them to be educated. Achieving change strictly requires actions to improve their access to economic resources and a secured livelihood (Shunmugasundaram, 2014).

Their participation in public life has to be enhanced by removing the hindrances, extreme housework responsibilities, and raising their social awareness through access to information. Recuperating significance of female taskforce requires their enhancement of managerial capability within each and every sphere of life.

Equal participation of women in opinionated process in every society and community can be achieved by enabling them to articulate their needs and their matter of concerns. Elimination of poverty, ill health, and illiteracy is essential to promote their potential in skill development, education, and employment (Kumar, 2017).

Women have to be assisted in realizing their rights and discriminative actions against them, including those related to reproductive health. Women have the capability to bring in ahead of conventional livelihood and achieve self-reliance economically, which has to be improved further. In political regard, equal involvement of both males and females in making decision balances the amalgamation of social structure, and it is required to reinforce the functioning of social equality as well.

Women participation in labor market ensuring identical pay for same work pertaining to women's livelihoods with secure incomes should be enforced in order to remove the gender gaps.

"Before, I was a wife who cowered before her husband and brought no income to the family. Now I am a community leader, an income earner and equal partner in my marriage. WE have helped me find my voice, my power, my community."

– Rosalba, WE Member, Guatemala

Women empowerment by definition says that it is a process to improve the women's control over finance, human, and intellectual resource of the country. Women empowerment in a country can be deliberated by the involvement of women in the areas, which are male imperious. Women empowerment is the most talked about word today. It is creating a buzz globally with more women coming and working in all the spheres along men. They are taking up charge of professions, which at one point of time was considered to be gray area for them to gain recognition and are setting benchmarks, being veterans.

They have started acknowledging a sovereign attitude either being as a homemaker or a corporate. There is an increment in independent to be in charge of over their lives and now are independent in the area of their own education, career, and lifestyle. The considerable upsurge in the statistic of employed and engaged women has made them financially independent, which, in turn, has made them confident to lead their own life with their own identity (Rawanda, n.d.).

The importance of women empowerment at work is now realized and recognized by the business leaders of the world. Their participation in driving innovation and boosting corporation's profit cannot be denied. The business world is setting clear goals for diversity or implements initiatives making the workplace indispensable.

13.2 WAYS THROUGH WHICH CORPORATE CAN PROMOTE WOMEN EMPOWERMENT AT THE WORKPLACE

- *Persuade and support women to make the rise:* Organizations can encourage women to achieve success by actively persuading them to take more challenging jobs and making them confident.

- *Endorse an unbiased parental leave policy:* An unbiased leave policy can remove the intricacy of women back to the workplace after child-birth. This will empower women while respecting the long-term goals and desires of women.
- *Generate unique defined functions for female employees:* Women task force is offering unique and different perspectives to the organizations today. They are now taking more challenging tasks and opportunities, which were once confined to male task force only.
- *Avert sexual harassment:* The organizations are now practicing zero tolerance to sexual harassment against women in the organizations today. They are making the place more safe and comfortable where women can optimize their competencies.
- *Generate more flexible job options for women employees:* Women face problems when they decide to start a family. They start playing low-career-oriented roles because of their family obligations. To sort out these issues, organizations should implement more flexible job options such as flexi time, compressed week, telecommuting, etc., to ensure financial and professional continuity.
- *Equity:* The study shows that many organizations are biased when it comes to salary. Women are paid lesser than their male counterparts. Organizations need to think on the salary equity, that is, equal work equal experience equal pay, works and matters a lot. Organizations should be encouraged to think about the gender equality and make policies sufficing them.
- *Women should be given leadership positions:* Organizations need to diversify the leadership position by giving opportunities to women to reach to the top positions. In.Corp is proud to diversify the leadership position and executive position in a bid to women empowerment. It has the work force ratio, which constitutes 30% men and 70% women, whereby women make up to 60% to the top management position.

While working outside in different professions, women conjointly take charge to thump parity betwixt to their professional life as well as their household work. They play the role of professionals as well as of mother, wife, and daughter simultaneously with remarkable ease and harmony. They are also giving strong support and cooperation to their male counterparts with all enthusiasm of alliance and to deliver all feasible co-operations in meeting the deadlines and set targets.

Empowering women is confined not only to cities and towns, but also in even small villages and towns; it is seen that they are now forging their viewpoints with a roar in the society. Women strongly deny in becoming a second contrivance to their opposite gender. World is changing and is witnessing attitudinal change in society and corporate with regard to women. Women are being treated with proper respect, dignity, fairness, and equality because it has been realized that their presence in the corporate or at home cannot be ignored. Using the competencies of the female counterparts by giving them proper and fair opportunities will boost up financial and social improvement of the world.

13.3 BENEFITS OF WOMEN EMPOWERMENT

Empowering women builds confidence in their competence to have a more substantial and decisive lives. It helps in removing their dependency and creating their own identity. Some of the benefits that women empowerment can have are the following:

- It will help them to bulge their lives with self-respect and sovereignty.
- It will boost their self-esteem.
- It will help in developing their own identity.
- It will help in developing a strong and healthy society.
- Women will be able to handle different positions in the organizations with efficiency, ease, confidence, and pride.
- It will help them to become financially independent.
- Empowering women contributes to economic improvement of nation by enhancing gross domestic product.

In spite of women being inured specific and significant position in every religion, there are many unethical customs and traditions, which are being followed as a norm since decades. Women are granted special recognition in different creed and caste; many unethical and inhuman norms and rituals are taking place with females since eternity. However, with the change in time and attitude, a number of positive changes are seen, and gradually, the patriarchal society is eroding giving room for sociopolitical rights to women (Fadia, 2014).

Swami Vivekananda quoted, "There is no possibility for the welfare of the world unless the state of women is enhanced; it is unrealistic for a flying creature to fly on just a single wing."

To make an effort for the modernization of society and growth and success of any organization or nation as a whole vital to bring women force in the mainstream and give them opportunity to equally contribute for the success, a perfect balance needs to be made in both rural and urban societies without being biased toward men (Women Education in Modern India, n.d.).

The consecutive governments of our country have employed various legal and constitutional rights for women to empower them to lead a purposeful and meaningful life. The Parliament of the country has also passed various bills to protect women from injustice and discrimination.

Another most significant law passed in the wake of Nirbhaya case is the Juvenile Bill, 2015, where the immature age appealing reprimand for crime at present is abridged commencing from 16 years of age, which was earlier 16 years.

It is essential for every human being to possess the confidence, self-worth, as well as the freedom to have choices in private and professional life alike, but unfortunately women as a being still need to realize their potential and worth in this male-dominated society. Gender bias is wrong and unreasonable in so many levels as it curbs individuals from striving toward their best and living a life free from domination and fear.

Women empowerment defines liberating the women from the immoral clinch of orthodox mindset of just being a commodity which can be made use of. Each and every woman should be endowed the freedom to decide about their what their priorities are? They should not be compelled to live like other's way. It means replacing patriarchy with parity (Lagarde, 2016).

Empowering women is an elevated and talked-about subject across the globe, but what is its significance? It is to know why there is a denial in equality and confidence in the social structure. We are in the so-called educated, modern, 21st century generation; still, women are constantly fighting for their rights. A docile housewife can turn out to be a magnificent entrepreneur or a corporate leader; then what is the need of this women empowerment concept?

We cannot deny the hard truth fact that women in our country India have made a substantial growth till now, currently first time our Minister of Finance is also a women, but still we have to strive to all the obstructions of patriarch social structure. Many filthy rituals and male-centric regulation and rules still exist in the era of artificial intelligence and Internet of things.

Today's women have also attained monetary autonomy, which has given them confidence to survive and also make others dependent on them. Women are responsible for their identity. They are successfully taking up diverse professions to prove that they are not second in any respect.

Empowering women in the factual and practical terms and not in speech, books, and journals can be attained, provided a massive transformation in conceptual behavior in the social structure with respect to women fraternity, in treating them with due respect, dignity, fairness, and equality. Eventually, change needs to be initiated from within (Anju Malhotra, n.d.).

India is called as Bharat Mata right, not Bharat Pita; then, how on Earth did India stand fourth among not safe place for women according to a recent survey, which also says that 54% of children are being sexually harassed, that is, one out of two children are being sexually harassed at their early ages.

Well, times have changed; century moves like clouds on the clear sky, but still is the world any better for women. Centuries of human history stand witness to women discrimination, you love her as a mother, you love her as a wife, and you love her as a sister, but why not as your daughter. Their births are considered meaningless, unwanted, and even some are killed before the birth. There are many women who have been forced into prostitution, who are imposed to dowry, and who are subjected to domestic violence and marital rape.

Life is a beautiful story of sharing and giving but done by women most of the time. Education plays a major role in empowering the women folks making them responsive about their rights and duties. Government has passed so many acts for the betterment of women, but then, think about the 33% women reservation bill, which was proposed a decade ago; maybe that was the reason we had to spent a lot of time to find a piece or robe as justice for Nirbhaya; this is due to procrastination in judicial proceedings and the existence of many shortcomings.

Financial independence is not the only solution for all evil but is certainly a step forward. The future must hold more good times for the women folk. Right from the Jhansi Rani, the brave heart, to Arunima Sinha, awardee of Padma Shri (first disable climb the Mount Everest), Hima das, and PV Sindhu who won gold medals in sports and many more (Kakkar, n.d.).

Women have contributed in all spheres of human activities yet under-privileged. "To awaken the people, it is women who are to be awakened because once she moves forward, the family moves, the village moves, and the entire nation moves"—famous saying by Pandit Jawaharlal Nehru.

Women empowerment means giving women all their rights they are supposed to get in the family, society, school, college, and country equal to men.

We all hear about women empowerment in our daily lives but only few were seriously concern about this. Our world is growing at a rapid pace, but today also the position of women needs to be improved as there are several challenges in the way of empowering women.

Lots of crimes against females in society, challenges related to education, health, and safety of the women, and child marriage all are hurdles for the growth of women in society. It is hard to believe that even after lots of acts, projects, and campaigns for this, the women are not enjoying the freedom for which they have rights from the very beginning. Because there is need to change mindset of people toward women (Kumari, 2014).

In past few years, rape crimes against women in the national capital have been very devastating. Swami Vivekananda said that "It is not possible for a bird to fly only on one wing." So, in order to be a fully developed country, India has to empower its half HR. Everyone should try on their side to respect women and treat them equally by supporting girls and women in crisis, mentoring a girl close to our home, helping a new mom, by respecting our mother, wives, teachers, and every single woman.

The recent report of the United Nations Educational, Scientific and Cultural Organization in 2015 revealed that there are 481 million women who lack basic literacy skills. The challenges grow multifold; adolescent girls face a number of risks where they have to quit their education in the school level or at the intermediate level (5 Challenges faced by women entrepreneurs, 2018).

Women empowerment relates to the process where women are able to develop and recreate what they are capable of and what they can do, which is denied in past. Empowerment deals with the recreation of women by using the utmost power and skills to strive a better future. It helps women to step forward with confidence in each and every step of life.

Women empowerment helps in better decision-making. Women are no less than men. Women are ahead in every stream. They are the biggest competitor as they are the best managers. Women is nowadays ahead in all the fields such as homemaker, entrepreneur, mother, manager, professional, etc.

Women empowerment also relates to the competency to accelerate the right to plan and control, managing the assets, income, and their own time to raise their economy, and also, they are the one who manages the risk in the family and corporate too. Many rights are reserved for women under women empowerment.

It helps women to enable their self-esteem in political, social, economic, and health status. In case of upbringing of a child, we see that always women are solely responsible, but the child is of both the parents, so women empowerment empowers women to enjoy the right to up bring their child with equal efforts of her husband. Women have the right to speak (Maggie Vlazny, 2018).

To empower woman, edification is the imperative tool, by the help of which a woman is able to gain understanding, expertise, and poise. These help women contribute wholly in the growth process. Universal Declaration of Human Rights affirmed that everyone has the right to education; it is not about women empowerment; the essence of the topic will be achieved when the other genders of the society support each other to share the equal respect and rights in the society. There is a big question mark that arises.

Government of India has implemented many schemes such as Sukanya Yojna, Beti Bacho Beti Padhao, etc. But what the use of those schemes where in reality there are so many fights we being women are fighting alone, whether in our own house, our offices, a street, or else we can say deep inside mentally (https://indianyojana.in/sukanya-samriddhi-yojana/, n.d.).

Attentiveness in the society regarding the constitutional rights and values of women and also their contributions in the growth of the nation needs to be developed in a number of spheres. People speak about gender equality in our society, but accepting equality in reality has become very difficult for all of us. And the perfect word that distance the equality gap is "EGO." Ego is among both the genders in developing or molding our society in a new way.

We can also see that illiteracy is mostly covered by women, whereas women are the real teachers in every house. The real meaning of women empowerment would be in progress when we will allow every woman to get proper education and also by allowing them to live their life freely and also to take decisions of their own. We have seen that in today's progressed world, women are holding pinnacle positions in their career. It India, they are holding the positions such as President of the country, speakers in the Lok Sabha, cabinet leader, leading sportswomen, CEOs, CFOs, CXOs, etc. Seeing this all, why do people think that women are not capable as compared to men? It is actually the old mindset that people are carrying alone with them, and this needs to change.

Nevertheless, there is a great gender disparity in the society across the world. Women are still facing so much of mayhems: acid attack, rapes, molestations, dowry killing, forced to do prostitution, etc. But still you have seen us surviving happily because of the inner strength that every woman has within herself. It is being said that India ranks fourth among most perilous nation in the world for women.

13.4 GOVERNMENTAL SCHEMES AND BENEFICIAL POLICIES FOR EMPOWERING WOMEN

The statutory policies and framework under which women development initiated from the early 1990s in the country, but the implementation was effective from the late 1990s. In today's scenario, the government has more than 34 schemes for female happiness and well-being operated by different social central and state organizations.

13.5 FOCUS ON HR INDUSTRY

The issue of empowering women has become an important concern across the globe. Many bureaus of the United Nations mentioned in their conclusion of research conducted that demographic disparity requires paramount precedence. It is subjected that women now do not have to wait to ask for their rights of equality. Women have always been officially encouraged to participate in the workforce.

If you visit government schools and government banks, it is not surprising to see a lot of women. In addition, as research suggests, the complete picture is not rosy for India, and that is despite the laws that have been brought by the government to ensure gender pay parity across all professions and age groups.

Women empowerment in the profession interprets that they also have power over themselves about their lives. It signifies that they are free to have plans and programs for their personal life, enhance their competencies, and self-sufficiency. Women empowerment is developed when the strengths and power exhibited at the workplace are established, honored, and used.

In the corporate particularly in HR department, the number of women is increasing with rapid rate. This scenario divulges the improving stature of women in corporate world in the present century. As per some findings of important research, women are holding more than 65% of the HR jobs across the globe. This means they are ahead of men in this field.

Women are more persistent than men in real life; nevertheless, there are definite qualities of women, which are keeping them to the lead in the competition, particularly their interpersonal skills. At the workplace, women are blessed to have the ability to quickly be acquainted with and recognize employees upon whom the company can bet upon. In the past 10 years, we have addressed a lot of women doing great in every field of work or industry (The Gender Policy Statement for the Education and Human Resources Sector., n.d.).

For many women, work is fulfilling, and there are plenty of opportunities for promotion and career development. Women in the workplace 2018 found that although companies report that they are highly committed still they are not getting their deserving recognition for their work.

There should be women-friendly working environments, where staff can leave meetings on time in order to pick up children, and important calls are not scheduled at times when women with families are unavailable; it can also benefit men who may not be wedded to the long hour's culture and who want a better work-life balance.

They have strong ability to understand people efficiently with their interpersonal competencies and skills. The power of female employees in the workplace is now clear. When it is about to set benchmarks for work related initiatives through diversity, the idea of being "indispensable" in the workplace has never been more important.

The points why females are doing wonders in the domain of HR are as follows:

- Comprehending skills.
- Multitasking skills.
- People management skills.

Organizations are continuously trying to innovate their HR policies and create a culture for their employees to thrive. One of the key aspects that HR managers are focusing on is the gender diversity and equality. No doubt women have made tremendous progress in corporate world, yet globally only around 39% of the total workforce accounts to women.

Even today, across the globe, female professionals continue to face challenges in making strides in their organizations. In order to make an overall difference, the workforce at any organization has to be diverse.

While focusing on equality, it is essential for organizations not to compromise on the basic needs for women such as ensuring safety security and comfortable working environment. HR policies have a great impact on influencing gender diversity and empowering woman.

- Zero tolerance to abuse and harassment.

- Encourage female employees to identify their talents and showcase their leadership by giving them high visibility initiatives.
- Policies designed around work from home, flexible working hours, women safety, etc.
- Connect with senior women leaders to share their motivational stories and journeys of life.

Special programs and policies certainly help in empowering women, but other than this, women should support each other also. However, sometimes, it is seen that in the rat race to achieve success and recognition, women often tend to bring down their own. Women in power should also use the opportunity to build an environment where other women can grow. A true gender diverse culture can be created only if women are empowered, and a culture of inclusiveness and equality is created.

It is the year 2019 and women already hold 60% of jobs in HR at various industries. HR has proven to be the most helpful platform for providing women a chance to outcast their talent and skills. No other industry, be it marketing, finance, or operations, has provided such wide exposure to women than the HR industry. And, it is growing day by day. In future, we are going to see a rise in this ratio of women in HR industry.

As a good HR, a candidate needs to be patient, polite, regular, influencer, multitasking, unbiased, and creative, which can be found completely in a woman. Thus, women are more preferred in the profession than men. Men for a profession are more interested in marketing operations than HR, which are completely different professions.

Many people think that the reason why women are more preferred in HR is their soft skills. They believe HR as a career would will give them an opportunity to become prominent than other sectors.

Sky and European broadcaster has all sorts of proposals intended to growing diversity in corporate.

When it comes to higher leadership position or higher pay, women has to face difficulty at that place, so at last, we can say that women are core competency in any firm, and the way they can manage HR no one can do.

In the cooperate world, the number of women employees is going up day by day. We can observe that they are continually creating their best existence in every department as sales, marketing, finance, etc. However, in a survey, it was found that women are holding diverse jobs in different segments, and it is coming to be almost 60%. It clearly shows "Women empowerment in human resource management" (15 Biggest Challenges Women Leaders Face And How To Overcome Them, 2018).

We can understand it with the help of the following points why they are making their best existence in the HR management.

- *Employee management ability:* As per the different research studies, it has been cleared that women are more focused and have more concentration power than men usually have. They have the ability to manage the people well. They are able to measure the employees' worthiness for the organization. They are good in counseling and best observer of performance. They actually know how to tackle the issues of the individual employee or a group.
- *Multitasking ability:* Actually, HR is a profession in which an employee must be a multitasking and multiperformer as they are required to do different tasks at the same time. They are liable for the recruitments, training process, employee engagement, operations, designing the strategies, etc. for the welfare of the organization. Usually, women are the best in this field as they handle all the things very politely and patiently than men.
- *Understanding ability:* Usually, women are more understanding than men. They can easily understand the requirements of the employees through their intense understanding skills. Naturally, they are soft spoken and have well communication skills so that they are much able to crack the problems emerging in the organization. Therefore, we can say that they have good ability to create the best work environment in the organization.

These are some points that describe the title as "Women empowerment in human resource management." There exist a belief and the research data, which clearly say that women are much better than men in this industrial segment.

Empowering women is the crucial and inevitable practice nowadays, with females holding various positions in the workplace. Women nowadays profess a different identity whether they are working outside their homes or living in their homes. They are self-reliant and are able to make them self-dependent in taking vital decisions in the area of their education, profession, and their life.

Empowering females in HR sector (Is your company recognizing "Her" enough?, 2019):

HR department is the most crucial in several HR jobs women are preferred more especially in corporate offices, but this is opposite in industrial relations and manufacturing units where male are preferred more due to trade unions and many other factors.

Some of the practices the organizations can adopt to bring equality in the workforce are the following:

- Persuade female employees to lead and initiate.
- Promote fair parental leave policy.
- Prevent sexual harassment.

There is a remarkable amplification in the professional females; some of whom are occupying massive posts in different work areas, enabling them to gain financial independence. This so-called revolution in the women stature around the social structure has become a means to support their decisions, direct their lives, and put up their own idiosyncratic individual distinctiveness. They have effectively taken up different careers that they are second to none. Few best examples in today's era where women are doing better than or equal to men are Kirthiga Reddy (Head, Facebook India), Mary Kom (Olympic Boxer), Indra Nooyi (Board Member, Amazon), and Hima Das (Sprint Runner); these are just a few names of women who have proved themselves to be better than the men at their respective professions.

However, while carrying out their professions, women balance professions and household very well. They are undertaking a number of roles, such as role of a daughter, mother, wife, and sister with exceptional consonance, ease, and once given chance to work. Women work and perform with a firm sense of belonging to exhibit all required assistance to their opposite gender (Dutta, 2018).

Women empowerment can not only be observed in the urban working or nonworking women, but also in the rural areas, women are making sure that their voices are being heard loud and clear irrespective of their age, class, etc.

Rather, it is a reality to a certain extent that women nowadays are not facing disparity in work and at home today, but still a huge number of women confront harassment and are abused at homes, society, and workplace in different ways: emotionally, sexually, and mentally.

The importance of women empowerment at work is now realized and recognized by the business leaders of the world. Their participation in driving innovation and boosting corporation's profit cannot be denied. The business world is setting clear goals for diversity or implements initiatives making the workplace indispensable.

Ways through which corporate can promote women empowerment at the workplace are the following:

- *Persuade and support women to make the rise:* Organizations can encourage women to achieve success by actively persuading them to take more challenging jobs and making them confident.
- *Endorse an unbiased parental leave policy:* An unbiased leave policy can remove the intricacy of women back to the workplace after child-birth. This will empower women while respecting the shared vision and needs of women.
- *Generate innovative responsibilities and positions for women employees:* Women task force is offering unique and different perspective to the organizations today. They are now taking more challenging tasks and opportunities, which were once confined to male task force only.
- *Avert sexual harassment:* The organizations are now practicing zero tolerance to sexual harassment against women in the organizations today. They are making the place more safe and comfortable where women can optimize their competencies (2018 Study on Sexual Harassment and Assault, 2018).
- *Generate more flexible job options for women employees:* Women face problems when they decide to start a family. They start playing low career oriented roles because of their family obligations. To sort out these issues, organizations should implement more flexible job options such as flexi time, compressed week, telecommuting, etc., to ensure financial and professional continuity.
- *Equity:* The study shows that many organizations are biased to salary. Female work force is paid lesser even though they are working equally in the same position with similar skill sets. Organizations need to think on the salary equity, that is, equal work equal experience equal pay works and matters a lot. Organizations should be encouraged to think about the gender equality and make policies sufficing them.
- *Women should be given leadership positions:* Organizations need to diversify the leadership position by giving opportunities to women to reach to the top positions. In.Corp is proud to diversify the leadership position and executive position in a bid to women empowerment.

13.6 CHALLENGES IN THE HR SECTOR OF EMPOWERING WOMEN

HR being the most crucial department for any organization, the challenges faced by the female employees are numerous. As with authority and power, responsibility and accountability also join hands.

- *Implementing policies:* Introducing and implementing new policies to the organization is an important challenge as the other colleagues or the management of the organization would not accept the changes or would anticipate doubts on the successful implementation of the same.
- *Decision-making:* From hiring an employee to appraisal and layoffs, the decision-making roles for a female HR are questioned. Females are being blamed for being biased and emotionally unsound.
- *Working in manufacturing units:* Females are considered weak in their mechanical skills, visiting the shop floor, dealing with labors, and handling their grievances.

Working in the HR department, one has to take care from the basic need to the self-actualization needs. The satisfaction and acceptance level is less in comparison to the dissatisfaction forces (Fuhl, 2018).

Leading ladies such as Leena Nair and Arundhati Das faced challenges and disagreements but had a tremendous success journey.

Women empowerment will only be achieved in the truest sense when there is a change in the thought process of the society with regard to the male counterpart, thinking of women as equal, and giving them with utmost fairness, respect, dignity, and justice. Yet, some parts of the country steeped in orthodox mindset, denying women with their rights in regard to their lives, marital status, sexual orientation, attire, and carrier.

The very practice followed in the society, that is, giving equal power and rights to women, helps in constructing a happy, prosperous, generous and improved place to live in. With increase in happiness for society and world, it is the women who make a difference. Women should be given equal chances in all the aspects of the society because on gender we cannot judge someone's capability, and it has also been seen that working women are actively participating in the success of the

We always know that women all over the world are a victim of inequality in this gender-biased society. No matter how much we promote and fight for equality, there is always the "women" component in the society who lack in awareness of their basic rights. This is a harsh reality of the society which nobody is ready to accept.

But as we all know that everything has a positive as well as negative aspect, similarly women empowerment in our society is gaining pace with the advancement and inclusion of education among the women community.

Therefore, we can conclude that women if given opportunity can do anything and everything with full dedication and willpower. In the above example, we saw that women have capabilities for which they are large in numbers, and the respective companies were benefitted because they hired women as salesperson, which, in turn, is a step toward women empowerment.

KEYWORDS

- **women empowerment**
- **development**
- **gender discrimination**
- **inclusive growth**

REFERENCES

(n.d.). Retrieved from https://www.indianetzone.com/39/women_education_modern_india.htm: Women Education in Modern India

(n.d.). Retrieved from https://indianyojana.in/sukanya-samriddhi-yojana/

(n.d.). Retrieved from http://hdr.undp.org/sites/default/files/2015_human_development_report.pdf

15 Biggest Challenges Women Leaders Face And How To Overcome Them. (2018, February 26). Retrieved from https://www.forbes.com/sites/forbescoachescouncil/2018/02/26/15-biggest-challenges-women-leaders-face-and-how-to-overcome-them/#7c33f8404162

2018 Study on Sexual Harassment and Assault. (2018). Retrieved from http://www.stopstreetharassment.org/our-work/nationalstudy/2018-national-sexual-abuse-report/

5 Challenges faced by women entrepreneurs. (2018, April 14). Retrieved from https://savvywomen.tomorrowmakers.com/wise/5-challenges-faced-women-entrepreneurs-listicle

Barbara. (2015, March). Retrieved from https://www.theparliamentmagazine.eu/articles/opinion/lack-education-making-women-powerless

Dutta, P. K. (2018, October 15). *Sexual harassment at workplace explained.* Retrieved from https://www.indiatoday.in/india/story/sexual-harassment-at-workplace-1368055-2018-10-15

Fadia, K. (2014). *Indian Journal of Public Administration.* Retrieved from http://www.iipa.org.in/New%20Folder/13--Kuldeep.pdf

Fuhl, J. (2018, Jan. 16). *What to expect in HR in 2018.* Retrieved from https://www.sagepeople.com/about-us/news-hub/challenges-facing-hr-professionals-2018/

http://hdr.undp.org/sites/default/files/2015_human_development_report.pdf. (n.d.).
http://hdr.undp.org/sites/default/files/2015_human_development_report.pdf. (n.d.).
http://hdr.undp.org/sites/default/files/2015_human_development_report.pdf. (2015).
https://indianyojana.in/sukanya-samriddhi-yojana/. (n.d.).
https://mhrd.gov.in/sites/upload_files/mhrd/files/statistics/AISHE2015-16.pdf. (n.d.).
Is your company recognizing "Her" enough? (2019, July 30). Retrieved from http://womenofhr.com/
Kakkar, A. (n.d.). *Why Is Financial Independence for Women Important*. Retrieved from https://www.jaagore.com/power-of-49/why-is-financial-independence-for-women-important
Kumar, G. R. (2017, March). Retrieved from https://www.thehansindia.com/posts/index/Hans/2017-03-20/Women-employment-and-empowerment-in-Indian-Economy/288084
Kumari, V. (2014, May). *Problems and Challenges Faced by Urban Working Women in India*. Retrieved from http://ethesis.nitrkl.ac.in/6094/1/E-208.pdf
Lagarde, C. (2016, November 14). *Women's Empowerment: An Economic Game Changer*. Retrieved from https://www.imf.org/en/News/Articles/2016/11/14/SP111416-Womens-Empowerment-An-Economic-Game-Changer
Maggie Vlazny, L. (2018, Oct. 8). Retrieved from https://psychcentral.com/lib/women-and-self-esteem/
Malhotra, A.; Schulte, J.; Patel, P.; Petesch, P. (n.d.). Retrieved from https://www.icrw.org/wp-content/uploads/2016/10/Innovation-for-Womens-Empowerment.pdf
Shunmugasundaram, M. (2014). Retrieved from https://www.researchgate.net/publication/280218999_WOMEN_EMPOWERMENT_ROLE_OF_EDUCATION
Shettar, R. M. (2015, April). Retrieved from http://osrjournals.org/iosr-jbm/papers/Vol17-issue4/Version-1/B017411319.pdf
Rawanda. (n.d.). Retrieved from http://www.rw.undp.org/content/rwanda/en/home/operations/projects/poverty_reduction/youth-and-women-employment-.html.
The Gender Policy Statement for the Education and Human Resources Sector. (n.d.). Retrieved from http://ministry-education.govmu.org/English//DOCUMENTS/GENDER%20POLICY%20STATEMENT%20FOR%20THE%20EDUCATION%20AND%20HUMAN%20RESOURCES%20SECTOR.PDF
Women Education in Modern India. (n.d.). Retrieved from https://www.indianetzone.com/39/women_education_modern_india.htm

CHAPTER 14

Emotional Intelligence in the Era of Industry 4.0

STEPHEN MCKENNA[1*] and ANUBHUTI GUPTA[2]

[1]*Curtin University, School of Management, Bentley, WA 6102, Australia*

[2]*Amity University, Greater Noida, Uttar Pradesh 201308, India*

[*]*Corresponding author. E-mail: agupta1@gn.amity.edu*

ABSTRACT

The unprecedented boom of artificial intelligence (AI) and machine learning (ML) like most disruption advances is energizing as well as frightening. It is energizing to consider all the innovative developments, from scheduling our calendars to make medical appointment and diagnoses, but it is frightening to think about the social and individual ramifications. As ML keeps on developing, we as a whole need to grow new aptitudes so as to separate ourselves.

It has been quite some time realized that AI and automation/robotization may disrupt business environment and related workforces. Driverless vehicles will compel more than 5000 vehicle drivers to look for some other types of work, and automated generation lines like Tesla's will keep on consuming assembling employments, at present they are twelve million and decreasing. Be that as it may, this is only the start of the disturbance. As AI enhances, which is going on rapidly, a lot more extensive arrangement of occupations will be affected. So, in the present scenario, it has become highly difficult to develop, gain, and maintain emotional intelligence (EI) in the era of AI which is creating highly volatile, uncertain, complex, and ambiguous environment that has become new challenge for human resource department to develop new strategies and methods for sustaining EI in the era of AI. This chapter explores different issues with EI when AI is proliferating and a model is proposed that can be adapted by an employee for gaining and maintaining EI in the era of AI that can help employees to become emotionally stronger in this highly competitive world.

14.1 INTRODUCTION

Though artificial intelligence (AI) fundamentally enhances the surrounding environment in various ways, but are serious issues as for the planned impact of AI on or and the organizations and their employees. There are desires examining a colossal number of jobless people in theaccompanying decades—basically in light of the impact of robotics and AI systems. More or less, the entire financial ecosystem has been going through a period of radical change in markets, organizations, training, government, social welfare, and work models that will be seriously affected.

14.2 JOBS AT RISK

Jobs that are repetitive could be effectively robotized; and it may slowly make various jobs obsolete. As we can take example of jobs concerned with call center/customer care center, operations, classification of document, discovery and retrieval, and content moderation are day by day more relying on innovation and computerization and are lesser dependent on human work. The equivalent is valid for jobs identified with task and backing of assembly lines and manufacturing plants: people are being supplanted by active robots that are able to explore the space, find and move objects, (e.g., items, parts, or apparatuses) or perform complex collecting activities.

AI demonstrates to be compelling in taking care of many progressively complicated tasks—which need handling of various signals, huge unstructured data and accumulated knowledge continuously. A remarkable case is the driverless cars which can catch and "comprehend" the roads and its elements; they can "see," choose and act continuously, toward very much characterized advancement goals.

Subsequently, we should start by changing our meaning of being "savvy," until this point in time, being "more intelligent" than other individuals was estimated by evaluations and test scores. The brilliant individuals were those that contended well, and got the most astounding scores by committing the least errors. The new smart will be resolved not through challenge and high test scores, rather it will be estimated by the nature of your reasoning, listening, relating, working together, and learning. We should concentrate on building up those aptitudes that require correspondence, cooperation, imagination, basic reasoning, and high commitment. Aptitudes that meet the enthusiastic needs of other individuals! "The new savvy will be tied in with attempting to beat the two major inhibitors of basic reasoning and group joint effort: our self-image and our apprehensions. Doing as such will make it simpler to see reality all things considered; instead of as we want it to be. To put it plainly, we will embrace humility. That is the means by which we people will include an incentive in a universe of smart innovation."

As sympathetic AI machines and human robots venture into the world and endeavor to understand emotions of human beings, it is the time to put the term "Mechanical" to excess.

We are quickly moving toward the day when human and robot similarity abilities would be recorded as one of the looked for after occupation necessity (accepting one figures out how to get a new line of work which needs human intercession insider savvy). This advances a legitimate inquiry: In the regularly developing universe of machines adapting continually, where does the new expectation to absorb information for us as people initiate?

Nowadays, there are many suffix attached to intelligence and this is resulting in diversification of technology. Humans, on the other hand, have genetic ability to have emotional connect.

14.2.1 EMOTIONAL INTELLIGENCE (EI)-PLAYING THE LEAD

"The most vital factor in Communication is hearing what isn't aforesaid." Those are the genius words of Peter Drucker that it is important to understand emotions and perceiving the same. Deciphering and understanding one's own feelings and grasping the mood of others is wherever the ubiquitous "Emotional Intelligence comes into play!" It is proved many times that leader having robust EI competencies are extraordinarily roaring not solely in scaling an ascending profit curve; however, conjointly in making a raining trend within the worker rate of attrition. The clarity of thoughts, words, enthusiasm, and also the inner motivation of a pacesetter casts its positive influence on the team. A transformational leader brings expansive advantages to the organization. Juggling between addressing massive teams and influencing people, a pacesetter carries the responsibility of guaranteeing that the message is obvious and is capsule in a remarkable type. Communication is *so* key for effective leadership. Each word of appreciation and motivation from a pacesetter causes an enormous ripple and builds the self-image of the people, the team, and also the organization as a full. Being acknowledged for one's work is so a basic human need promoting an appreciative culture among the team leads to high motivation and positive progress.

14.2.2 EMOTIONAL QUOTIENT (EQ) IS THE KEY FACTOR

In this madness around accounts of AI assuming control over all that we do, it turns out to be critical that pioneers assist individuals by understanding that EQ will assume a flat out basic job in characterizing our future jobs and occupations. Simply take a model, while any boot can convey a specialized discourse, the passionate validness which is an essential component in interfacing with crowd must be finished by people with high EQ. People will keep on being profoundly social creature with the need to have an enthusiastic association with things!

14.2.3 CLOSENESS IS THE MANTRA

Working projects nowadays are prevalently in a worldwide virtual condition with layers of unpredictability underneath them. Unobtrusive yet significant factors, for example, peace promotion, cooperation, time zone varieties, and between social affectability represent a consistent test. This prompts a habitual need to improve the group's passionate jargon to remain associated regardless of the separation. Venture directors with high enthusiastic insight are vested with the capacity to change the whole undertaking atmosphere and move the task to an effective completion.

14.2.4 EMPATHY—IS VERY IMPORTANT

By placing yourself from one's point of view, you feel the spirit!
"Sympathy" requires a high level of receptivity to an individual's condition and needs. Some of the time, the least difficult of arrangements supplant the most mind boggling procedures. Close by the appeal of the promotions and industrialism instigating store format, what about a certifiable grin?

Underneath the articulations like "Client experience" and "Client Delight," lies the intrinsic human should be euphoric and wishing the equivalent for other people. Seeing "Individuals" as the genuine benefit unfurls new business measurements. The mix of fulfilled clients and persuaded workers means a successful business.

14.2.5 MAKE EMPLOYEE'S EI STRONGER

As feelings and sympathy saturate innovation with "Counterfeit Emotional Intelligence," it may be the time we center around the characteristic "enthusiastic insight" we are given with. People need to continually upsurge their enthusiastic knowledge to configuration considerably further developed passionate viewpoints in the innovation to come and furthermore to make a universe of concordance for them. HR and pioneers have duty to make a situation for their employees so that they have stronger EI.

14.3 OBJECTIVE OF THE CHAPTER

This chapter is written with an aim to explore the ways that how HR can play a role for maintaining EI among employees in the era of AI, when many people are under stress of losing their jobs to AI.

Here, in this paper, a model has been proposed for HR and co-corporate people that how they can sustain EI in the Industry 4.0.

14.4 LEADERSHIP THEORY AND IMPACT OF TECHNOLOGY ON IT

In this part, it has been highlighted how various advancements in technologies contributed toward different theories of leadership and how AI might be affecting the human resources and corresponding leadership role and how HR leaders should come up with recent disruptions due to the introduction of AI.

The main machine age drove the path for the board as a control, as the framework of the processing plants required somewhat association, and guidance along with sensible vigilance. In the early start of the modern unrest, formal form of leadership was not being formed up (Bolden, 2004). Be that as it may, an overall view called Great man theory began to flow amidst the 19th century, which comprised of the considering extraordinary pioneers as brought into the world with their gifts, where great administration is beyond the realm of imagination to expect to instruct. In the early 1990s, the first authority initiative hypothesis was developed, called Trait theory.

Trait theory intended to distinguish explicit normal qualities of fruitful pioneers and recommended that these qualities were inborn. The theory was supplanted with speculations on various administration styles and conduct, concentrating on what pioneers do instead of their acquired attributes.

The response to what kind of leadership fits various circumstances were given the assistance of contingency hypothesis, which developed during the 1940s, is still regularly utilized today. As per contingency hypothesis, administration ought to be adjusted to the circumstance (Goffee & Jones, 2000). With greater accentuation weighing on situational attributes as the determinant factor for good initiative, the perspective on administration drifted from concentrating exclusively on the pioneer's conduct and intentions to likewise incorporate the supporters (Bolden, 2004; Goffee & Jones, 2000).

Today, innovation is encountering what Brynjolfsson and McAfee (2016) portray as the time for machines based on computers and the mechanical advancement it brings. The PC and the web improve correspondence and make it simpler to unite individuals, information, and aptitudes).With new and updated technological support, the speed of work at organizations has increased and also it has a similar effect on the kind of innovation that is taking place at different organizations (Tapscott, 2014).

As indicated by Pearce and Manz (2005), leadership is, therefore, made in the transaction with others as well as developed inside by people in an association.

14.5 HR AND LEADERSHIP ROLE IN THE ERA OF AI

This part introduces the writing on AI's potential ramifications for the future position of authority. To begin with, it portrays the future position of leadership as including the observing and direction of AI. Leadership which introduces work culture conducive for AI has been introduced, just as how these conceivable outcomes and difficulties make the requirement for the pioneer to have the option to screen and guide AI. Second, it portrays what is to come pioneer's obligations as moving toward the social parts of the working environment, where the influential position incorporates rousing, supporting, and empowering representatives. It mirrors the significance of accentuating milder qualities, for example, relational abilities and inventive reasoning when AI replaces specialized aptitudes.

14.6 HR FOR MOTIVATION, CREATION, AND SUSTAINABILITY

Nowadays, AI is able to accomplish all physical errands that by and large no one but people can perform face and voice recognition, and hence computers utilization is increasing gains (Hirsch, 2018).

Companies manufacturing cars have likewise made considerable progress in making self-driving vehicles with the assistance of a computerized programming associated with cameras and sensors (Brynjolfsson & McAfee, 2016).

Another model is AI emulating people. A bank in Southeast Asia utilizes an AI robot to communicate with clients. The robots can comprehend and converse with clients through facial and language-handling methods and react with the assistance of both inherent information and new learning created as a matter of fact (Finch et al., 2018).

Despite the fact that AI is showing signs of improvement at perceiving passionate states, Brynjolfsson and McAfee (2017) contend that people are yet need to reflect and change mental states. As indicated by two, the psyches and feelings of people are not maintaining consistency when it comes to AI to repeat as people are social to their inclination. People are consequently expected to make inspiration, influence and solidarity, among representatives. They also put that it will along these lines become increasingly significant for people to make arrangement and responsibility among representatives in the working environment. Leaders should be those guiding and empowering their subordinates.

Dewhurst and Willmott (2014) also proclaim AI to give imperative sources of info; yet, people to have a relative preferred position with regards to rousing individuals and indicating emotions. Past writing, subsequently, recommends that the pioneers should put more accentuation on what is impossible by AI when it is completed by skilled and intelligent employees. As in one of the Southeast Asia's banks that has employed intelligent robots to interact with customers; duties for staff have turned out to be progressively intricate and moved toward higher worth client commitment (Finch et al., 2018). Consequently, leaders should search for abilities in correspondence and inventive reasoning when contracting workers (Plastino & Purdy, 2018).

As Tapscott (2014) contends, when the innovative skill increments in associations, the pioneer should concentrate on starting and keeping up connectivity with team members to make work-learning experience fruitful.

14.7 EI–AI PROPOSED MODEL

In this case, the advantages of an affiliation increase in view of those lower costs, by then salary and work may thus create in a robotizing firm continue to express that examinations of a large number of associations in different countries has been shown that this advancement happens in around half of the associations that experience benefit increase.

The Oxford University/Deloitte Study furthermore shows the manner in which that various new openings will be made the same number of new businesses and new openings spring up, close by jobs requiring the capacities that machines cannot facilitate. Anyway, it is like manner communicates that creation sure new openings overshadow those being abstained from requires preparing a workforce to help the high-capable spots that advancement is presumably not going to rule. This has noteworthy implications for HR and its workforce planning agenda, asking HR to look at the structure of occupations and occupations, especially those that require reliability, creative mind, and energetic knowledge aptitudes that cannot be imitated by a motorized machine. The most significant thing in the present situation is the means by which representatives will oversee AI in their work and it will be their key execution pointer.

An examination in Harvard Business Review as of late expressed a legitimate issue "we will in general spotlight on what robots can do, more than how we will work with them." Representatives should work near to robots and, if you like, PCs will be seen as associates. In any case, this suggests redescribing work and our character at work. This needs association aptitudes and shared participation to help the upsides of an extraordinarily mechanized endeavor and will see HR working near to managers to become adjusted to this new novel activity in the workplace. Notwithstanding that HR ought to find a good pace its affecting instrument compartment and help capacities in habits that will be totally startling from any used already.

However, clearly it may be the start of EI working with AI, the two working pair can have drastically momentous outcomes to benefit the entire human race and to result into our common working lives all the more empowering and improve our money-related status. So, what do HR specialists and pioneers prerequisite for this "new world?"

It is a given they ought to be cautiously insightful and have strong definitive arrangement capacities or restructure aptitudes! In any case, overall, we require a proactive procedure that grips robotization and comprehends the points of interest this may result into. This got together with a wealth of flexibility, versatility, and eager expertise will enable us to coordinate the executives and people through these monstrous changes.

Walt Disney has once expressed: "We continue pushing ahead, opening new entryways, and doing new things, since we are interested and interest holds driving us down new ways." So, pioneers must guarantee this new changing world with AI ought to be managed all interest and this is the main route dread of AI future can be dealt with.

Along these lines, this part proposes a model for HR experts and for pioneers for this new universe of AI for supporting EI among their representatives with the goal that they can perform well in this difficult "new world." This is a five stage model (the name of the model has been given *Mental Model*) which is shown in Table 14.1.

TABLE 14.1 Steps in Mental Model

Step No	Name of the Step	Description
1	Change the mental frame	Try not to battle the technical advancement and innovation, e.g., AI can improve results and also can lessen cost—thus, do not battle with innovation. Embrace the disruption in the respective field and try for making it productive and reciprocal
2	Strength Weakness Opportunity Threat matrix	Take AI as a part of your system. With AI you have to beat your competitors and with it only you have to proceed not without it. Do SWOT analysis of the system and employees with AI
3	Gap analysis	After SWOT analysis, proper gap analysis should be done for analyzing the gap for their employees for EI in reference to AI
4	Training	According to gap analysis, HR should impart training to the employees
5	Collaborative approach	The HR should train their employees in such a way that they treat AI and machines as their counterparts, not rivals and they should think about the means to improve productivity, quality of the system together. And, this is called collaborative approach

14.7.1 CHANGE THE MENTAL FRAME

- *Do not fight the change.* AI is here to stay and will contain timelines and costs. Instead of being resistant to the technology, embrace change and how it can transform the way you operate.

- *Look inward.* Explore yourself and what is your true potential, like never before. Do your SWOT analysis, and focus on a natural understanding of what you can do best.
- *People matter.* Do not put yourself in a back seat; instead, connect with your coworkers, and consider where you can collaborate to create the "new" and the "meaningful."

Try not to battle the technical advancement and innovation, for example, AI can improve results and also can lessen cost—thus, do not battle with innovation. Embrace the disruption in the respective field and try for making it productive and reciprocal.

14.7.2 SWOT MATRIX

Take AI a part of your system. With AI you have to beat your competitors and with it only you have to proceed not without it. Do SWOT analysis of the system with AI (Table 14.2).

TABLE 14.2 SWOT Matrix

Combination of Human and Machine	Opportunity	Threat	Description
$S^H S^M$	$S^H S^M O$	$S^H S^M T$	Evaluation of strengths of human + machines in the environment of threat and opportunity
$S^H W^M$	$S^H W^M O$	$S^H W^M T$	Evaluation of strengths of human and weakness of machines in combination in the environment of threat and opportunity
$W^H S^M$	$W^H S^M O$	$W^H S^M T$	Evaluation of weakness of human and strength of machines in combination in the environment of threat and opportunity
$W^H W^M$	$W^H W^M O$	$W^H W^M T$	Evaluation of weaknesses of human and machines in combination in the environment of threat and opportunity

14.7.3 GAP ANALYSIS

HR should focus on employees as well as machines strength and weakness all together and then give training accordingly.

14.7.4 TRAINING

After analyzing the gap, proper training to the employees should be given to strengthen their weakness and hence will help in enhancing their EI.

14.7.5 COLLABORATIVE APPROACH

The HR should prepare their representatives so that they treat AI and machines as their partners, not adversaries and they should consider the way to improve efficiency, nature of the framework together. What is more, this is called collective methodology.

For some time, it has been realized that AI will influence workforces and markets. Automated creation lines will keep on dissolving producing occupations. Self-driving vehicles will constrain drivers of trucks, prepares and transports to search for elective types of work that can use their aptitudes.

At the point when AI propels, which is occurring quickly, a more extensive number of occupations will be affected, including those that require certain degrees of psychological capacity. A portion of these are occupations that, until a couple of years back, nobody could consider having been done off without the inclusion of a talented work power, for example, educating, drug, money-related exhorting, promoting, and business counseling.

These occupations may appear to be unpredictable; however, there are different errands which robots/AI may perform superior to people. In all actuality, numerous gifted occupations are close to celebrated handling, which incorporates gathering information, deciphering the outcomes, deciding a practical arrangement, and executing this plausible arrangement.

14.8 AI AS THREATNING AGENT FOR SKILLED EMPLOYEES

Highly trained workers, having years of training, can command high rates cause of following skills:

(1) Performing routine activities fast and effectively.
(2) Experience and decision-making for a feasible solution.
(3) Astuteness for helping others in implementing that feasible solution.

14.9 HUMAN TOUCH IS STILL FUNCTIONAL

AI might be capable of diagnosing advanced technological or medical issues and provide solutions for reaching a satisfactory conclusion. An employee is still fit for motivating and decision-making jobs.

Also, AI may be capable for diagnosing any disease and suggest a course of medication more effectively as compared to medical practitioner; however, it requires a medical practitioner to engage with a patient, which can comprehend the patient's life actual parameters like his financial background, family support, etc., and may help the patient in determining the best treatment.

It is these capabilities of man that may become a lot of and more valued over consequent decade. EI skills such as influencing, persuading, social understanding, and sympathy can become differentiators as AI and ML take over.

High EI will be needed for everyone who needs to excel in the job as automation increases.

14.10 EI AND ITS REACTIONS

Computerized reasoning and ML may before long exceed expectations people regarding the over two referenced capacities. It will change the abilities necessity for workers who need to support and exceed expectations in their professions as new innovation has changed the scene.

As we become increasingly acquainted with machine innovation through our own cell phones, trusting and drawing in with machines gets simpler and our favored decision much of the time. People are restricted and regularly one-sided, regardless of whether deliberately or unknowingly; we have our inclinations and our feelings sway upon our basic leadership.

We can never keep completely side by side of the considerable number of changes in our specialized topic and have a limited arrangement of encounters, propensities, desires and bits of knowledge. We get pushed, get sick, need time away from work to eat and drink, exercise, rest and participate in an entire host of different exercises that are imperative to us as people.

14.11 EI SOON BECOMES MAIN SKILL SET

Abilities in EI will before long become compulsory for workers in the event that they need to support in this profoundly aggressive world brimming with mechanical headways.

There are a few endeavors to carry these abilities into the instruction framework. They are viewed as something that is significant in the early years in nursery and elementary school; however, presently cannot seem to turn into a fundamental piece of human advancement all through, and past, essential, optional, and tertiary instruction.

EI preparing in association is regularly observed as a jolt on to administration and the board programs. Superior to nothing, they can be wide and oversimplified in their conveyance, coming up short on the profundity of utilization to achieve manageable, implanted change through an intensive investigation of the development.

ML and AI both are having the capabilities to enhance our lifestyle at economical cost. Embrace the disruption as it is economical to everyone.

Encouraged advancement of passionate mindfulness, getting the board and articulation through preparing and training achieves transformational change.

Psychological insight can be accustomed to realize change in working with feeling to comprehend their importance and effect upon execution. These progressions are in social aptitudes, sympathy, correspondence, confidence, relational connections, and social duty, to give some examples.

These are genuine abilities for the 2020s and past as AI turns out to be increasingly settled.

14.12 WORKING WITH AI AND EI

Here are three recommendations around working with AI and EI.

Following are the suggestions for adopting EI for AI:

(1) Do not fight against technology: AI and ML have the capabilities to better our lives and lifestyle with minimum cost. We should embrace this update in our industry and try to make it beneficial for everyone.
(2) Evaluate your own abilities in networking with, guiding and assessing people: Perceive how feelings support everything that we do. Use enthusiastic data as an information source that takes into account more critical thinking and basic leadership, and to grow progressively

bona fide connections. Know your qualities and restrictions with regards to EI.

(3) Investing in developing your EI: The simplest technique is to move your psychological model to that inessential in your activity profile and to begin organizing on how you can enhance overseeing, affecting, and identifying with others.</NL>

Push somewhat forward by getting out preparing and training from individuals with top to bottom understanding of working in the field of enthusiastic knowledge.

That you can give and that what you can perform superior to anything any machine, regardless of how canny it is, is identified with the individuals around you. Start to put resources into and build up these passionate capacities similarly that you have the more specialized pieces of your activity.

In the event that one is great audience, help or pioneer, at that point he may have a significant job to do as computerization disturbs the eventual fate of work.

Hence, Human Resource Department should train its personnel like that they takes machines as a part of the system and work together.

14.13 DISCUSSION AND CONCLUSION

Extended bits of knowledge, consolidating human info and AI, are the key driver to arrive at exchange and to shape a well-off society. We are on the cusp of mechanical adjusting not in the scarcest degree like each, we have seen your time as of late in mankind's history. Mechanical progress is requiring an extra significant focus on innately human aptitudes—fundamental considering, passionate knowledge, and regard decisions. At interims the stage economy, development is interfacing individual everything being equal and limits with extra openings. It is also enlivening crafted by each gifted and incompetent masters at interims the measure of prepared correction, making the planet work for everybody.

In this time, where AI application is predominant, it is significant for bosses to assist their workers with sustaining their EI as it a significant factor prevailing in this focused condition.

The proposed model, for example, Mental Model has been proposed for HR proficient which will help them for supporting EI among their workers. Individuals ought not to fear innovation and should regard it as their partners. This methodology will help individuals to support and keep up their EI.

KEYWORDS

- **artificial intelligence**
- **automation**
- **disruption**
- **emotional intelligence**
- **machine learning**

REFERENCES

Bolden, R. What is leadership? Technical Report, University of Exeter: Exeter, July, 2004.
Brynjolfsson, A. & McAfee J. *The Second Machine Age; Work, Progress, and Prosperity in a Time of Brilliant Technologies*; Norton & Company: New York, 2016.
Brynjolfsson, A. & McAfee, J. The Business of Artificial Intelligence, What it can and cannot do for your organization, *Harvard Business Review Digital Articles*. [Online] 2017, pp. 3–11. https://hbr.org/cover-story/2017/07/the-business-of-artificialintelligence (accessed April 30, 2019).
Dewhurst M. & Willmott, P. Manager and machine: The new leadership equation, *McKinsey Quarterly.* [Online] 2014, 3rd Quarter, no. 3, pp. 76–83.
http://ludwig.lub.lu.se/login?url=http://search.ebscohost.com.ludwig.lub.lu.se/login.aspx?direct=true&db=bth&AN=102111149&site=eds-live&scope=site (accessed Apr. 28, 2019).
Finch, G., Goehring, B., & Marshall, A. (2018). Cognitive innovation: Top performers share their best practices, *Strategy & Leadership,* Vol. 46, no. 1, pp. 30–35.
Goffee, R. & Jones, G. (2000). Why should anyone be led by you? *Harvard Business Review*, Vol. 78, no. 5, pp. 63–70.
Hirsch, P. B. (2018). Tie me to the mast: Artificial intelligence and reputation risk management, *Journal of Business Strategy,* Vol. 39, no. 1, pp. 61–64.
Pearce, C. L. & Manz, C. C. (2005). The new silver bullets of leadership: The importance of self- and shared leadership in knowledge work, *Organizational Dynamics*, Vol. 34, no. 2, pp.130–140.
Plastino, E. & Purdy, M. (2018). Game changing value from artificial intelligence: Eight strategies, *Strategy & Leadership,* Vol. 46, no. 1, pp.16–22.
Tapscott, D. *The Digital Economy Anniversary Edition: Rethinking Promise and Perilin the Age of Networked Intelligence*, 2nd Edn, McGraw-Hill Education: New York, 2014.

CHAPTER 15

Importance of Succession Planning in Empowering SMEs For Inclusive Growth

ANSHIKA SHARMA[1*] and PREETI TEWARI[2]

[1]*IFTM University, Moradabad 244001 Uttar Pradesh, India*

[2]*Lloyd Law College, Greater Noida 201306, Uttar Pradesh, India*

[*]*Corresponding author. E-mail: anshi1986@rediffmail.com*

ABSTRACT

With the growing trends and developments in the industrial sector, it has become the need of an hour to have the perfect individuals in the acceptable job at the best time and at the correct places. The sustainability of an organization entirely depends on enough people or appropriate people who can rapidly occupy the key positions when it becomes vacant. It sounds like a commonsensical. However, despite everything, it is not occurring as adequately as it must. The organizations and all the more altogether pioneers and administrators, inside this specific situation, necessitate looking with augmented analysis at the need to oversee worker progression effectively to ensure that they have the perfect individual in the opportune spot, imminent authority, the board and specialized ability to convey government business. Small- and medium-enterprises (SMEs) play a vital responsibility in the inclusive growth of the economy of every nation. SMEs comprise over 80% of the sum number of business undertakings and bolster technological improvement. SMEs, at a standstill overwhelm the monetary panorama of the majority nations and encompass being perceived in causative astonishingly to their GDPs. Solitarily, the most prominent rationales for the breakdown of small-scale enterprises are the absence of succession planning (SP). This chapter focuses on the magnitude of SP in empowering SMEs for its wide-ranging development in the economy. The chapter also draws light on the rise of small business enterprises as well as the succession process to be followed in these organizations. The end results of this section demonstrate that the majority of the SMEs include no succession plan set up and the capacity and capability of the heirs are not deemed in succession planning. The chapter suggests to facilitate SMEs ought to build up a prescribed arrangement for progression, express the persona of a descendant, and endow with fundamental company executive education, supply instruction /monitoring to the incumbent chief executive officer and all stakeholders to help him in planning succession and guaranteeing that proprietors are agreeable after disentanglement.

15.1 INTRODUCTION

Succession planning (SP) was first presented by Fayol who acknowledged whether succession planning needs were disregarded, organizations will not be set up to make crucial advances. Succession planning empowers an organization to prepare for the nonappearance, departure, destruction, retirement, or end of an individual. It suits the intelligibility of culture and the improvement of significant aptitudes for an organization. Succession planning has been used as a formalized method for overseeing changes in authority for over 50 years. At first, it is used to imagine smooth advances at the top level in organizations; it has formed into a system that various organizations see as a reason for key positions transversely over limits and levels. While operational definitions change, the central importance has proceeded as before during the time as the strategy of succession planning has created.

Succession planning for some time acts as a function of HR system that has been talked about in expert conferences and inside expert diaries and periodicals. In the late 1990s, human resource experts and leaders knew about the requirement for dignified vital arranging of workforces, particularly, the unadulterated socioeconomics of the open administration condition, at all stages. The unavoidably evolving employees, impelled by the unfaltering mass migration of baby boomers toward withdrawal of work, and merging commonly with an expanding interest for open administrations, denoted the nonrational workforce the board requirement for legitimate arranging and advancement of mindful systems in the zones of enrollment, maintenance, and succession planning.

Succession planning grasps not just distinguishing qualified and inspired contender for advancement to official positions, yet in addition building up their abilities that would set them up to execute viably in initiative situations.

Succession planning procedure incorporates three fundamental parts. The primary segment is the choice of competitor dependent on past understanding and foundation in the midst of development of ability pool. Through this, one ensures that every crucial role has elective prospective successors and every ability has various latent advancement ways. The pattern within the extent of succession planning is growing which envelop every role not simply top administrative place. It has demonstrated that organizations reflecting junior levels have a healthier benefit. If there should arise an occurrence of excluding the whole organization, it is imperative to distinguish basic positions that are basic for the organization, office, distribution, work entity, or group to accomplish the vital job end results.

Recognizing an improvement preparation and transcribe is an obligatory piece of the procedure. The arrangement ought to be there to survive custom fitted toward human being needs and curiosity of the successor. Best improvement strategies incorporate 360° input, official instructing, tutoring, organizing, work tasks, and activity knowledge. The third part of the SP procedure changes the executives and procedure the board, including capacity and subforms like a vital perspective on SP, board duty, usage contemplations and so on. Progression plan ensures that organization approaches required human assets, quantitatively and subjectively. This arrangement will decide the advancement openings and formative needs of applicants and manufacture the executive's responsibility (Christie, 2005).

Along these lines, it is exceptionally fundamental to interface progression intending to business technique to get need sort of individuals with the required arrangement of aptitudes for what's to come. Nevertheless, this linkage has not been practiced in the real-world even in organizations with the best SP. The board duty like some other definitive wide program is fundamental for compelling use of SP. With no assistance, SP is not implementable paying little mind to whether the setup stage is advanced honorably. Mindfulness and correspondence are other contentions in the procedure of SP. It gives the idea that the finest position for both affiliation and person is that contention hoist and discussions about a direct subject to a clear system. Predictable evaluation is noteworthy in Procedural Management.

SP is used to delineate a large collection of activities that incorporate making game plans for specific changes in activity inside associations. The word SP has usually implied preparing for organization intelligibility at the chief executive officer (CEO) level, yet the present SP obliges authority congruity at all stages. Starting late, SP has been drilled even more routinely and intentionally in a noteworthy number of bigger and at stages far underneath senior power.

SP is the path toward diagnosing the focal necessity for academic capacity and authority all through the organizations after sometime and preparing individuals for current and prospective work commitments required by the association. This characterized as a method for distinguishing basic administration positions, beginning at the degree of venture administrator and manager and stretching out growing to the most astounding roles in the organization.

It additionally portrays the administrative roles to give most extreme adaptability in parallel administration moves and to guarantee that as people accomplish more prominent status, their management aptitudes will expand and turn out to be increasingly summed up in connection to add up to hierarchical targets instead of absolutely departmental destinations (Carter, 1986).

SP is not an end in itself or an add-on activity. It frames a subset of more extensive organization ways to deal with workforce planning which tries to guarantee when all is said in done that the ideal individuals are in the correct spot at the opportune time to accomplish fruitful business results. Like workforce planning, SP includes inquiries regarding the changing idea of work and the sorts of jobs that are probably going to develop instead of concentrating exclusively on the present jobs which may not be required later on.

The focal point of SP is to guarantee a progression of up-and-comers who have what it takes, learning and ascribes to go after opportunities in basic jobs when they emerge, as opposed to taking a gander at the total staffing requirements for whole work families over the organization. Different methodologies that add to incorporated workforce planning incorporate endeavor vigorous and far-reaching statistic examination, and the execution of activities, for example, directed enlistment programs, maintenance techniques, performance management strategies, knowledge management frameworks, and learning and advancement mediations. Succession management systems, which spotlight on creating representative capacity, empower organizations to react to change. On the basis of literatures, the researcher has identified the following framework which includes key areas to effective SP (Figure 15.1).

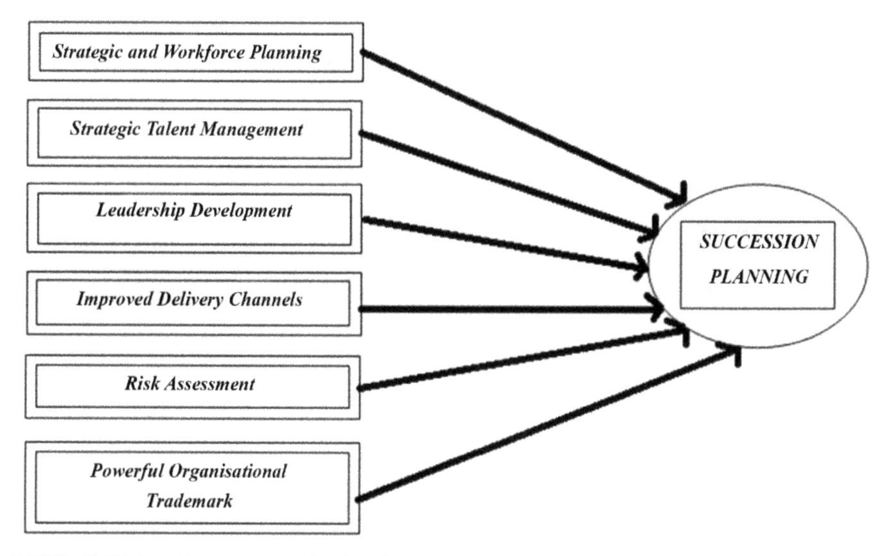

FIGURE 15.1 Key areas to effective SP.

15.1.1 EMANATION OF SUCCESSION PLANNING

The area restrains a portrayal of investigation hooked on SP and the management since been announced from an authentic point of view. The historical backdrop of SP and the management has three periods. A clarification for every timeframe, the cause for study, inclines, and correlated the exact research pursues. The verifiable setting of SP and the management in this investigation is partitioned in three eras. The primary era cover ups from 1960 to 1980. In this period "The Growth of Succession Planning" has been discussed. The subsequent period, 1980 to 1990, is top depicted as "Rising Trends and Advancement of Succession Planning." The last time frame, "The Succession Planning Ahead," discussed about 1990 till present.

Although, few discussions subsist on the sources of progression while an exploration theme is the main part of observationally support investigations of succession happened in the middle of 1950s to the middle of 1960s. In this timeframe, the essential point of convergence of research was on CEO progression and the board improvement. At that point, human resource arranging comes up with mode in the middle of 1970s to middle of 1980s. Now inquire about consideration moved concentration to anticipating the HR needs of a whole organization as opposed to concentrating exclusively on the senior administration group. Exploration in SP started vigorously during the 1980s (Kesner and Sebora, 1994). From that point forward, SP has developed by way of the floods of progress in working environment and human resource management (HRM).

15.1.2 THE GROWTH OF SUCCESSION PLANNING BEFORE 1980

Alongside the rise of HR, during this period, a considerable lot of the compositions with respect to individual components are extremely compelling. Several discussions arise apprehending the inception of an investigation into succession. The formal executive development programs in huge organizations announced the normal components in 1947. The Grusky's exploration was the first to distinguish the key factors in the succession condition, to build up a research replica, and to test a speculation.

In this timeframe, the most fundamental topic for research is on CEO succession which is an occasion for an official improvement. The research in this period is classified under two heads: Psychological/organizational hypothesis and connected speculations. In human science and authoritative speculations, the inescapable interests were coordinated toward building up whether a change in administration decides and are influenced by framework level execution, and assuming this is the case and in what aspect they were influenced. Tactical planning, HRM, and organization development (OD) exploration and speculations added to the progression framework. Tactical planning, the noteworthiness of individual factors in key definition and execution, and the use of human asset data in business arranging were dissected in HRM works of literature. SP was asked about as a correlative system in which the stockpile was facilitated with the solicitations of the organization. The administration of a succession system is the central mediation point in the move from an OD perspective of individual change to a perspective of long-broaden hierarchical level change.

During the 1960s, explore subjects consolidated the birthplaces of succession, generally standing out succession from inside from outside, the association between the pace of succession and authoritative size, the association between succession repeat and post-progression execution, and the association among succession and execution. During the 1970s, rising investigation inspected progression repeat, the relationship of the style of successor and succession, and the association between the directorate and CEO succession.

Asbury explored trade and business rehearses to do with staff organization at the official point in 1947. The research considered 53 organizations and recognized 5 basic components in their official arrangement for official advancement which also include organization research, selection, assessment, improvement of official level ability, and stock control. The research additionally presumed that the board and line the management inclusion were two basic aspects in fruitful executive succession and development planning.

In one of the study, the "state of the art," and distinguished the five regular components in managerial advancement programs. They incorporated a characterized association arrangement, routine evaluation, set up replacement tables, the advancement of sky-scraping possibilities, work revolution, and preparing programs. Trow (1961) analyzed the extent of succession into top positions in 108 little organizations. It discovered organizations that readied for succession were more averse to encounter budgetary trouble during executive turnover. Likewise, it saw a solid relationship between getting ready for succession and benefit of an organization (Zaich, 1986).

The normal components along with the investigations are the meaning of organization purposes and preparations, assurance of total labor, necessities for arranging time frame, appraisal of in-house abilities, and assurance of gross human resource prerequisites to congregate organizational objectives, and enlargement of an activity chart and projects to accomplish targets.

15.1.3 RISING TRENDS AND ADVANCEMENT OF SUCCESSION PLANNING IN 1980S

SP is the subject that pertain a more prominent consideration, all through this time of exploration, than earlier years. The purposes behind the expansion of SP study as the consistent branch of 20 years of examination concerning succession. Various scientists started to find out the effect of both the extent and nature of the planning.

Various researchers contemplated the methodologies and phases of SP. They underscored jobs of erudition and the advancement of the board occupants in anticipation of succession. During this period, many of the scientists coordinated strategic human resource literature and SP. They noticed that succession ought to be intended to coordinate chiefs with systems, and plans ought to be explicit when graphing situational moves and timing.

In 1982, Carnazza provided details regarding his inside and out meeting research including 15 organizations with succession/substitution planning programs. This is one of the most as often as possible cited research of this period. It accepted the motivation behind succession/substitution planning projects which guarantee the expansion of an adequate supply of qualified people to occupy future opportunities in key administrative and expert roles.

The reason for the examination was to explore how organizations accomplish the goals of succession/substitution planning programs. The discoveries of exploration incorporated in accompanying:

- The bigger is the organization, larger the formal succession/substitution planning.
- Organization should perceive that significant time is fundamental, maybe up to 5 years for a succession/substitution planning system to be actualized completely and for the organization to start to accumulate the normal advantage.
- A representation of succession/substitution planning, which shaped the reason for their self exploration discoveries, are particulars of the techniques expected in order to accomplish two fundamental goals of succession/substitution planning program: choosing the situation to be secured, which included a procedure of distinguishing basic roles, and the individual to be incorporated, which incorporates the way toward supervising essential administrators?
- The embodiment of succession/substitution planning is connecting individual capacity to job need.

In one of the examinations during this period, the specialist has investigated the administration's succession and advancement arranging practices around at that point. The investigation perceived four one of a kind approaches to manage the board succession and advancement arranging: casual, decentralized, brought together and coordinated. As shown by the investigation, the elements impacting the assurance of the approach are association structure and the administration style, size, and abundance of the organization, and the organization's advancement rate. They in like manner explained the six methodologies used in executing the tasks: (1) senior administration contribution, (2) data necessities, (3) evaluation, (4) the board survey, (5) formative strategies, and (6) instructive preparing. They related every approach to managing the administration succession and advancement arranging with the used framework. They underlined the criticalness of making and installing a program that fits each organization's traits and culture.

Among a few sorts of research on SP, one circulated in 1986 rose up out of the examination for his own one of a kind doctoral paper submitted in 1984. This was the fundamental research done on SP as a structure and its relationship with organization execution. The centrality of this investigation of SP is twofold. The succession system as "the rules and techniques that structure the setting for a typical progression occasion (i.e., a modification in occupation incumbency), including official improvement and position practice." Secondly, its investigation is the first to try to see SP and its effect on authoritative exhibition or results. Frequently, past research thought about the association of one of the components raised by CEO progression and organization execution, yet not the association of SP and authoritative execution.

15.1.4 THE SUCCESSION PLANNING AHEAD: 1990S TO THE PRESENT

During the 1990s, nobody appeared to bring up issues about the significance of SP. As of late, business organizations, yet a wide range of organizations, including educational foundations and government workplaces, understood the need for SP and its implementation that fit in their own organizations. Hence, look into this period extended to other business organizations, for example, learning establishments, government, nonprofit organizations, healthcare, and independent ventures. The themes hold inside the general shade of SP has turned out to be different.

Other nondoctoral dissertations likewise center basically around CEO progression. For example, CEO selection, succession, and organization performance. The reason for its investigation is to offer a progressively absolute calculated replica of the connection amid CEO-related issues and organization performance. It manufactured a few speculations. Among them, the subsequent speculation identified with SP is that a CEO who has a particular heir as a primary concern will head a higher performing firm.

The most important research discovered during the literature audit is in 2001. The research inquired about the succession management frameworks and human resource results. The one reason convoluting the investigation of the impact of succession anticipating HR was the incredible inconstancy of planning practices crosswise over firms. It contended that an arrangement might be incredibly straightforward, accommodating just the expansion of reinforcements and probable successors to the most superior-level administrators. The study found that the other outrageous might be extremely official as well as a recorded standard and methods for management succession at all stages.

A portion of the attributes, for example, line-men involvement, nonpolitical succession criterion, the validity of SP personnel, survey, and criticism, and powerful data frameworks, influenced the presentation of HR. The impediment of the announced research was that no circumstances and logical results connections or course of the setback of factors existed. This area contained an audit of the exploration of SP and the management from an authentic point of view. With 30 years of foundation, no solitary investigations are straightforwardly identified with this examination. One of the studies inquires about on the predecessors, outcomes, or possibilities of a CEO succession result. At that time no critical modification in research points and factors has happened. With respect to the investigation into SP, those likewise will in general examination the factors of SP as opposed to SP itself. At the end of the day, examine on SP legitimately researches components or factors of SP and recommends those elements or factors may be helpful when actualizing a SP program. The estimation instrument of the results for SP has turned out to be changed from exclusively relying upon the monetary presentation of an enterprise to learn HR adequacy and speculators' response.

15.1.5 SUCCESSION PLANNING IN SMES

The controversy of SP in small-scale enterprises is not contemporary to the literatures. In spite of the fact that the significant spotlight is on the business transfer. SP is the place where the enterprises plan for the prospective exchange of ownership. Truth be told, it happens when the company proprietor wishes to depart from the company in any case needs the business to proceed. The rationale behind this is to move responsibility for firm to any of the relatives instead of closing down the business altogether. Handler (1994) audits the exploration to date on progression in the field of family business management. Five surges of research are featured: (1) progression as a procedure, (2) the job of the author, (3) the perspective of the people to come, (4) complex degrees of examination, and (5) attributes of powerful progressions.

Sambrook (2005) investigated a portion of the contentions in deliberate progression related to finding and effectively creating head successors to guarantee the endurance and development of small enterprises. The researcher distinguishes different crucial factors incorporating enlisting representatives with potential, thinking about the work/profession thought processes of potential successors and methods for moving hierarchical and individual/implicit information from the proprietor administrator to the successor, regardless of the fact that whether an inner worker or another buyer. With the assistance of the basic model, the researcher has distinguished three sorts of information moves and two levels of progression. The researcher offers new bits of knowledge into SP and the model gives a structure to creating successors.

For most family-possessed organizations, antecedents do not wish to mull over progression since they dread the loss of intensity and status. It could be noticed that there was a call for to maintain the business in the family since it is a key determinant for family-run firms all-inclusive. In little firms, it is commonly the business person who is exclusively in charge of the administration of human assets. Regularly in light of the fact that little firms cannot legitimize full-time human asset experts because of restricted size and assets, and still numerous business people do not see HRM in little firm as an exceptionally refined procedure requiring authority. While being little, this errand will not present quite a bit of an issue for generally business people. Anyway, in one of the examination, the creator expresses that when a firm develops and it's a number of representatives increment the multifaceted nature of HRM extend.

As of late, it was discovered that while the coordinators of SMEs need to pass ownership to their successors (progression), the last dynamically want to be diminished of this weight and cheers when the business is offered to outside individuals (move). These outcomes are reliable with the examination for explicit cases inside the UK. Their investigation focuses on the picked leave methodologies of innovative family firms. They locate that both the accessibility of a ready successor and the particular individual and family esteems included affect the left course on the improvement of SME and the job of SME area in advancing monetary and social advancement by making open doors for business. Studies also indicate that innovation has a genuine effect in a large portion of ventures and in all parts of the economy, while organizations and undertakings keep on experiencing extensive changes. Utilization of these advancements is disquieting the measures of business, achieving essential differences in endeavors. The discoveries thought that cutting-edge organizations are impractical without the assistance of innovation, which is significantly affecting the tasks of small- and medium-scale business. It is pronounced to be fundamental for the endurance and improvement of economies all in all. SMEs are attracting consideration created and creating nations just as experiencing significant change nations.

Elżbieta Roszko-Grzegorek (2008) found that to investigate business move as a vital objective and potential wellspring of changes in SMEs in Poland. The significance of this investigation is recognized by the way that it offers scientists and professionals observational confirmations that connection the succession process with a vital direction. In this paper the researcher contended about the point of business progression which ought to supplant existing innovative assets, however, it should prompt the auxiliary reestablishment. With this examination, the researcher attempts to clarify that business move/progression in SMEs is not seen by respondents as a procedure that requires long haul arrangements. The business move still happens fundamentally inside the family and is not respected from the crucial perspective. The outcomes propose various fascinating qualities of effective changes. To start with, in regard to potential successor it is demonstrated in the experimental examination that it ought to be a relative and this is by all accounts the most significant viewpoint shown by test organizations. Pushing toward key objectives of the sample organizations' expansion of offer development, augmentation of the organization's worth and furthermore money related freedom of the family firms are of exceptional premium.

Obadan and Ohiorenoya (2013) analyzed the procedure of SP in private venture undertakings in the lodging business in Benin City. For those that participate in SP, the investigation analyzes the procedures required just as the elements that record for this basic administration conduct. The discoveries in this examination demonstrate that the vast majority of the independent company ventures have no progression plan set up and the capacity and ability of the successors are not considered in SP. In any SP, as the successor is being prepared to take over, arrangement should also be put in place to make the founder or entrepreneur who is disengaged comfortable after the disengagement. This will alleviate the fear of losing the grip or ownership of the business. This examination recommends that independent venture tries should develop a proper course of action for progression, grant the character of the successor, give the fundamental business the board guidance, give preparing/checking to the occupant CEO and all partners to help him in planning succession and guaranteeing that proprietors are agreeable after separation. In the study entitled *Factors Influencing Strategic SP for Small- and Medium-Family Enterprises* the author identify the factors which determine successive strategic succession plans for family SME businesses in Nairobi Central Business District. There are three research questions that have been addressed by this study: What role does effective leadership play on successive family succession strategy? What role does gender play in determining successive family succession plans? To what extend does successor attributes factors affect family succession plans? In this paper the author has used descriptive survey design to ascertain and describe the characteristics of the variable of interest in this situation. The population of interest in the study was SMEs operating in Nairobi County Central Business District (Ashanda, 2015).

15.1.6 IMPORTANCE OF SUCCESSION PLANNING IN EMPOWERING SMES

A large portion of SMEs over the world are endorser monomaniacal, in addition to this, it subsequently turns out to be basis for these organizations, all things considered, any organizations headed for legitimate SP on all the significant stages in different offices just as the top generally level. The majority of SMEs do not endure the succession starting with one age then onto the next age because of the absence of appropriate arranging and it is discovered that regularly the arrangements and arranging comprise not been embraced in favor of mental reasons, especially the dread of ancestors clashes on possession or stakeholding concern.

Numerous SMEs proprietors has established and fabricated their organizations exclusive of really thinking about what would occur whilst the ideal opportunity for them to resign inserting their organizations in danger. Organizations neglect to get ready for the opportune and compelling SP particularly the positions of authority are regularly observed to be found napping prompting disturbance to ordinary business exercises thusly loss of a piece of the pie. Thus to maintain a strategic distance from such circumstance SP ends up basis for SMEs in light of the fact that it empowers the organization to distinguish gifted workers and give the vital tutoring and preparing to create it for potential advanced stage with more extensive duties and executes them easily.

Succession planning is often a deliberate way of dealing with:

- Structuring a solid ability team/viable initiative channel to guarantee headship permanence.

- Counseling and establishing probable descendants within the organization in a manner so as to fit in their qualities.
- Analyzing and enrolling the pre-eminent possibility for classifications of roles at vital stages inside the organizations.
- Pondering HR toward the ability maintenance and advancement procedure capitulating more prominent degree of profitability and viably threat management.

Consequently, SP requires a key methodology for the business procedure which empowers handling of intense subject matters to settle on the best choices about the eventual fate of promoters and their enterprises. Few decisive advances which should be remembered while functioning a powerful SP technique are as follows:

- Coordinating business system and effort management.
- Determining gaps and irregularities in procedures and individuals.
- Talent classification and talent management
- Fostering succession management policy
- Enforcing SP
- Assuring to assess and screen constantly to implement compelling SP.

Consequently, while implementing a compelling SP procedure various basic components which assume urgent positions are:

- Sponsors/owners/company pioneer ought to acquire actually engaged with the procedure as it entails duty.
- Sponsors/owners/company pioneer should have responsible for creating successors as probable pioneers.
- Member of staff at all levels must be focused on self-improvement and development in the organization.
- A long-term vision should be remembered because accomplishment depends on company efforts.
- Succession ought to be connected toward vital arranging and interest later on the development of the company.
- Succession planning consists of a coordinated HR processes and ought to incorporate preparing, advancement, and performance appraisal at ordinary interims.
- SP ought not to be just upward yet it ought to likewise be horizontal.

Although implementing the SP approach in small- and medium-scale enterprises sponsor/possessor must think about leave system which is best reasonable: an ancestors move, advertising of enterprise totally and leaving, or shutting down the business in order to stay away from superfluous clash. It ought to likewise empower SMEs in structure solid money related safety for the sponsor/company owner and their family similar to single biggest resource and a noteworthy wellspring of individual value and confidence for their inclusive growth.

15.1.7 SUCCESSION PLANNING PROCESS IN SMES

Succession planning is an unpredictable procedure. It includes numerous variables and develops over a significant lot of moment. Exploration on SP in SMEs is generally affirmed the significance of SP wanting to the endurance of the organizations. Besides SP is not a standard upgearing exchange wherever the organization is offered to the most elevated proposal; basically the beneficiaries acquire the family business as of the originator and need to maintain the business in an expert manner.

The problem of SP in SMEs is not novel to the literatures. Nonetheless, the significant spotlight is a business move. In spite of the fact that SP is an essential event in the company's life cycle researchers' put almost no regard for the particular aspects that upgrade organization ability to endure and how folks interface with the procedure itself. As such, the procedure of company succession ought to consider the idea of the focused setting where procedure happens. Thus, thought of succession as a step-by-step process is inevitable.

So as to give a response to the expressed above mentioned inquiries the hypothetical examination is isolated into various components devoted as

- obstructions during succession process,
- assortment of successor,
- succession planning.

15.1.8 OBSTRUCTIONS DURING SUCCESSION PLANNING PROCESS

Succession is the utmost decisive problem confronting every family business. It includes difference in authority starting with one age then onto the next to guarantee the coherence of family responsibility for company. Succession within SMEs is novel; the aftereffect of the double character of these organizations. In that capacity, Succession is an undeniable component in the organizations' life cycle—the more youthful age succeeds the more established one so as to guarantee coherence. Succession planning is risky for SMEs and sky-scraping disappointment velocities are frequently announced. No arrangement or getting ready is an issue regularly referenced in succession literatures. The absence of SP as the principal motivation behind such a significant number of first-generation family firms companies do not make to the subsequent age.

Few preventing features fundamentally worried about genuinely stacked issues and the various interests of partners. The powerlessness to give up by the original organizer is broadly noted in the prose on succession and an inhibitor as it prompts a specific degree of inflexibility. The exact research of trial organizations, it is demonstrated that still there is a conviction in which the legitimate guidelines are excessively confounded to effectively finish the succession process.

15.1.9 SUCCESSION AS A PROCESS

Concerning succession, researchers have contended either an animating or slowdown sway on the organization advancement and improvement. Succession can similarly be a wellspring of the fundamental various leveled and managerial energizing of the enterprise that could display a trademark bit of space of SMEs. SMEs stimulate business advancement and enable successors to direct break up divisions in like manner keeping up a key good way from conflicts. Researchers dependably underline the essentialness of acceptable foreseeing succession to avoid succession problems getting the chance to be inhibitors of endurance and improvement. The one of a kind job that succession plays in the present cycle of the business and the endurance of the SMEs regard as succession as an improvement driver for the SMEs anyway exactly when opportune arranged and implemented. Be that as it may, in the observational research exhibits that there are insignificantly understood in associations about the progression procedure. For the most part, the timing is agreed by the respondents gathering to the succession method. Despite the way that the fitting time allotment be 5–9 years to plan the succession process is between.

15.2 SUCCESSION PLANNING AND TACTICAL OBJECTIVES

The strategic management and succession plans empower organizations to:

- determine management capacities and routine guidelines,
- guarantee congruity in the organization performances,
- personality extraordinary possibility designed for superior administration positions,
- fulfill yearning of representatives meant for professional success.

In the course of arranging procedure, succession plans steer activities in order to upgrade the nature of administration. The SP could assemble an upper hand through the prevalent improvement of successor's authority ability. Indicators of a fruitful succession process are observed to be subject to the occupant's eagerness to move to one side, the successor's readiness to assume control over, accord among relatives on the most proficient method to transmit SP.

The proper fit between the successor skills and necessities of business which is likewise identified with the planning of the succession process. The achievement of succession is anyway a mind-boggling blend of various partners' fulfillment discernment with the procedure and firm key objectives. Various investigations underscored to facilitate succession that must be foreseen elongated ahead of time and oversaw as a planned procedure. The absence of SP is a significant explanation behind the sky-scraping transience in family business. The establishment to potential research on succession in the family-firms presents commitments in five classifications:

- company methodology,
- family firms,
- successors determination criterion,
- successors preparation,
- association among proprietor author and successor.

Despite what might be expected, in the meadow of family firms, succession is viewed as tactical. For example, a family business that goes from age to age. On this ground, the connection between a family and a business is seen as a wellspring of firms. In any case, succession is frequently separation and upper hand perceived as basic for organizations survival. In this manner, in this field, the administration of succession is regularly seen as a feature of the organizations vital arranging. Lamentably, in an experimental examination over 35% of respondents demonstrated that they do not design succession yet and just somewhat over 2% composed succession plan. The outcomes demonstrate that organizations yet philanthropic insufficient regard for SP.

15.2.1 ASSORTMENT OF SUCCESSOR

The succession procedure depicted in the family-firms organizations literatures incorporates three basic advances that are related to successor choice. The initial pace is to get ready posterity for the position of authority at an untimely age before fusion of the family firm. The subsequent advance is to incorporate posterity in family business in various situations. The next step includes the posterity assuming control above the power of the family-business company.

With respect to criterion, the literature emphasizes a universal inclination toward meaning of progressively target procedure of determination. Social generalizations, for example, demographic request are steadily trailing their role. Successors are always chosen based on their encounters and aptitudes and promise to the enterprise. The choice in picking the oldest for the pioneer is frequently supported by the dissimilarity among their situation in the family and their situation in the organization.

In the literatures, it is additionally brought up the family's trust as definitive to invigorate partners to perceive the successor's job in the firm. In spite of the fact that in the ongoing writing, the generalizations are losing their capacity in this example see that over 75% of defendants need probable successor to be an individual from a family.

The subsequent concern respects the successor preparation. The degree is to characterize successor ideal way of encounters. Researchers concur the procedure of intra-family move is an elongated one. It begins in adolescence and is portrayed by two basic focuses. The primary is the point at which successor enters the industry on a permanent premise. The next is when the successor acquires the initiative. Moreover, researchers firmly suggest a long run administrative practices outer the family firms. It empowers the successor to gather involvement in a differing situation and to build up the personality with management style. The improvement of an administrative mover in the family enterprises and in the various places that entails associating with various partners in viewed as necessary. At long last, ongoing commitments call attention to the importance of being prepared as a business person at scholastic echelon as significant for the inter-generation succession achievement.

15.2.2 CORRESPONDENCE AMONG INCUMBENT AND SUCCESSOR

The association among proprietor and successor is an additional matter which is frequently discussed in the literature. The idea of the association is usually viewed as vital designed for the problem of the family-business company succession. Its disposition is important to decide procedure, planning along with the viability of progression.

With this viewpoint, one of the investigations has described the vibrant relationship alongside succession process and recognized following phases:

- possessor administration,
- guidance and improvement,
- fraternity,
- authority transmission.

Customarily, the two jobs ought to commonly modify their conduct along with the procedure and perfectly with its condition of development. It has been likewise demonstrated that the relationship is important, yet in addition roles' mutual perception. Besides, a few commitments demonstrate that the decision of "venturing out of intensity" is not a simple. There could be numerous reasons, for example, the dread intended for the eventual fate of the firm, in favor of their own dignity and character, for the prospective failure of regard both family and the network, and the absence of faith in the successor aptitudes. Because of it, the exact research demonstrates the occupant favors much of the time to choose who to decide for the potential successor However, various arrangements have been proposed to defeat senior's protection from the difference in authority, for example, helping his/her to end up mindful of his/her conduct or energize his/her to use on his/her experience to begin another enterprise.

Sharma (2004)'s survey of the literature adds some fascinating bits of knowledge to the broad-spectrum depiction of the succession process. In one of the research study, organization succession disappointment in family-business companies assume urgent job. The real contention is that succession disappointment in family-business companies is capable of frequently followed backside to the aberrant among a hierarchical past and future.

15.2.3 BUSINESS PREMEDITATED PURPOSE

The inter-generational relocation process itself might impact the company's technique. Models will incorporate how much new age is acknowledged and their capacity to actualize their own vital motivation. In the investigation of the family firms system, it must look at inter-generational case in points of procedure. With a center grants assessment of succession practice, this is a significant component of the system in the family firms. Despite of the fact that the small- and medium-scale organizations, be significant in favor of monetary development of nation. Ongoing writing indicated out that they have face distinctive management issues, for example, general management of the organization, the executives of abilities, and furthermore the board of succession process.

In this chapter, aspects identified with the management issues and succession, the management issues assume an urgent role. When getting some information about key objectives and their significance for the OD accompanied such outcomes; the three most significant objectives to those respondents are an amplification of offer development, a boost of the organizations worth, and furthermore money related freedom of the family.

As pointed out by the consequences of a research completed on family firms, the principal management issues arranged by significance are:

- to give monetary freedom to the family;
- to boost deals development;
- to boost the company's worth;
- to give advancement and development of the organization;
- new interests in the organization and its items and administrations; and
- to verify the life span of the organization.

15.3 METHODOLOGIES TOWARD SUCCESSION PLANNING

An organization way to deal with SP will impact the thought of the framework's adequacy. No single or straight approach is recommended, given any methodology has taken must be pertinent to the agency specific situation and requirements. Albeit each workplace should consider those conditions applicable to their own organization to decide suitable SP arrangements and procedures, the next might be considered.

Key advances may include:

1. devising the procedure
2. securing tactical assimilation
3. evaluating the existing state of affairs
4. identifying as well as evaluating high potential
5. execution: scheduling and enterprise improvement
6. incorporate Opportunities for Feedback and Regular Review
7. evaluation: a few contemplations.

Every parameter is talked about in the accompanying areas. The areas mean to give various key inquiries to think about when structuring and actualizing SP.

15.3.1 DEVISING THE PROCEDURE

Inquiries to consider include:

- Have they characterized a business case for SP? Do they have the planned initiative quality required inside the organization?
- Is the procedure "possessed" by senior administrators and bolstered by line managers?
- Have they thought about how information emerging from the procedure will be overseen and whether innovation will be utilized to help the program? Consider the degree to which the HR zone will deal with the procedure.
- Have they guaranteed the procedure is straightforward and considered the degree of detail at which the succession management approach will be imparted over the organization (being mindful so as to keep up the privacy of individual subtleties)?
- Have they manufactured criticism frameworks to staff into the procedure? These might incorporate conveying the procedure in all terms just as nourishing back explicit data to singular workers as fitting.
- Does the methodology center on the key improvement of capacities?
- Are the results of the procedure quantifiable after some time?
- How will they assess the methodology as far as authoritative effect?

15.3.2 SECURING TACTICAL ASSIMILATION

From the business and workforce planning structure, recognize the critical roles to organization prosperity just as those liable to rise as basic over the medium to the long term. Consider the workplace socioeconomic and repercussions of varying socioeconomics on the contribution of initiative competitors.

Inquiries to consider include:

- Is the methodology adjusted and incorporated with other human asset forms? For instance:
  - o Are the distinguished capacities consolidated into development planning?
  - o Is the procedure coordinated with the performance management system?
  - o Have they tended to the development needs of managers who will encourage the procedure and mentor people?
- What are the few of particular leadership capabilities that will enable the organization to be more compelling than its (potential) rivals? Which ones are not quite the same as today? Consolidate future business needs, basic achievement factors, qualities, techniques, and anticipated difficulties in this investigation.
- In taking a medium- to long-term view, is the succession the management procedure initiating at any rate one to two levels beneath the objective jobs to manufacture the required capacity? That is, would they say they are taking a gander at future ages of pioneers and not simply current "feeder gatherings" (in the event that they exist) to key role?

15.3.3 EVALUATING THE EXISTING STATE OF AFFAIRS

- Perform a peril evaluation of prospective departure as of obtainable basic jobs.
- Draw on a statistic investigation spilling out of the more extensive workforce planning structure.
- Project future staffing prerequisites in critical roles, taking a gander at the interior and outside components (counting enrollment and maintenance examples) and recognize most pessimistic scenario situations. From this, decide the degree of any pending initiative lack

by anticipating necessities; inward versatility and wearing down throughout the following five years, for example, consider the current 'bench strength'.

- Analyze the hole between the current ability for key jobs and future necessities, and recognize methodologies for shutting the hole. Systems could incorporate, for instance, interior ability development, outer enlistment to target specific quick aptitudes gaps or exceptional projects to select and create authorities. The money-saving advantage contemplations of these methodologies would be surveyed.

15.3.4 IDENTIFYING AS WELL AS EVALUATING HIGH POTENTIAL

- Clarify the obligation regarding recognizing and evaluating potential. Most performance management frameworks center around discussion between managers and their staff about what performance is expected and how the staff is performing. Succession management, similar to career development, is not in every case best-taken care of by prompt managers. Effective succession management requires efficient contribution by managers all through the performance management framework.
- Define what "high potential" signifies inside the workplace setting and inside the setting of the recognized basic hierarchical roles (e.g., the abilities, learning, and individual characteristics vital for superior).
- Focus on the exact identification. So as to limit subjectivity however much as could be expected and to guarantee a careful methodology, utilize numerous approaches to survey plausible along with recognize representatives who might conceivably occupy the distinguished positions. Utilize the performance management framework as a beginning stage for evaluating performance, potential and developmental needs. Appraisals should be exhaustive and proof based.

Important data could include:

- obiographical information
- ocurrent performance
- oobserved behavior
- oadaptability
- o360° feedback and formal appraisal results
- ointerviews to decide career inclinations and self-observations

- oassessment of the probability of remaining with the workplace
- obehavioral meetings to decide the past performance in testing circumstances
- oviews of the scope of ranking managers performance and the overall estimation of specific attributes, (this could incorporate the perspectives on the administrator once-evacuated where this framework is set up in the organization or a senior level audit body)
- opsychometric testing or other outside evaluation devices.

On the off chance that it is consistent with the general methodology taken by the organization, consider furnishing people with chances to self-nominate and to express an enthusiasm for seeking after positions of authority dependent on their own inclinations and targets; they would then be able to be evaluated by the data parameters recognized previously. So as to add to the accomplishment of assorted variety objectives, offices could likewise empower articulations of enthusiasm from a various scope of representatives.

15.3.5 EXECUTION: PLANNING AND UNDERTAKING IMPROVEMENT

- Kinds of tasks or know-how which might be tendered at the same time as quickened advancement openings focused against future business need instead of explicit employments; join presentation to a scope of testing encounters as a key component.
- Consider whether to assign specific arrangements of obligations as "development role" inside the organization, given that fewer levels and more extensive ranges of control make it harder to orchestrate formative assignments for individuals with high potential.
- A further issue to consider is whether to utilize a "co-manager" system to slide older managers into retirement while getting ready new pioneers, to add to the exchange of learning and judgment.
- Establish separately customized development plans. Development plans can be founded on results from 360° criticism or the outcomes from an improvement focus or other demonstrative instruments, and can consolidate factors, for example, singular ability necessities, foreseen job difficulties, required authoritative information and comprehension, and individual variables.

In SP, development plans normally include:

- challenging work-based experiences chosen by senior pioneers as an development procedure (counting work pivot, unique assignments and cross-practical inclusion including exceptional activities or teams);
- exposure to the vital motivation and to senior authorities of the organization;
- well-focused on preparing (e.g., official improvement projects or formal abilities preparing where fitting);
- self-development methodologies; and
- use of senior mentors.

15.3.6 INCORPORATE OPPORTUNITIES FOR FEEDBACK AND REGULAR REVIEW

- Clarify particular duties. The organization ought to recognize the looked for after abilities, be straightforward and predictable in actualizing the presentation the management framework, give access to formative assignments, and give direction and criticism. People are in charge of their profession and responsible for gathering formative destinations and picking up and exhibiting new abilities. It is imperative to take note of that while organization and manager backing and contribution are basic to the effective performance of a development plan, the individual must assume extreme liability for gathering formative objectives and keeping up the nature of their presentation.
- Another issue to consider is that development plans are frequently conceived however not implemented. An elective methodology might be for the person to make a rundown of improvement needs related to their supervisor (and director's administrator) and mentor and make it the subject of ordinary audit.
- Identify how candidates will be considered responsible for their improvement. Think about how regularly to survey and catch up on development plans and whether reward structures are lined up with advancement in gathering formative objectives. Think about whether those workers who have been offered access to concentrated improvement openings will be offered proceeded with access if their presentation is not kept up for the time being: will they be offered the most testing assignments if performance is peripheral in the present performance cycle, despite the fact that they are attempted

development action and perhaps confronting a precarious expectation to absorb information?

15.3.7 EVALUATION: A FEW CONTEMPLATIONS

Set up clear time spans for actualizing and assessing of the methodology and its results. These issues should be considered in advance by both the organization and the individual(s) concerned. For the workplace, assessment of results could be made as far as whether hierarchical hazard erstwhile decreased otherwise limited. For the entity, this possibly will incorporate contemplation with reference to the level of capacity growth and showed modification during recital and conduct into the working environment.

Monitor the progression managing framework itself—intermittent assessment at this level could incorporate evaluating progress on individual improvement designs, the level of inclusion of current pioneers or senior officials and the extent of the interior to outer arrangements. Casual audits could be planned quarterly as opposed to depending on an enormous scale yearly process.

15.4 CONCLUSION

In the present focused business condition, it is more basic than any other time in recent memory to have the perfect individuals set up all through the organization. Moreover, on behalf of this to be agile sufficient to quickly restructure the capacity ought to another industry prospect rise or one of the significant performers out of the blue pulls back. Nowadays, 60% of the organizations with high impact talent management research do not have strategy for SP and just around 1/4 encompass an endeavor extensive loom. Substantially more unusually, organizations through an undertaking extensive loom have recently it for a typical of 1.6 years, speaking to the shocking improvement into this critical bit of talent management. The call for SP for empowering SMEs is emerging to stretch out past the attention on the management and positions of leadership which is ending up increasingly significant as organizations find a way to manufacture superior and high contribution workplaces wherein basic leadership is spread out and activity is disseminated all through with empowered as well as varying workforce. The reason for the research is to investigate the ascent and hugeness of succession management in SMEs and how the associations are executing put into practices and thusly checking their comprehensive development in long run. The research in like manner include getting data about SP and the board advancement in SMEs and in endorsing best progression making courses of action for these undertakings for their wide-extending development in the economy. The confinement of the exploration is related to the literatures which show enormous absence of the study on the evaluation of SP and the management improvement programs in SMEs. Subsequently, literatures open on the present point are constrained lying on a worldwide perspective instead of an Indian perspective. On the basis of this, the present study grasped to find why SMEs do not get by past the instigators and owners and their significance for its development. In this manner, there is the prerequisite for SP in SMEs.

KEYWORDS

- **succession**
- **succession planning**
- **tactical goals**
- **SMEs**

REFERENCES

Alcorn, P.B. (1982): Success and Survival in the Family-owned Firm. McGraw-Hill, New York.

Ashanda, S. M. (2015): Factors Influencing Strategic Succession Planning for Small & Medium Family Enterprises. Thesis and Dissertaion.

Babu, G. R., Harinarayana, CH. and Chandraiah, M. (2014): Technology development in MSMEs. Indian Streams Research Journal, 4(10).

Bennedsen, M., Nielsen, K. M., Perez-Gonzalez, F., and Wolfenzon, D. (2007): Inside the family firm: the role of families in succession decisions and performance. Quarterly Journal of Economics 122, 647–691.

Blaskey, M.S. (2002): Succession planning with a business living will. Journal of Accountancy, 193(5), 22–23.

Brockhaus, R.H. (2004): Family business succession: suggestions for future research. Family Business Review, 17(2), 165–177.

Carter, N. (1986): Guaranteeing management's future through succession planning. Journal of Information Systems Management, 3(3), 13–14.

Churchill, N.C. and Hatten, K.J. (1987): Non-market-based transfers of wealth and power: a research framework for small business. American Journal of Small Business, 11(3), pp. 51–64.

Diamond, A. (2006): Finding success through succession planning. Section Management, 50(2), 36–39; 4 (10), 2140–2149.

Elżbieta Roszko-Grzegorek (2008): The role and importance of succession planning in Polish family firms. Acta Universitatis Lodziensis, Folia Oeconomica 224.

Erven, B.L. (2004): Management Succession Issues in Family Business. Available at http://www.famiz.com/Orgs/Cornell/articles/real/erven.cfm.

Friedman, S. D. (1984): Succession systems and organisational performance in large corporations. Unpublished doctoral dissertation, The University of Michigan, Ann Arbor.

Friedman, S. D. (1987): Leadership Succession. New Brunswick, NJ: Transaction Books.

Garman, A.N. and Glawe, J. (2004): Succession Planning. Consulting Psychology Journal: Practice and Research, 56(2), 119–128.

Greer, C.R. and Virick, M. (2008): Diverse succession planning: lessons learned from the industry leaders. Human Research Management, 47(2): 351–367.

Groves, K. S. (2007): Integrating leadership development and succession planning best practices. Journal of Management Development, 26(3): 239–260.

Hall, D. T. (1986): Dilemmas in linking succession planning to individual executive learning. Human Resource Management, 25(2), 235–265.

Handler, W.C. (1990): Succession in family firms: a mutual role adjustment between entrepreneur and next generation family members. Entrepreneurships Theory and Practice, pp. 37–55.

Handler, W.C. (1994): Succession in Family Business. A Review of the Research Family Business Review, SAGE Publications, 7; 133.

Handler, W.C. (1994): The succession experience of the next-generation. Family Business Review, 5(3), pp. 283–307.

Härtel, C. E.J., Bozer G. and Levin, L. (2008): Family business leadership transition: how an adaptation of executive coaching may help. Journal of Management and Organisation, 15(3), 378–391.

Harvey, M. and Evans, R.E. (1995): Life after succession in the family business: is it really the end of problems? Family Business Review, 8(1), pp. 3–16.

Hunte-Cox, D. E. (2004): Executive succession planning and the organisational learning capacity. Unpublished Doctoral dissertation, Washington, DC: The George Washington University.

Ibarra P (2005). Succession planning: an idea whose time has come. Journal of Public Management, 87(1), 18–24.

Ibrahim, A.B., Soufani, K. and Lam, J. (2001): A study of succession in a family firm. Family Business Review, 14(3), 245–58.

Ibrahim, A.B. and Ellis, W. (2004): Family Business Management: Concepts and Practice. 2nd Edition, Kendal/Hunt, Dubuque, IA.

Janjuha-Jivraj, S. and Woods, A. (2002): Successional issues within Asian family firms: learning from the Kenyan experience. International Small Business Journal, 20(1), 77–94.

Johnson, L.G. and Brown, J. (2004): Workforce planning not a common practice. IPIMA-HR Study Finds, Journal Public Personnel Management, 33(4).

Karaevli, A. and Hall, D. T. (2003): Growing leaders for turbulent times: is succession planning up to the challenge? Organizational Dynamics, 32, 62–79.

Kesner, I. F., and Sebora, T. C. (1994): Executive succession: past, present & future. Journal of Management, 20, 327–372.

Lansberg, I. (1988): The succession conspiracy. Family Business Review, 1(2), 119–143.

Martin, L. (2001): More jobs for the boys?: Succession planning in SMEs. Women in Management Review 16(5): 222—231.
Miller, D., Steier, L. and Le Breton-Miller, I. (2003): Lost in time: intergenerational succession, change and failure. Journal of Business Venturing, 18(4), 513–531.
Obadan, J.A. and Ohiorenoya, J. O. (2013): Succession planning in small business enterprises in EDO State of Nigeria. European Scientific Journal, 9(31), 64–76.
Poutziouris, P. (1995): The development of the family business, Proceedings of the Institute for Small Business Affairs Workshop on Family Firms. Manchester: The University of Manchester Business School, Manchester.
Rhodes, D. W., and Walker, J. W. (1984): Management succession and development planning. Human Resource Planning, 17, 157–173.
Rothwell, W. J (1994): Effective succession planning: ensuring leadership continuity and building talent from within. New York: AMACOM.
Rothwell, W.J. (2002): Putting success into your succession planning. Journal of Business Strategy, 23(3), 32–38.
Rothwell, W. (2005): Effective succession planning: Ensuring leadership continuity and building talent from within. 3rd edition, New York: Amacom.
Rothwell W. J (2010): Effective succession planning, ensuring leadership continuity and building talent from within. 4th edition, New York: American Management Association.
Sambrook, S. (2005): Exploring succession planning in small, growing firms. Journal of Small Business and Enterprise Development, 12(4), pp. 579–594.
Sharma, P., Chrisman, J.J., Pablo, A.L. and Chau, J.H. (2001): Determinants of initial satisfaction with the succession process in family firms: a conceptual model. Entrepreneurship Theory and Practice, 25(3), pp. 17–35.
Shen, W., and Cannella, A. A. (2003): Will succession planning increase shareholder wealth? Evidence from investor reactions to relay CEO successions. Strategic Management Journal, 24, 191–198.
Stavrou, E. (1999): Succession in family business: exploring the effects of demographic factors on offspring intentions to join and take over the business. Journal of Small Business Management, 37(3), pp. 43–61.
Steele, P (2006): Succession Planning. White Paper (Revised), Regis Learning Solutions.
Stinchcomb, J.B., McCampbell, S.W., and Leip, L. (2009): The Future is Now: Recruiting, Retaining, and developing the 21st Century Workforce. Washington, DC: U.S. Department of Justice, Bureau of Justice Assistance.
Walker, J.W. (1998): Perspectives: do we need succession planning any more? Human Resource Planning, 21(3), pp. 9–11.
Wortman, M.S. (1994): Theoretical foundations for family-owned business: a conceptual and research-based paradigm. Family Business Review, 7(1), pp. 3–27.
Zaich, L. L. (1986): Executive succession planning in select financial institute. Unpublished doctoral dissertation, Pepperdine University.
Zajac, E. J. (1990): CEO selection, succession, compensation and firm performance: A theoretical integration and empirical analysis. Strategic Management Journal, 11(3), 217–230.

CHAPTER 16

Recent Financial Reforms: A Case Study on Developing and Deploying an Income Prediction Model for a Personal Loan Portfolio

GAURAV NAGPAL,[1*] ANKITA DHAMIJA,[2] and DIVYANSH GUPTA[3]

[1,3]*Birla Institute of Technology and Science, Pilani, Rajasthan 333031, India*

[2]*Lingaya's Lalita Devi Institute of Management & Sciences, Delhi 110047, India*

[*]*Corresponding author. E-mail: gaurav.nagpal@pilani.bits-pilani.ac.in*

ABSTRACT

The core business of business-to-business-to-consumer financing organizations is in the field of corporate lending, housing finance, consumer finance, and asset management. The objective behind this project is to build predictive model for Personal loan Portfolio and deploying it at the frontend afterward. Although with the help of this model, it will be easy to predict income of future as well as the existing customers in Personal loan portfolio. This model was further deployed on the cloud-based server of the company with the use of an application programming interface. Hence, when any new customer applies for a loan, the model will be triggered, and the predicted income of the customer will be generated. Hence, this predictive model will help in checking the financial health of the customers. This study touches upon Data Pulling and making it available for analysis followed by Data Preparation; Model building with the help of Machine Learning Algorithms; Out of time testing; Model Testing; Model Deployment, and Model re-calibration. Due to the sensitivity of the data worked upon and the exact model worked out, this chapter does not disclose the name of the organization where this model was deployed, and also does not mention the technical output of the model.

16.1 INTRODUCTION

The business-to-business-to-consumer (B2B2C) lending organizations may have many live partners which include small fintech startups, large digital wallets, international electronics companies, and new-generation tech-first nonbanking finance companies (NBFCs). These partners provide products which include student loans, consumptions loans, healthcare loans, personal loans, and many more. Some of the live partners may be repeat borrowers which signal the commitment of such organizations to build long-term relationships.

The way such organizations operate, the partners act as front-end demand aggregators and solve the problem of the customer while these organizations act as the credit underwriter and balance sheet partner.

Being a fully fledged NBFCs, such organizations have many departments, that is, operations, credit, collections, data science, legal, compliance, IT and many more.

(1) *Operations:* The operations department is responsible for smooth and efficient working which results in profitability. The work of operations department is to oversee daily operations activities of different partners which include application collection, application verification. They are also tasked with doing checks on the information provided pertaining to the loans.
(2) *Collections:* The collection team is responsible for, that is, controlling defaults in loans. They are also responsible for controlling NPA and recovers dues from written off portfolio of the respective zone. This includes setting up strategy for collection of amounts from the clients, loan takers. They are responsible for maintaining the health of the portfolio to its best.
(3) *Data science*: The data science department mostly deals with the data that is collected and use this data to provide valuable insights and helps in policy-making process. This data is also used in developing predictive models for different portfolios. Data science department also deploys the different predictive models. Also, data science department creates different types of reports for collection team, operations team, etc.

16.2 PROBLEM STATEMENT

Digital framework is meant to deliver personalize financial production with a simple experience to customers, through multiple digital channels in a transparent and secure manner to disrupt the Indian financial service market. This requires a strong foundation of automation and data science. Data science is a multidisciplinary blend of data cleaning, data analysis, modeling/statistics, and prototyping.

Fintech aggregators try to minimize information they seek from customers during the application process. Given this, creating a model to predict customer's income or predicting future delinquency status of customers accurately becomes sacrosanct. Also, models were generating application risk code developed for different product type like consumption, student, personnel, etc.

Personal loan portfolio is one of the major portfolios at such organizations. This portfolio covers broad spectrum of loans like loan for medical expenses, marriage, family need, investment, etc. Also, this portfolio has a very broad range of loan amount which can vary from a few thousand rupees up to several crores' rupees. So, we can say that this is a very diverse portfolio with different types of customers.

Being the Fintech aggregators, these organizations try to minimize the information they seek from customers at the time of loan. Thus, there are many customers who do not provide their income or the provided income by them varies significant from the actual income. So, it becomes very important to have an estimate of income of all the customers. This estimated income can be used as their actual income for those customers who did not provided theirs at the time of application for the loan and can be compared for those customers who have provided their income in the income statement.

So, for finding the estimated income, predictive modeling is used. It employs different type of processes like data cleaning, data exploration, and many types of fitting and boosting methods which will eventually result in best estimate.

After the model development phase, this model is going to be deployed on the cloud-based server of the company [which in this case is Amazon Web Server (AWS)].

Deploying the predictive model is basically making it accessible to relevant people and departments in the company.

Deployment of the model includes developing a web application programming interface (API). This web API is used as a container in which the developed model is stored. The predictive model is calibrated in such a way that it pulls the data of each independent variable used in the model with the help of an URL from the cloud-based server where the information of the customers is stored. Whenever a customer applies for a loan, the model will be triggered. The deployed model then predicts income for that customer.

16.3 OBJECTIVE

The idea to develop an income prediction model occur because earlier it was thought that income based from customers records were correct. But, it was seen that the default rate was high among the high earning customers which did not make any sense. So, it was concluded that reliable estimates of a customer's income are very important in today's world. The company wanted to put a structured framework, so that the dependable income estimates can be made.

To achieve this objective, the following steps were formulated to make an income prediction model.

16.3.1 END-TO-END MODEL

- Data pulling and making it available for analysis followed by data preparation.
- Model building with the help of machine learning (ML) algorithms.
- Out of time testing.
- Model testing.
- Model deployment.
- Model recalibration (if required).

16.3.2 METHODOLOGY

This section will broadly explain the processes which are going to be used in the developing as well as deploying the predictive model. The work of this project is mainly done on the four softwares, that is, MS-Excel, Python, SQL workbench with the database as Amazon Athena, and the testing is done on R-Studio software.

The main software used to write the model development as well as deployment script is Python. Python is an interpreter, high-level programming language for general purpose programming. It supports multiple programming paradigms including object-oriented, imperative, and functional and many more. Python was considered best for code writing as it is an open-source language with a vast compatible functional library. These libraries provide wide range of functionality like graphical user interfaces, web frameworks, multimedia, databases, networking, test frameworks, and automation. Some of the important functional libraries are pandas, NumPy, train-test split, and many more. During the development phase, these libraries were extensively used.

Apart from that, R-Studio was used for model building. Model building means testing the uniform dataset which was created after treatment in Python, with the ML algorithms. The reason to use R-Studio was that it provides better estimation results in comparison to Python in-built libraries.

16.4 MODEL DEVELOPMENT

This is the process for developing the predictive modeling. The main motive of this whole exercise is to determine if the variables in already existing data can influence and provide some good predicted value of the independent variable. Model development includes various steps and each step involves many processes.

The steps that will be used in this model development are as follows:

- Data pull and data preparation.
- Data cleaning.
- Data exploration.
- Model building with ML algorithms.

16.4.1 STEP 1: DATA PULL AND DATA PREPARATION

This is the first step in the process of estimating income. Data of customer is being pulled from different sources with the help of information provided by them at the time of application. This data contains many types of variables relating to customer, which are as follows:

- *Bureau variables:* These include various types of data of customer like bank-related data, debt-related data, different types of scores (like CIBIL), and many more.

- *Demographic variables:* These include variables which are socio–economic for a customer like age, gender, city, state, contact information, and many more.
- Other variables that are used are like reason for taking loan, number of dependents, work place information, etc.

After accumulating all the required information, the data is being prepared by merging and joining all the data variables, so that it will become one uniform dataset.

16.4.2 STEP 2: DATA CLEANING

Now, after the data that has been prepared contains different types of formats, spelling mistakes, nonuniform population in several variables, information present in different typography, that is, lowercase and uppercase which can create duplicates for a single observation (e.g., *Mumbai, MUMBAI, mumbai* all will be consider as different values but, they are same) and many more. So, this dataset cannot be used for model development as model building is only possible on uniform dataset which do have a single format. Many types of technics are there which can be used to remove these errors and make the data uniform.

1. *Removing typo errors and correcting typography:* After inspecting the prepared dataset, there are many variables which do have typo errors like spelling mistakes, insertion of special characters (e.g., &, !, „ *, and %) and presence of lower- and uppercase. So, if these types of characters are present, they will increase the number of values in a variable. These errors are removed from the dataset with help of codes and other different techniques.
2. *Data imputation:* This method is used to impute the blank or missing population in the dataset with the use of appropriate value. These values can either be mean, median or anything else depending on the trend shown in the dataset.
3. *Finding an outlier in the variables:* Outliers are the points that are inconsistent with the rest of the dataset as in a way these points are very distinct from the rest of the points; thus, they can bias or skew any analysis performed. So, it is very important to remove these points from the dataset. Thus, the performance will remain unbiased at the time of prediction or estimation.

4. *Creating binary for different variables:* These are newly created variables on the existing variables. These existing variables do contain some type of repeating values, so all the repeating values are defined in a single class (like making one repeating value as 1 and another repeating value as 0). So, it becomes very easy to use these variables at the time of data exploration.

16.4.3 STEP 3: DATA EXPLORATION

This is the next step is model development after the data is cleaned and have been uniformly populated. In this step, usually trends and some type of correlation is tried to achieve among the dependent and the independent variable. Thus, in this step, many types of functions are used.

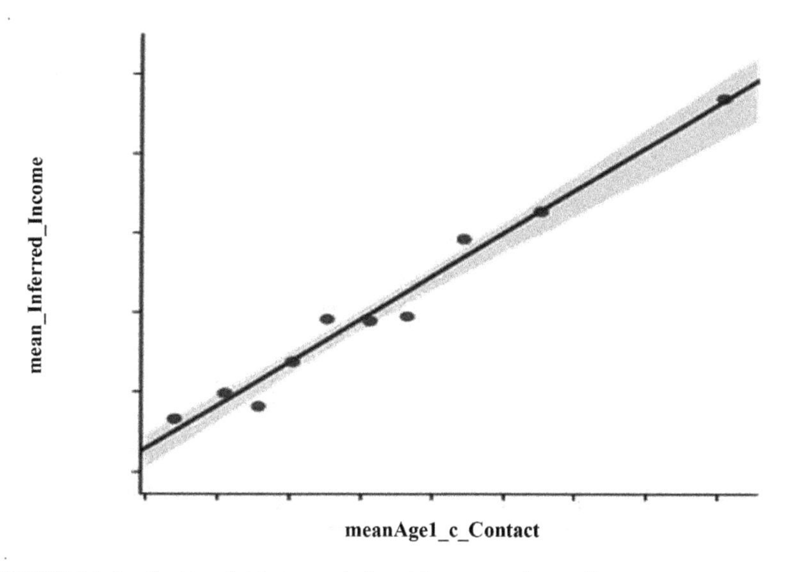

FIGURE 16.1 Scatter plot between inferred income and age of customer.

Figure 16.1 is the example of type of trends which are generated for dependent (inferred income) and independent variables (age of the customer) in data exploration step. In this trend, there is a very good correlation between dependent and independent variable. But, all the correlations are not that good.

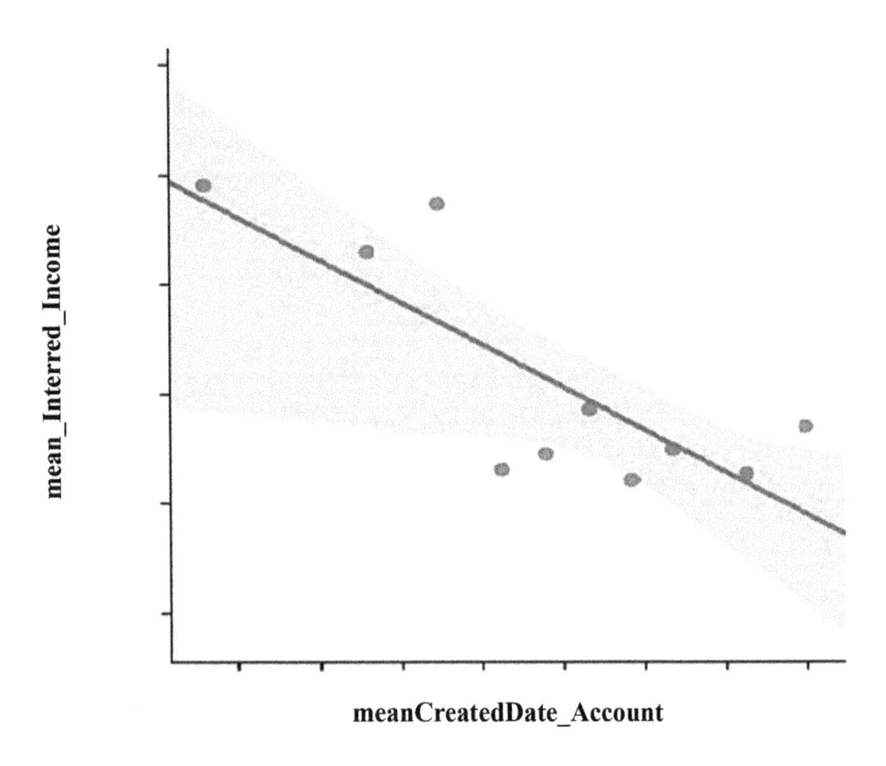

FIGURE 16.2 Scatter plot between inferred income and age of account.

For example, in Figure 16.2, the linear trend is not the best as there is no direct correlation. Thus, the different types of trend like polynomial, power, exponential, and logarithm are tried and then the one which shows best correlation is used in the model development.

The best fit and trend is tried to observe and the function which can provide, is used in the next step, that is, model development.

16.4.4 STEP 4: MODEL BUILDING WITH ML ALGORITHMS

This is the final step in model building/development process. In this step, different types of techniques and methods are employed to find the optimal method which can provide the best result on training dataset. ML is a very important tool in model development. It has a wide variety of tests like random forest regressor, generalized linear model (GLM), gradient boosting method, ada boosting method, and so on, which are employed to predict the estimated income on the test or experimental dataset. For building this model, ML algorithms of random forest regressor, and GLM were giving the best results.

Random forest regressor is one of the most effective algorithms for predictive analysis. This is an ensemble algorithm which reduces high variability by using bagging technique. It determines the statistical relationship between two or more dependent variables, where change in dependent variable is associated with, and depends on, a change in dependent variable. The way it works is it reduces the variance of the model by the way of averaging the prediction.

GLM is a very effective tool in solving problems relating to logistic regression. This model is in a form of a simple linear equation with an independent variable, a dependent variable with a coefficient multiplied, and a constant intercept. But in our dataset, the dependent variables are not linearly correlated with the independent variable, so there is a function called link function which transforms the dependent variables, so that they can be linearly related to independent variable. The link function used in model building is called logit function. While developing the income prediction model for personal loan portfolio, random forest regressor and GLM are used but the best results are achieved through GLM technique.

16.4.5 STEP 5: MODEL DEPLOYMENT

Model deployment requires developing a web API. This web API can be then hosted on a cloud which then can be automated, so that the model developed can run uninterrupted and can be used any time.

The first process is to develop a web API. API is short for API. Web API are the tools which are used to make information and application functionality accessible over internet. This web API will work as a container in which the developed model can stored. For this model, deployment flask API is used.

What is Rest: Rest is the acronym for representational state transfer. It has emerged as the standard architectural for web server and web APIs. It is a way of providing interoperability between computer systems on the internet. Rest was originally designed to fit the HTTP protocol that the World Wide Web can use.

Flask API: Flask is a Python-based framework that enables us to quickly build web application including managing HTTP requests. The advantage of using flask is comparison to other web framework it tends to be written on a blank canvas, so it is well suited to a contained application.

Flask maps the HTTP request to the Python functions, the internal flask server is like http://127.0.0.1:5000/, it tries to check if there is a match between path provided and the defined function. After that flask runs, the contained Python script and display the obtained result on the browser.

16.5 WRITING MODEL DEVELOPMENT SCRIPT

From the earlier stage of predictive model development, not all the variables which are pulled during data pulling and preparation step are used in the later stage of model development. There were some important variables which were affecting the characteristics of model. So, the first step is to identify those variables.

These variables are going to be pulled from the database which is present on the cloud with the help of URLs. The data is received from these URLs is in JSON format which cannot be directly used as the input for the model development script. So, this data is then stored in a data frame rather than a JSON file.

16.5.1 FOR NUMERIC VARIABLES

As we know from the model development stage, the predictive model takes uniform input, that is, all the variables should have some value. The predictive model will give error, if we try to input a variable which has NULL value. So, the data cleaning step which we used earlier has to be used now. The data cleaning should be done in the same way and in the same order as it was done earlier. So, first all the typographic errors have to be removed and if the special characters like (! @, #, & etc.) are present should be removed. All the upper- and lower-case alphabets have to be converted in a uniform format, that is, either all lowercase or all uppercase. After this step, missing value imputation is performed; the value should be the same as used in the model development process. After this step, outlier treatment is performed, if there are outliers present in the variable, they are removed, thus the variable become unbiased at the time of prediction or estimation. Now, the next step is to find out if any of the variables used in the predictive modeling uses a different feature or trend, that is, if the variable is showing better trend with polynomial, power, exponential function in comparison to linear function. Hence, these variables are then converted into their cross-ponding function.

16.5.2 FOR CATEGORICAL VARIABLES

Categorical variables are those whose values are not exactly numeric, their values can be form of alphabetical or combination of alpha–numeric. So, they are not directly used in the model development process. Instead, they are used to create new variables (binary flag variables), this new variable is defined in such a way that if the variable has influencing value, it is assigned as 1 otherwise it is assigned as 0.

After this step, we have converted every variable in the form of a number or an integer. Thus, they are now ready to be use as an input for our GLM model.

16.6 CREATING GLM ALGORITHM IN DEPLOYMENT SCRIPT

The predictive model, we developed earlier for personal loan portfolio, give the best result by using GLM model. The GLM model is implemented in R rather than used in Python. Even though Python has its own in-built GLM model, it does provide different results in comparison to the GLM Model in R. So, we have to implement the GLM model of R in Python.

GLM model can be shown as linear equation. It can be represented as shown above.

The Yi gives us the predictive coefficient. By raising Yi as the exponential power, it will give us the predicted income.

16.7 RESULTS AND DISCUSSIONS

For the model development part, best results were obtained from the GLM algorithm.

The graph in the figure (Figure 16.3) is obtained for the predicted and the actual income for the training dataset. We can see that there is a very good correlation between the predicted and the actual income which is a very good indicator and proves that dataset is fitting perfectly on the training dataset.

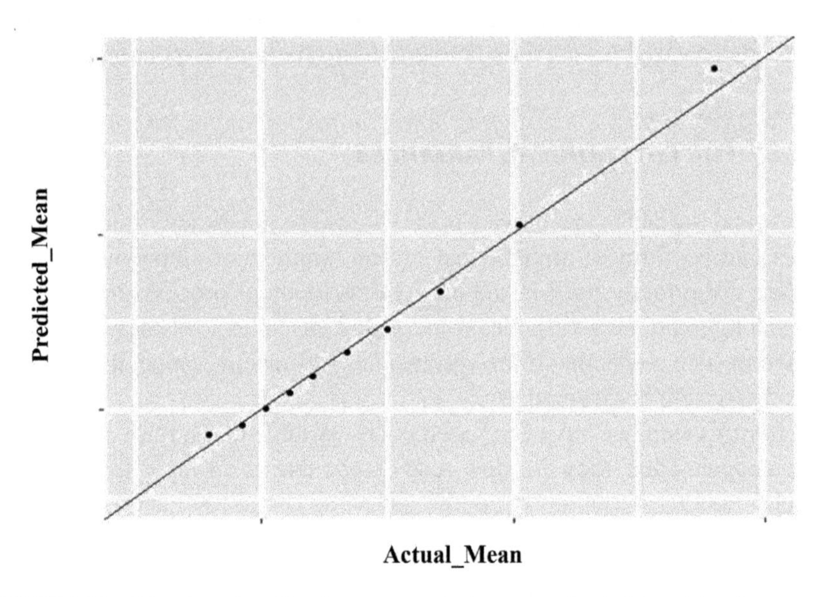

FIGURE 16.3 The plot of predicted mean vs actual mean on the training dataset.

Figure 16.4 is the graph showing the correlation between the predicted and the actual income on the test or experimental dataset.

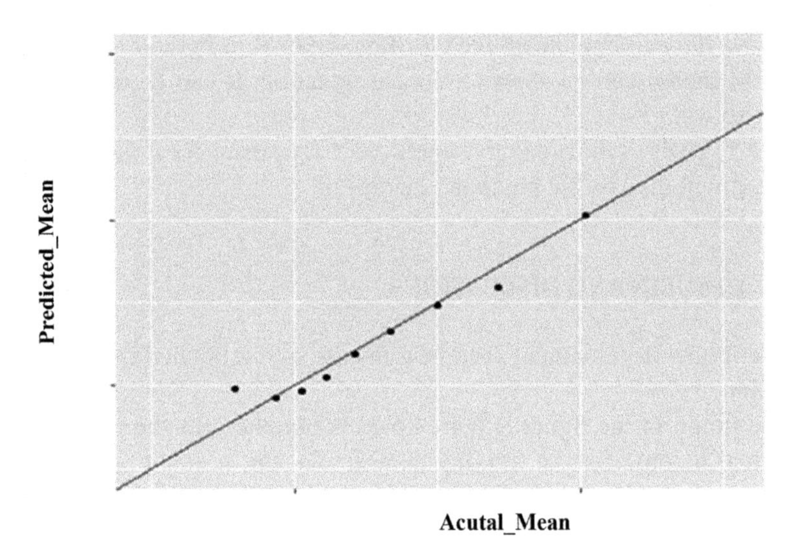

FIGURE 16.4 The plot of predicted mean vs actual mean on the experimental dataset.

This graph also follows the same trend as the graph of training dataset which proves that the model that has been developed is very fair in predicting the income of the customers.

These two graphs showed very good prospects for the model. Hence, after that, model was decided to be deployed.

The model is deployed on the company's cloud which is AWS. It is working on the company's internal network and is automated. Whenever a customer applies for a loan, the model is triggered to generate an income prediction which is pushed to the loan management system of company. Apart from the income prediction, model will generate a decile ranging from 1 to 10, which indicates if a customer is a low- or high-income customer. All the data underlying the model would also be populated in the loan management system by the API created for the purpose of the project. This will enable easy and effective periodic testing of the model.

Hence, it can be concluded that the model deployment solution is a complete tool itself, enabling income model prediction and testing.

KEYWORDS

- **model testing**
- **model deployment**
- **model recalibration**

REFERENCES

[Online] https://docs.python.org/2/library/math.html.
[Online] https://docs.python.org/2/library/warnings.html.
[Online] https://en.wikipedia.org/wiki/JSON.
[Online] https://scikit-
[Online] https://scikit-learn.org/stable/modules/generated/sklearn.metrics.r2_score.html.
[Online] https://www.pythonforbeginners.com/requests/using-requests-in-python.
[Online] https://www.w3schools.com/python/python_json.asp.
Building and Documenting Python REST APIs With Flask and Connexion, [Online] https://realpython.com/flask-connexion-rest-api/.
Designing a RESTful API with Python and Flask, [Online] https://blog.miguelgrinberg.com/post/designing-a-restful-api-with-python-and-flask.
Generalized Linear Model, [Online] https://en.wikipedia.org/wiki/Generalized_linear_model.
Generalized Linear Models, [Online] http://www.statsoft.com/Textbook/Generalized-Linear-Models.

Introduction to Generalized Linear Models, Penn State Eberly Collage of Science, [Online] https://onlinecourses.science.psu.edu/stat504/node/216/.
JSON, learn.org/stable/modules/generated/sklearn.model_selection.train_test_split.html
Mathematical functions, NumPy, [Online] http://www.numpy.org.
NumPy, [Online] https://en.wikipedia.org/wiki/NumPy.
pandas (software), [Online] https://en.wikipedia.org/wiki/Pandas_(software).
Pandas, [Online] https://pandas.pydata.org/.
Python JSON, R2 score from sklearn metrics, Random Forest Regression, [Online] https://turi.com/learn/userguide/supervised-learning/random_forest_regression.html.
Random Forest Regressor from scikit learn, [Online] https://scikit-learn.org/stable/modules/generated/sklearn.ensemble.RandomForestRegressor.html.
Random Forest, [Online] https://en.wikipedia.org/wiki/Random_forest.
SmythPatrick,CreatingWebAPIswithPythonandFlask,[Online]https://programminghistorian.org/en/lessons/creating-apis-with-python-and-flask#what-flask-does.
Train-Test Split from sklearn model selection, Using the Requests Library in Python, Warning control

CHAPTER 17

Effects of Career Planning and Development on Employee's Performance: A Case Study of Equity Bank Rwanda Public Limited Company

SINGH SATYENDRA NARAYAN[1*], SETH SHYIRAKERA MUNYANZIZA[1], AND SONAM RANI[2]

[1]*University of Kigali, KG 541 St, Kigali, Rwanda*

[2]*GL Bajaj Institute of Management, Greater Noida 201306, Uttar Pradesh, India*

[*]*Corresponding author. E-mail: sonamrani0908@gmail.com*

ABSTRACT

The chapter intended to measure employee performance through career planning and development, a case study of Equity Bank Rwanda PLC (Public Limited Company) in the period from January 2017 to January 2019. The research tool used by the researcher for data collection is primary data and secondary data. The research design is a survey research design and the data collected from both sources used to interpret the variables of the articles. The population of the study involved Equity Bank Rwanda employees and the researcher used Microsoft Excel sheets to interpret data collected. The result of the research shows the positive relationship between career planning and development and the employee's performance with the Equity Bank Rwanda PLC. The researcher concluded that there is a positive effect of career planning and development on employee's performance within the Equity Bank Rwanda PLC.

17.1 INTRODUCTION

Nowadays the terminology of work has been changing, and due to these changes in work terminology, the job of career counselors and clients becomes more challenging. Moreover, post-modern society and globalization have a specific impact on careers.

In the age of globalization, all employees are affected by different work cultures; some organizations have common work culture. In order to search the meaning and purpose of life; an employee faces many problems related to work and life. The challenges faced by employees and efforts by the nation to overcome with the problem of unemployment and employment are common examples of universal issues that seem to affect many individuals from diverse cultures.

Equity Bank Rwanda as the partner in development joined the Rwandan market in 2011. It started its operations back in 1984 as a microfinance bank in Kenya and became a commercial bank in 2004. While the bank started in 1984; there were no clear career planning paths, employees' commitment and loyalty were the foundation of the company's performance. Thereafter, the senior management saw the need of promoting career programs to be able to develop, coach mentor, and retain key employees as competitors had started recruiting high performing and competent employees of the upcoming new listening; caring, and high performing East African Bank.

17.2 THE PURPOSE OF THE STUDY

The purpose to conduct this study is to find out the relationship between career planning and development and performance of individual in Rwanda.

17.3 OBJECTIVES OF THE STUDY

The four main objectives to conduct this study are as follows:

1. To establish how career planning and development influence the performance of an employee within Equity Bank Rwanda PLC.
2. To assess the influence of employee promotions along his/her career path in an individual's performance within Equity Bank Rwanda PLC.

3. To assess how career planning and development practices (like employee counseling, coaching, mentorship) influence the performance of employees within Equity Bank Rwanda PLC.
4. To determine how employee–employer career partnership can influence an employee's performance within Equity Bank Rwanda PLC.

17.4 RESEARCH QUESTIONS

This study answers the following research questions:

1. How do career planning and development influence the employee's performance within Equity Bank Rwanda PLC?
2. What is the influence of employee promotions along his/her career path on an individual's performance within Equity Bank Rwanda PLC?
3. How do career planning and development practices (career counseling, coaching, mentoring) influences the performance of employees within Equity Bank Rwanda PLC?
4. How does employee–employer career partnership influence employee performance within Equity Bank Rwanda PLC?

17.5 SCOPE OF THE STUDY

This study will be scoped to the content/domain and geographical scopes.

17.5.1 CONCEPTUAL SCOPE

The study domain of human resources management lies in assessing the relationship between career planning and development employee performance in Rwanda. It is chosen because the researcher would like to evaluate how to achieve expected employee performance within a company as a result of career planning and development practices adoption such as employee coaching, mentoring, counseling, training, and promotions.

17.5.2 GEOGRAPHIC SCOPE

Due to the shortage of time, the researcher covered only Equity Bank Rwanda PLC and not all banks of Rwanda.

17.6 SIGNIFICANCE OF THE STUDY

At the end of this research, the findings from it are to be useful to different organs and even individuals in the following ways:

About the government organs, the research findings will identify the picture of employee's career challenges in private sector which can reflect those that government entities are facing and these findings will act as an eye-opener to solve these challenges.

This research will inform Equity Bank Rwanda management, in particular, the points of strengths, weakness, opportunities, and threats as far as employees' career planning and development is concerned and this is to help together with the proposals that have been advanced by the researcher to improve employees' performance and individual productivity.

17.7 LITERATURE REVIEW

It is an in-depth study of work that has been done by other researchers on career growth that embrace both career advancement and career development (Figure 17.1). Career growth has always been a major concern to researchers, decision-makers, and human resource experts.

Irrespective to the size, sector, market or profile of the organization, career development is the key practice followed by every organization. To achieve better and meet out the future skill requirement, all the high performing organization watches out not only recruitment policies but also analyze and concern for employee development (Mwanje, 2010).

According to Greenhaus et al. (2010), career development is an ongoing process throughout an employee's work life. It plays a significant role for both organization and individual. It is an effective technique or process that helps the organization to retain and motivate its employees to develop the required skills for setting realistic goals and achieve their targets efficiently (Hall and Lorgan, 2009).

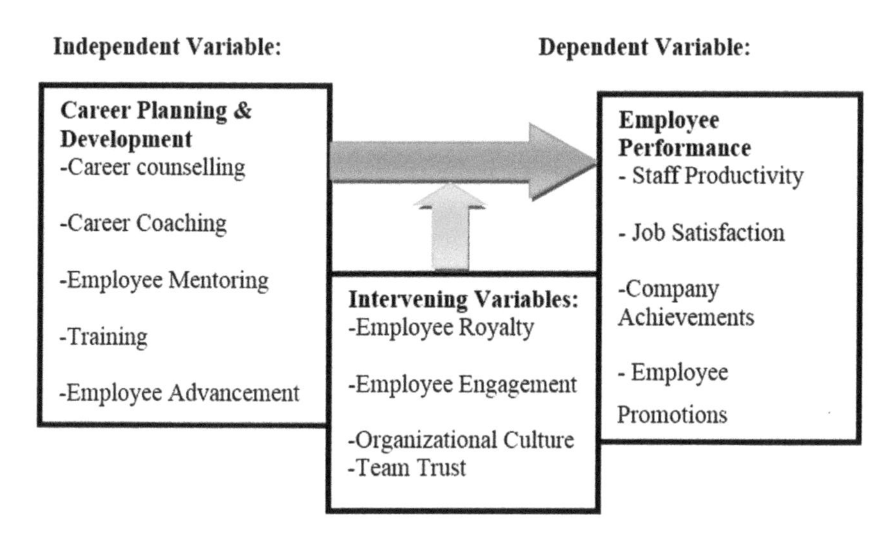

FIGURE 17.1 Conceptual framework.

17.8 CAREER ADVANCEMENT

Career advancement is the path of progress embrace merit of employee irrespective to race, gender, and ethnicity. It plays a significant role in employees' work life by providing them various long term benefits like job security, better pay, etc. It is a process that leads an organization to achieve high performance with a motivated workforce. It also provides opportunities to employees to improve their skill sets and get an education and undergo training programs to the overall development and all these help them to get promotions in the future.

17.8.1 EMPLOYEE PERFORMANCE

A significant role played by the workforce in any organization. We cannot ignore the role of employee performance in the success of the organization. Afshan et al. (2012) define performance as, "the achievement of specific tasks measured against predetermined or identified standards of accuracy, completeness, cost, and speed."

17.8.2 PROMOTIONS

An unbiased promotional opportunity for employees leads to job satisfaction. If an organization has fair promotion policies then employees show dedication and better performance.

In this case, public sector employees get promotions after a fixed time period so they are satisfied with their job in this aspect whereas in the private sector, promotion depends on the performance of employee.

17.8.3 REWARDS

It is something that an employee receives in return for good performance or for the attainment of service. Reward management is the process that begins with the formulation end up with the implementation of policies. The basic objective of giving rewards to employees is to motivate and retain talent in the organization.

The organization success depends upon the employee's performance and employee's performance depends upon the appreciation and reward received for their performance.

17.9 METHODOLOGY

17.9.1 RESEARCH DESIGN

During the research at Equity bank Rwanda PLC, the researcher applied the qualitative approach. The researcher showed the quality of respondents from Equity Bank Rwanda PLC by using various formats of questionnaires and interview guides. The research interpreted data using Microsoft Excel.

17.9.2 POPULATION OF THE STUDY AND SAMPLE SIZE

Information was collected and obtained from Equity Bank PLC staffs and the management of the Bank as planned by the researcher who was conducting this study (Table 17.1). The target population for this research was 12 employees including four Officers; three Senior Officers; two Assistant Managers; two Managers, and one Senior Manager of Equity Bank Rwanda PLC.

17.9.3 SAMPLING TECHNIQUE

Due to limited time; the researcher selected few employee randomly from the total population of the bank. The sample was obtained through random sampling technique among various strata identified and the researcher collected data at 100% rate among identified strata.

TABLE 17.1 Population Details within Equity Bank Rwanda PLC.

Positions	Population	Sample
Officers	4	4
Senior Officers	3	3
Assistant Managers	2	2
Managers	2	2
Senior Managers	1	1
Total	**12**	**12**

17.10 DATA COLLECTION TECHNIQUES

The researcher used a qualitative approach analysis, data collected from Equity Bank Rwanda employees was edited and coded and analyzed using Microsoft Excel sheets to make tables and graphical representation of research variables.

17.11 DATA INTERPRETATION

The findings were generated from responses to the general questions and research questions on the impact of career planning and development on employee performance.

17.11.1 RESPONSE RATE

A total of 25 questions were distributed among 12 staffs and were filled up and returned to the researcher with a 100% response rate.

TABLE 17.2 Gender Distribution

Years Range	Number of Staff	%
20–30 years	5	42
31–40 years	4	33
41–50 years	3	25
51 and above	0	0
Total	12	100

The researcher noticed that the bank has a younger workforce as 42% are between 20 and 30 years old and more 33% are between the ranges of 31 and 40 years old (Table 17.2). This explains why the involved population is more involved in career success through performance achievements.

TABLE 17.3 Staff Experience with Equity Bank Rwanda PLC

Years' of Experience	Staff Number	%
0–1 year	0	0
1–2 years	1	8
2–3 years	5	42
3–5 years	4	33
5 and above	2	17
Total	12	100

Research findings revealed that 42% of respondents are between 2 and 3 years' experience and 33% are between 3 and 5 years of working experience; this proves that involved employees are more ambitious in career advancement for success (Table 17.3).

TABLE 17.4 Gender Analysis

Gender	Staff Number	%
Male	8	67
Female	4	33
Total	12	100

Findings prove that the company employees are more men than women as among all the respondents 67% are men and 33% are ladies (Table 17.4).

TABLE 17.5 Respondents' Education

Education	Staff Number	%
A Level Certificate	0	0
Diploma	0	0
Degree	10	83
Master and above	2	17
Total	12	100

The researcher noticed that the bank resources are well educated as 83% holds a degree and 17% among respondents are at master's level (Table 17.5).

TABLE 17.6 Positions Occupied by Respondents

Positions	Staff Number	%
Officer	4	33
Senior Officer	3	25
Assistant Manager	2	17
Manager	2	17
Senior Manager	1	8
Total	12	100

Among involved population; 33% are Officers; 25% are Senior Officers; 17% are Assistant Managers; 17% are Managers, and 8% are Senior Managers within the bank (Table 17.6).

17.12 RESPONDENTS' DATA INTERPRETATION PER RESEARCH QUESTIONS

The researcher used questionnaire to collect data; these are related to career planning and developments and employee performance within Equity Bank Rwanda PLC. The researcher distributed a questionnaire among 12 employees occupying various positions and the responses were gotten from all involved staff. The researcher used various types of questions and below is a representation of respondents' feedback: Strongly Agree (SA), Agree (A), Neutral (N), Disagree (D), and Strongly Disagree (SD).

TABLE 17.7 Effect of Career on Employees Performance at Equity Bank Rwanda PLC

S/N	Questions	SA (%)	A (%)	N (%)	D (%)	SD (%)
1	I am motivated and committed to see the company succeed	58	25	8	8	0
2	Career advancement programs at workplace motivates and increase productivity of employees	33	50	0	17	0
3	Through career counseling, employees are able to have a positive sense of career achievement	33	42	8	8	8
4	Employees are given an opportunity to explore their personal concerns that may interfere with their job performance	17	42	25	8	8
5	At Equity Bank Rwanda employees' career development depends on performance	17	50	17	8	8

Above respondents views revealed that 82% of the staff involved are well motivated and committed to see the bank succeed, 8% were neutral, and 8% disagreed that they are committed to see the bank succeed 83% of the respondents agreed that career advancement programs of Equity Bank motivates employees and enhance their productivity, only 17% disagreed on that situation within the bank (Table 17.7).

The researcher noticed that 59% of respondents accepted that employees are given an opportunity to explore their personal concerns that may interfere with their job performance, 25% preferred to be neutral, and 8% disagreed about such practice at Equity Bank Rwanda workplace.

More than half of the respondents are in agreement that at Equity Bank Rwanda employees' career development depends on performance, this justified that 17% strongly agreed and 50% agreed that they witnessed this within Equity Bank Rwanda workplace. The researcher noticed that 17% of respondents were neutral on this and 16% of the total population disagreed on that situation within Equity Bank Rwanda.

Figure 17.2 is a representation of the respondents' personal view in terms of career planning and development and performance at the workplace.

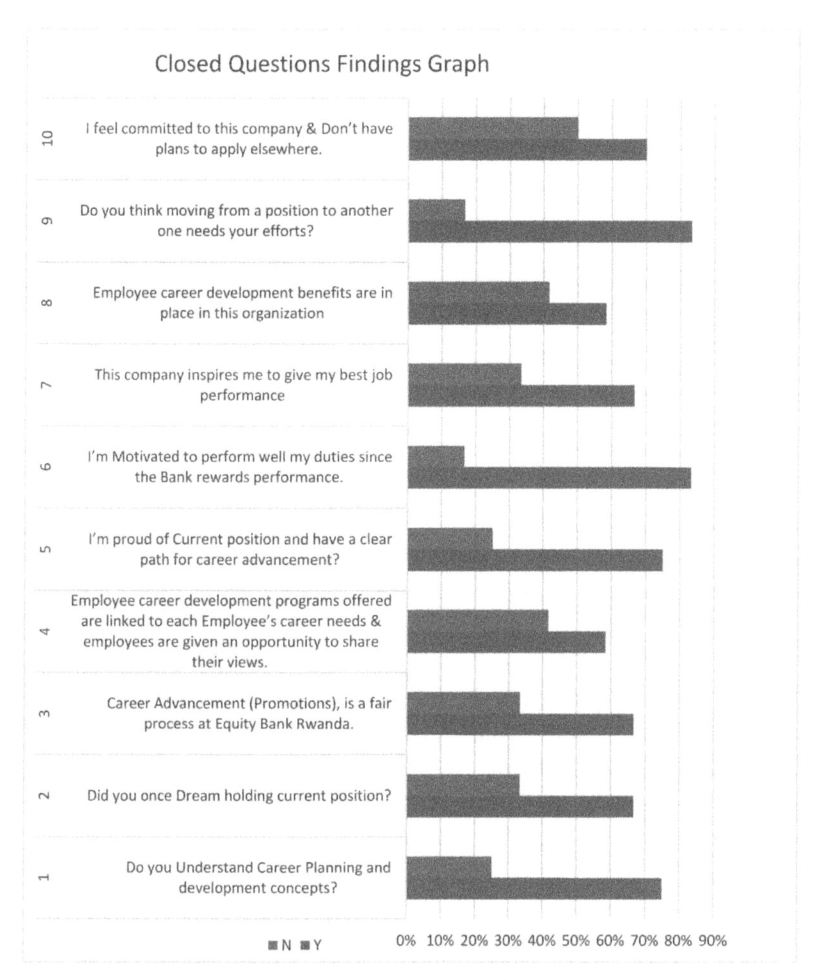

FIGURE 17.2 Career planning, development, and employee's performance.

To review the accuracy of the respondent understanding on career planning and development; the researcher asked all of them if they do understand such concepts and 75% of the respondents strongly agreed that they do and 25% agreed that they understand. This convinced the researcher that their responses will be appropriate to give the right interpretation of the status of career planning and development at Equity Bank Rwanda workplace.

On the side, 67% of the population strongly agreed that they have planned their current positions and 33% agreed on the same. Again 75% of the respondents said that they are happy with current positions within the bank.

This is well explained by the fact that 67% again strongly agreed that Equity Bank Rwanda promotions practices are fair and 33% agreed as well.

Findings reveal that 83% of the respondents strongly agreed that they are more committed to performing more since the bank rewards performance and 17% agreed on the same note. The researcher noticed that 58% said that the employee's career development benefits are in place in Equity Bank; indeed this is the source of their motivation.

The researcher noticed that 70% of staffs said they are not planning to leave the bank and other 30% said if other better opportunity comes they can leave and join another employer. This reveals how committed and happy are the bank resources and the efforts of the bank to advance employees' careers toward success.

17.13 FINDINGS

Data interpretation has led the researcher to identify the below findings:

- Career planning practices such as employees counseling, employees coaching, and employee mentorship gives a positive result on employees' motivation and commitment at workplace leading to employee's performance.
- Career planning and development strategy at workplace; is a tool Equity Bank Rwanda uses to enlighten employees' career paths.
- Employee–employer career partnerships make impact than one way efforts.
- Fair promotions practices at workplace are the source of employees' trust, morale, and good working environment.
- There is a positive relationship between career planning and advancement and employee's performance.
- Company's overall performance is a result of employees' performance.

17.14 CONCLUSION

The researcher noticed good and conducive work environment within Equity Bank Rwanda PLC and partnership of employer and employees in career building which impact on employee career advancement, individual performance, and overall bank performance.

Career planning and development is a key to success while in employment, it is a subject many scholars focused on and insisted that it should be care for as organization performance is a result of successful employees.

Hence, employees give their full dedication and work selflessly for the organization growth when they realized that the organization or employer have concern for their individual career growth and development rather than concern only for the organization's mission.

The researcher conclusion is that the outstanding performance of the bank; has a source in effective career planning and development strategies that the bank has put in place.

The researcher recommends further research on the role and of an effective succession planning program on employee commitment and career success within an organization.

KEYWORDS

- **career**
- **career planning**
- **employees**
- **career development**
- **employee performance**

REFERENCES

Armstrong, M. (2001), Human Resource Management Practice: Handbook, 8th Edition, and Kegan.

Arulmani, G. & Arulmani, S. N. (2004), Career Counselling.

Bernardin, J. (2010) Human Resources Management: An experiential Approach, Irwin McGraw-Hill.

Bohlander, S. G. (2004), Managing Human Resources.

Chris, D. S. (2009), Promotions as a Motivation Factor Towards Employees' Performance.

Diriye, A. M. (2015), Perceived Relationship Between Career Development and Employee Commitment and Engagement at Nairobi City County Government.

Greenhaus, R., Callanan J. and Godshalk H. (2000) Career Management in Public Organization (12th edn.), New York, England: Pearson Education Publishers.

Hall, D.T., and Associates (1986). Career Development in Organizations. (1st edn.), San Francisco: Jossey-Bass Publishers

James, A. & Van Esbroeck, R. (2008), International Handbook of Career Guidance.

Mare, J. G. (2013), South African Journal of Psychology 43, 2013.

Mwanje, S. (2010). Career development and staff motivation in the banking industry: A case study of the Bank of Uganda. (Unpublished Masters' dissertation). Makerere University, Uganda

Kakui, I. M. & Gachunga, H. (August 11, 2016), Effects of Career Development on Employee Performance in the Public Sector, Strategic Journal of Business Management, 3:307–324.

Parsons, F. (2009), Choosing a Vocation. Boston: Houghton Mifflin.

Sharma, M. S. & Sharma, M. V. (2014), Employee engagement to enhance productivity in current scenario. International Journal of Commerce, Business and Management, 3.

CHAPTER 18

HR Analytics for Competitive Advantage

SHIKHA KAPOOR

Amity International Business School, Amity University, Uttar Pradesh, 201313, India

**Corresponding author. E-mail: skapoor2@amity.edu*

ABSTRACT

Human resource (HR) analytics is the field of investigation that suggests to apply scientific procedures to the human resource department of an organization in the expectation of enhancing employee performance. In this way it shows the sign of enhancement in rate of return. HR analytics does not simply manage collecting data on employee efficiency; rather, it intends to provide insight into each procedure by gathering data and after that utilizing it to settle on significant choices about how to enhance these processes. It enables an organization to utilize the wealth of employees for making better decisions for improving organizational performance. HR investigation associates the business data with individual's data, which can help to set up vital associations later on. The key part of HR investigation is to give information on the effect of the HR department on the overall organization. The uncertainty in the environment makes it imperative for the HR professionals to use important qualitative and quantitative tools to design the HR process metrics. This will help in smooth transition from the transactional role to transformational role of organizations and become strategic partners in support of organizational growth and development.

18.1 INTRODUCTION

HR is to ascertain about the HR analytics for fostering the status of the HR profession for service industry that can go far to make India best suited for human capital investment. The acknowledgment of this guarantee relies on our individual and collective ability to ace the art and the science of HR analytics. The intensifying globalization of the job market combined with a regularly swelling deficiency of skillful staffs and advances in technological innovation has brought about massive changes to the recruitment practices throughout the world using HR analytics. Future investigations can concentrate on broadening the proposed hypothetical systems into an approved mode. Furthermore, measuring the ramifications of Evidence-Based Management practices to HR and organizational performance can likewise be explored. Recently, some of the authors who dedicatedly contributed to this field are Pfeiffer and Sutton and Briner, Denyer, and Rousseau. Most of the organizations focus majorly upon collecting and reporting data instead of outcomes hence analysis is the need of the hour.

Human resource (HR) analytics deals with investigation to the human resource department of an organization by applying scientific procedures in the expectation of enhancing employee performance and in this way showing signs of enhancement in rate of return. HR analytics is not only to manage collecting data on employee efficiency, rather, it intends to provide insight into each procedure by gathering data and after that utilizing it to settle on significant choices about how to enhance these processes.

A vast majority contemplated HR analytics as a very important element in fetching better return on investment for talent management. There are pre-existing wide and sophisticated usage of analytics in the domains like supply chain, marketing, finance, etc., where there are dependable metrics and the data which can be predictive of business decisions. As per the report of IBM (2009), the need is consistent analytical point of reference in HR profession to make positive and fruitful business results. HR analytics is to identify potential candidates to succeed in a role, the attributes of high performing employees and the probability of their termination.

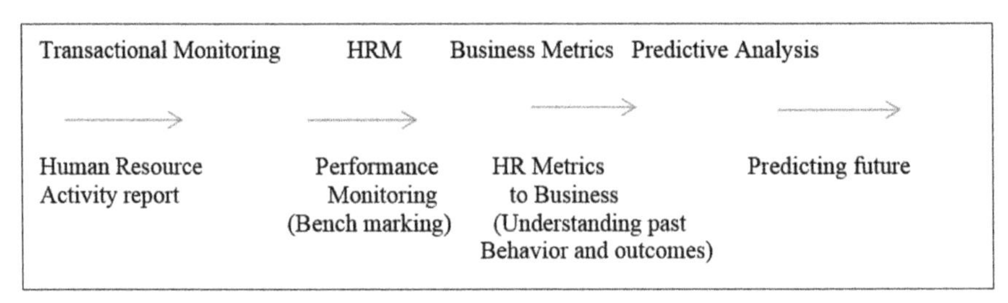

FIGURE 18.1 Human capital metrics.

The fundamental day-to-day issues of HR and larger people or employee-related matters of the organization can be done through the study of Big Data by applying the process of analytics. The Big Data analytics being applied to human resources of an organization and the various stakeholders with which it interacts in the external environs is giving rise to the importance of a field getting embedded in the HR realm called as the "Workforce Analytics" or "Human Resource Analytics." With the help of this we can now apprise management with a lot of conviction about emphasizing on what type of training programs and skill development needs to be strengthened.

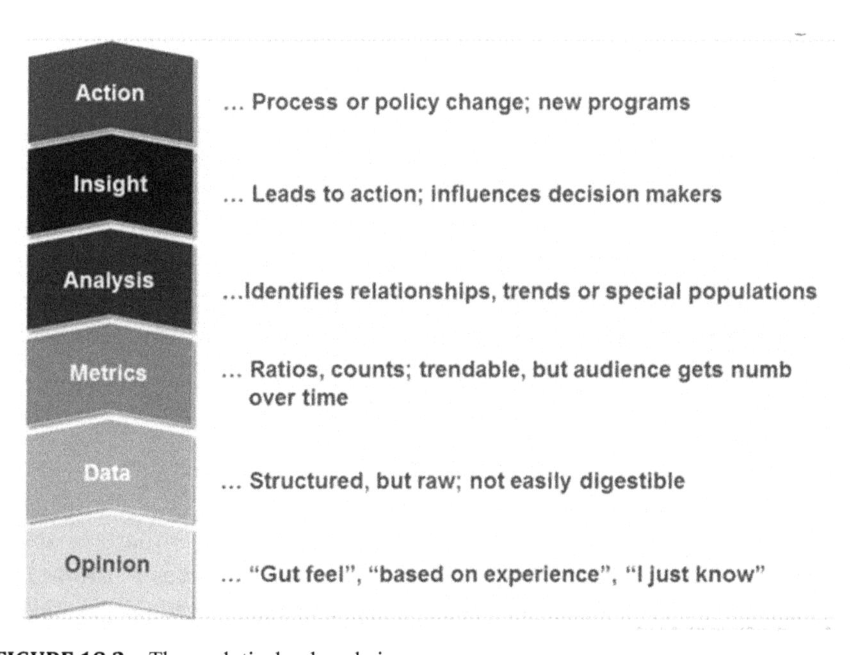

FIGURE 18.2 The analytical value chain.

A research was conducted by Institute for Corporate Productivity (2012) on the analytical practices and capabilities of HR. It was observed that most of the organizations are unrehearsed to deal with intensification of its data. It was also seen that HR organizations are proficient at collection and measuring the HR activities. But the capacity to measure conclusions/ recognize the aspects affecting the results to the maximum level is done by few organizations.

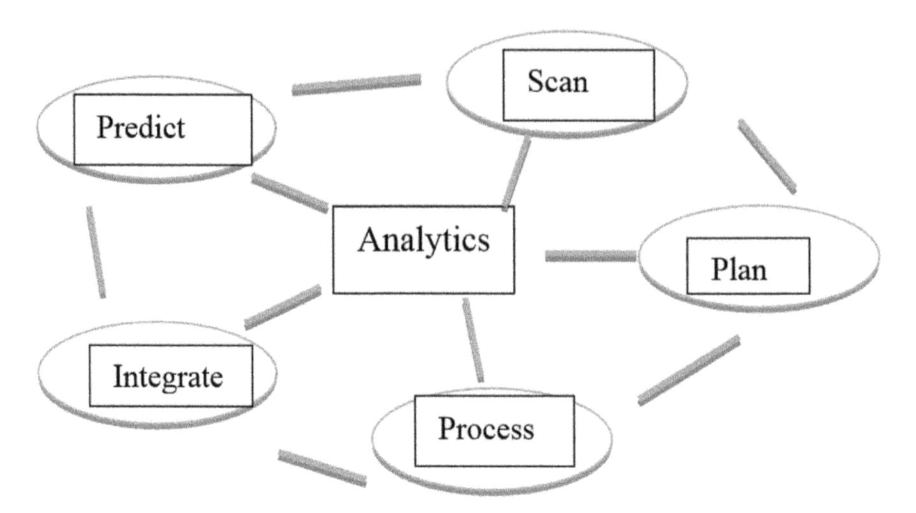

FIGURE 18.3 Connections through statistical analysis.
Source: Jac fitz-Enz (2010). The New HR Analytics: Predicting the Economic value of your Company's human Capital Investments.

The skeptics of HR analytics also claims that "the value of employees cannot be measured or predicted"—adds Hoffmann et al. (2012), describing the work force analytics or HR analytics is "a way of dealing people like widgets."

18.2 DEFINITION OF HR ANALYTICS

There are various opinions prevailing to HR analytics over the time span. Gustafson in 2012 stated that analytics in human resources is known as Talent Intelligence (Snell, 2011) HR analytics (Mondore et al., 2011), Talent Analytics (Davenport et al., 2010), or Workforce analytics (Hoffmann et al., 2012). HR measurement (Kapoor, 2015)

In the opinion of Bassi, "HR analytics is the application of methodology for improving the quality of people-related decisions, using HR metrics all the way to predictive modeling, for organizational performance improvement." Gustafson (2012), in his work, cited Hoffman, Lesser, and Ringo's descriptions of HR analytics as workforce analytics for expressing analytical activities and techniques used in a company/organization's workforce, that is, its employees. It aims by building a workforce for achieving business strategies with easy and simplicity and, thus, these techniques are helpful in getting the insights into how to organize, structure, and motivate the workforce. HR as a concept and its application has used statistical methods to track the manufacturing downtime, costs of labor and employee benefits, worker productivity, etc.

HR analytics measures impact of HR policies and practices on an organization's performance and therefore it is an impactful way to prove its value in organizations.

18.2.1 TOWARD PREDICTIVE ANALYTICS

The predictability in HR was analyzed by La Grange and Roodt in 2001. Ingham's analysis on predictive analytics is not about running statistical methods. In this regard, Ingham also cites conversation between Jac Fitz-Enz and David Creelman about predictive analysis. The Institute for Corporate Productivity, in its report (2012), argues that there has been an underuse of predictive analytics for measuring human capital—even by high performing organizations.

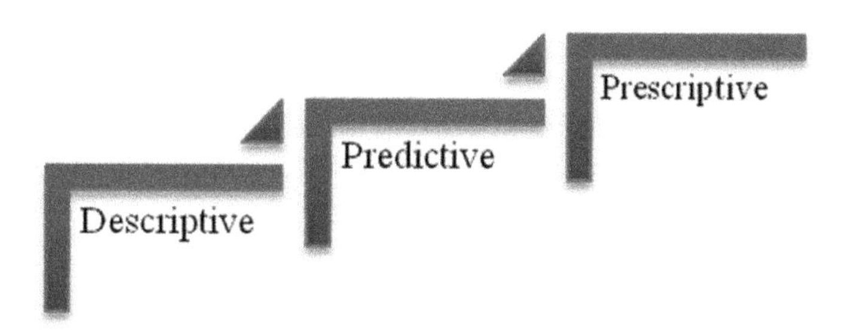

FIGURE 18.4 Fitz-Enz, Philips, Ray—three levels of analytics.

The three levels of analytics are explained by Fitz-Enz, Philips, and Ray (2012) as:

1. Descriptive analytics: give answers to questions such as "what happened" and "what is happening now";
2. Predictive analytics: prediction is related to the outcome of the organizations.
3. Prescriptive analytics: the most demanding and ultimate level of HR analytics "what is the best course of action?"—predictions and decision-making, and impact of those decisions.

LaValle et al. (2010) also describe another analytical capabilities into three levels as:

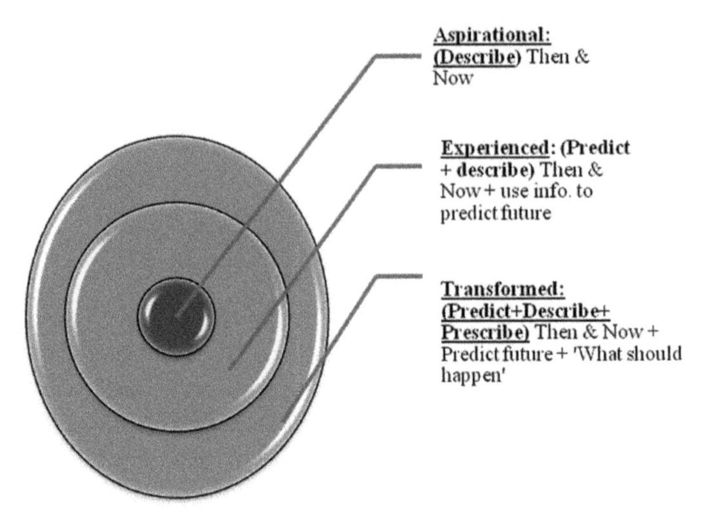

FIGURE 18.5 Levels of analytical capability.

HR analytics as an emerging area, competence, and capabilities can virtually act as a medium in all the roles that an HR professional can play in an organization by focusing toward its impact on the larger or smaller business outcomes. It is shown below in the Dave Ulrich's HR model in the following manner.

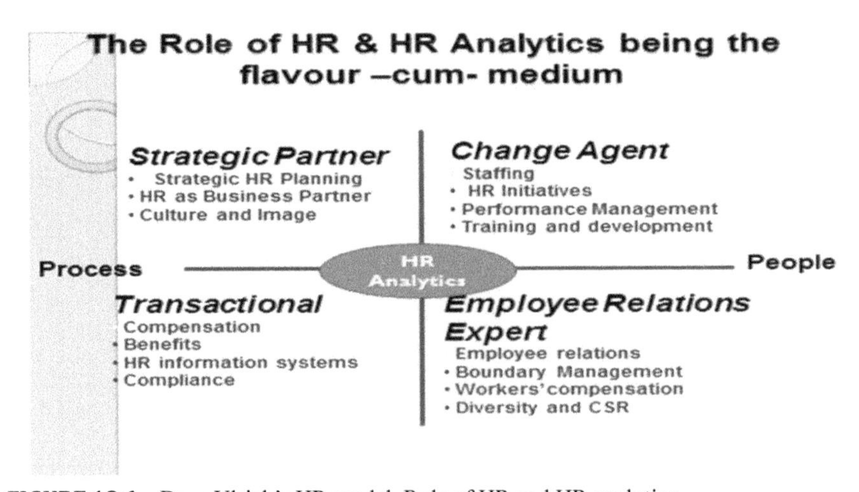

FIGURE 18.6 Dave Ulrich's HR model. Role of HR and HR analytics.

This flavor and medium of HR analytics can greatly enable in the effective delivery of the roles with a predictive edge both on the process and people perspective.

Further, the following is recommended, as a part of the organogram of HR function in a business enterprise, how HR analytics can contribute in the HR delivery model:

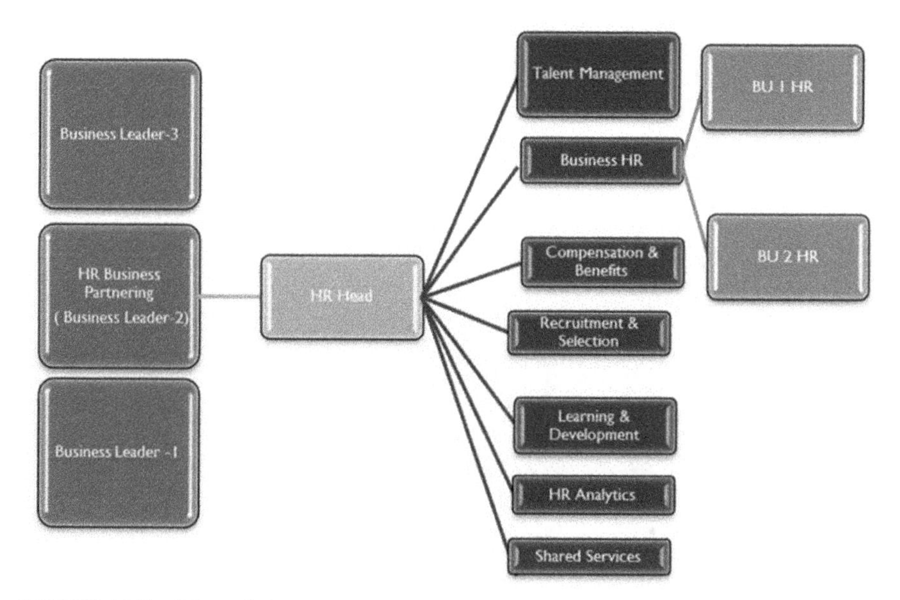

FIGURE 18.7 HR analytics to HR delivery model.

Finally, the HR practitioners have to now move from effectively rolling out various HR initiatives or processes to clearly showing insight into the business outcome. This will enable in accomplishing through the application of HR analytics.

18.3 KEY HR METRICS

Vokic (2011) is of view that the following indicators are for valuating individual HR activities like controlling of a particular program, activity, function, process, policy and this valuation is done by measuring HR indicators in a particular HR area/subarea as depicted beneath.

HR FUNCTIONS	EXAMPLES OF INDICATORS
HR planning	• Extra work hours per employee /year • Rate of replacement • Consultancies in a particular area / year
Job analysis	• Job description • Job analysis / Job cost • Job evaluation and Time required
Recruitment	• Number of applications/ recruitment service • Internal Employment rate (IER) • Selected candidates / recruitment source
Selection	• Employment cost/ selection method • Number of candidates interviewed, tested, etc. • Satisfaction of internal clients with the selection process
Performance Management	• Reliability of performance appraisal • Percentage of formally appraised employees • Average time required for performance appraisal
Compensation Management	• Number of existing benefits and promotions • Employee satisfaction with respect to salary, benefits, rewarding etc. • Average salary per employee
Training & Development (T&D)	• Training per hour per employee • Employee satisfaction with T&D programmes • Changes in behavior, knowledge, attitude, or change in performance with respect to Training and Development • Savings as a result of Training and Development programs.
Career Management	• Career management programmes and number of employees involved in it • Career management programmes and its cost
Health & safety Issues	• Average cost of work injury • Work injuries and time lost • Number of internal health & safety inspections

FIGURE 18.8 HR indicators in HR functions.
Source: Reprinted from Vokic (2011).

18.4 ALIGN HR ANALYTICS WITH COMPANY STRATEGY

Big Data insights play a central role in helping the organization identify the talent which is not only likely to succeed in future leadership position, but also to stay in the organization. In several recent global research studies conducted by analyst firms such as McKinsey & Company, AT Kearney, and SHL, it has been statistically proven that an organization can improve the odds of its success or influence positive outcomes by targeting their efforts around those individuals whose predicted career trajectories align with the company's business needs.

In the current state of affairs is prevailing in many Indian organizations, particularly those which have not fully automated their HR function and related services and even if automated they maintain and utilize people data in silo manner. These data are entirely disintegrated. To add to this, these data are not captured in a consistent manner. As for example, the data relating to Performance Management and Potential of employees remain in different formats and databases and those pertaining to compensation, payroll, health benefits, etc. are in different formats and even database types in many organizations. These data do not speak to each other, nor to other functions of business. So it is sitting on a huge pile which is simply mined for miniscule ad-hoc purposes to provide the management short-term view of people dimensions of the business. There is a constant challenge of consistency of data across. HR processes (transactional and transformational) and integration issues with other functional processes of the business. It is imperative for HR analytics to commence its journey with building data consistency breaking the silos of data across various subprocesses of HR. These data of HR need to speak to data of finance, accounts, marketing, or other function's requirements in a seamless manner (i.e., data integration) for yielding insights to the business leaders in decision-making for dimensions in HR Management. Added to these challenges in the human resource data space, there is a strong need for accessibility of data and its being efficiently mined at appropriate time to enable the management in taking informed HR decisions and evolve strategies.

The market space is today having many enterprise resource planning solutions and software applications which are offering management information off the shelf to transact business, with HR being no exception. All these are greatly enabling in simplifying and smoothening the transaction activities of the function of HR. But is it adding value to HR functioning in proactively predicting trends and patterns to tell what a good turnover of employees would be or what a bad turnover would be in terms of some concrete numbers or build up a scenario backed by logic or data or predict something with more certainty?

Linking HR data, using cause–effect statistics, to actual business outcomes, is becoming the increasing need of the times. HR analytics is one-third effective database management, data mining, application of business statistics with cross-functional business understanding, visualization and reporting and two-third application of logic, reasoning and predicting or forecasting trends about human resource process practices. For example, what to look for in lateral recruits to see that they are both best cultural fit in the new organization and develop fullest loyalty to groom their next phase of career based on the data gathered about them through structured manner and strengthen the relevant HR processes based on personality and behavioral profile analyses.

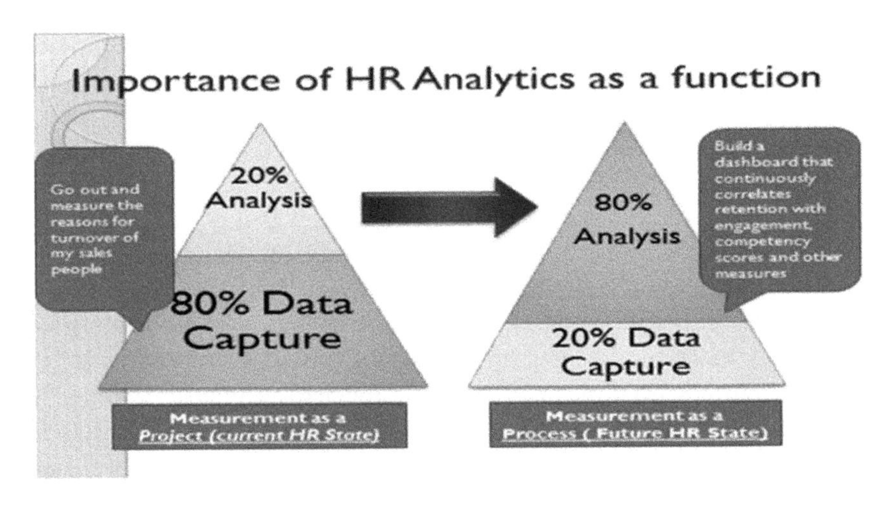

FIGURE 18.9 Importance of HR analytics as a function.

There is a value chain in HR analytics that can be depicted as below:

18.5 COMMON HR MODELS IN USE

Due to mounting interest in HR analytics, diverse processes and models have established a special space into the human capital investment arsenal, quoted by Fitz-Enz, Philip, and Ray (2012). This segment will discuss a few of these models that offer feasible and sustainable preferences for organizations who wish to venture in HR analytics practices.

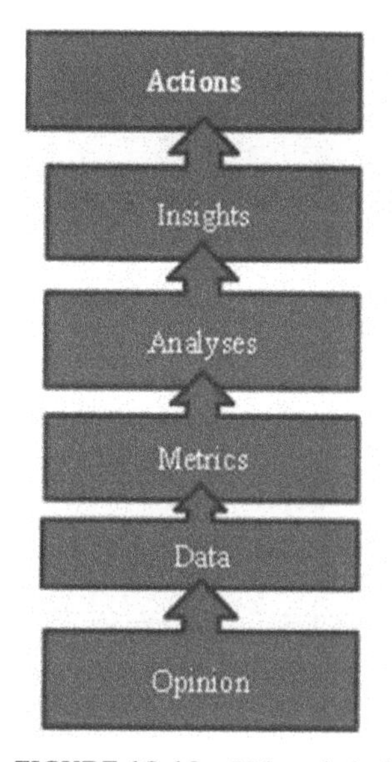

FIGURE 18.10 Value chain in HR analytics.

18.5.1 HC BRIDGE

This framework is one of the fundamental models used to address the task of linking business to HR initiatives. It was developed by Boudreau and Ramstad (2002, 2007) who used the symbol of a bridge to label the links between investments in HR programs and sustainable business success.

Three basic elements of success existing decision frameworks are efficiency, effectiveness, and impact. The anchor points as per the model are dissected into a set of linking elements. This helps in articulation of the framework with maximum clarity.

The model is used for inferences for HR practices and investment at the bottom. It is a planning tool for sustainable success.

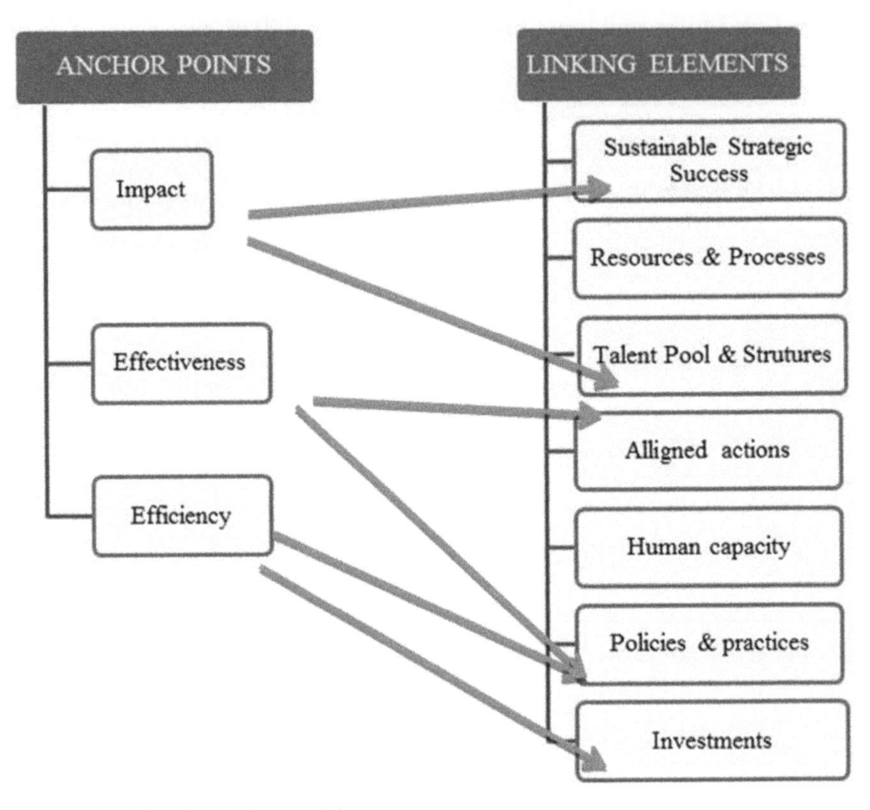

FIGURE 18.11 HC Bridge model.
Source: John W. Boudreau and Peter M. Ramstad (September 2004). Talentship and Human Resource Measurement and Analysis: From ROI to Strategic Organizational Change. https://www.strimgroup.com/wp-content/uploads/pdf/TalentshipHRMeasurementAnalysis.pdf

18.6 TALENT ANALYTICS OR HR ANALYTICS MATURITY MODEL

Josh Bersin, Principal Consultant, Deloitte Consulting, used research methodology (involved 480 large sized organizations) and found that only 4% companies achieved the capability to perform "predictive analytics" about their workforce. This was effectively done by utilizing the people data available in their respective organizations and developed effective linkages with other functions of the business. They are capable of scenario planning of human resources which is seamlessly integrated with the strategic planning and visioning of the business. These organizations can greatly forecast Risk Analysis and recommend ways for their mitigation. Further, as per their research, only 10% have done any significant "statistical analysis" of employee data which enables in development of making people model impacting the various HR processes such as Talent Acquisition, Succession Planning, Compensation Management, etc., which has strategic connection with various business decisions and a kind of cause and effect analysis for each action taken in the HR realm.

The remaining 84% of the surveyed organizations by them are on the lower end of dealing with data management and effective usage of analytics. They are still grappling with the routine jobs of standard and ad hoc reports to deliver standard operational metrics. These have no intelligent analysis of any kind which yields any meaningful information to the management for making dynamic HR policies or programs.

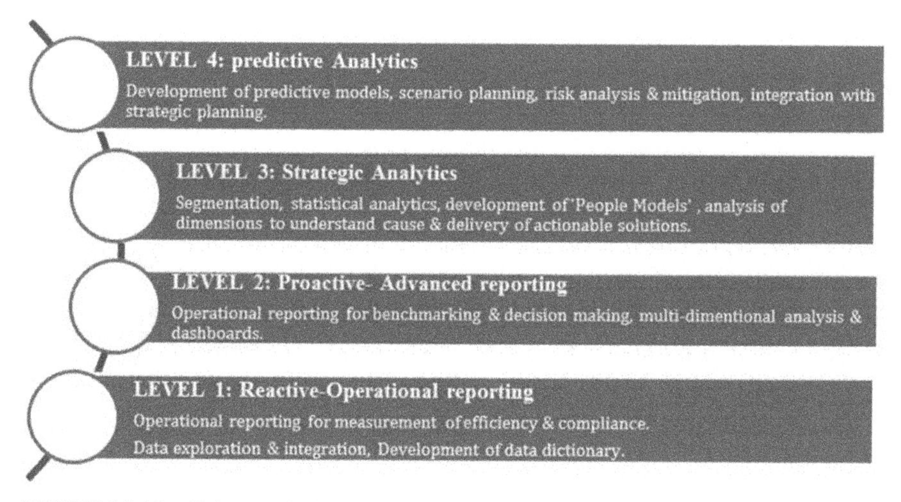

FIGURE 18.12 Talent analytics maturity model.
Source: Deloitte talent analytics maturity model proposed by Josh Bersin (2013).

This model defines the maturity organizations in terms of talent analytics. It begins with the first level or level 1 and goes up to level four. The level one is the reactive, operational reporting of HR data which moves up to level 2 wherein, the organizations become more proactive, this involves benchmarks and multidimensional dashboards. According to Bersin (2013), third level is "Strategic Analytics." In this level the statistical analysis, development of models and segmentation are used. The fourth level of maturity is predictive analytics. This involves scenario planning, predictive models, and integrates with the strategic planning.

At level four, the retention rate is 38% higher which involves three times the revenue per employee of first level HR organizations.

Going forward, some questions that can be answered using HR analytics are such as:

- What kind of investments in people has the greatest impact on our business?
- How should our people requirements adapt to changes in the business environment?
- Which business units/departments/individuals need attention?
- Which jobs will need to be filled as a result of the company's growth, and when?
- What will be the future causes of employee turnover?
- What are the effective counter actions?
- Is our particle measuring systems measuring what it is supposed to measure?

18.7 HR SYSTEMS BEING USED

As per Bersin's research, the average large organizations have minimum 10 diverse HR applications and respectively their core HR system is more than 7 years old. To bring this data together so as to make it useful, a lot of efforts and vigor are required. The real discipline of analytics requires a lot of skills in data analytics, visualization, statistics, cleaning, and problem solving. The organizations still struggle to find such HR professionals who possess these skills and bring them together.

The acronym "DELTA" has been used by Davenport et al. (2010) for defining the technology as very critical when it comes to mastering talent metrics. The essentials required for building analytical capability is distinctly explained using this model below:

18.8 SCOPE OF HR ANALYTICS

Diverse elements are required to perform the type of analytics that would depict some relationship between performance of the organization and HR practices. Systematically, at first, good metrics are required then possibly and importantly, well established analytic models and thereafter, valid measures of company performance.

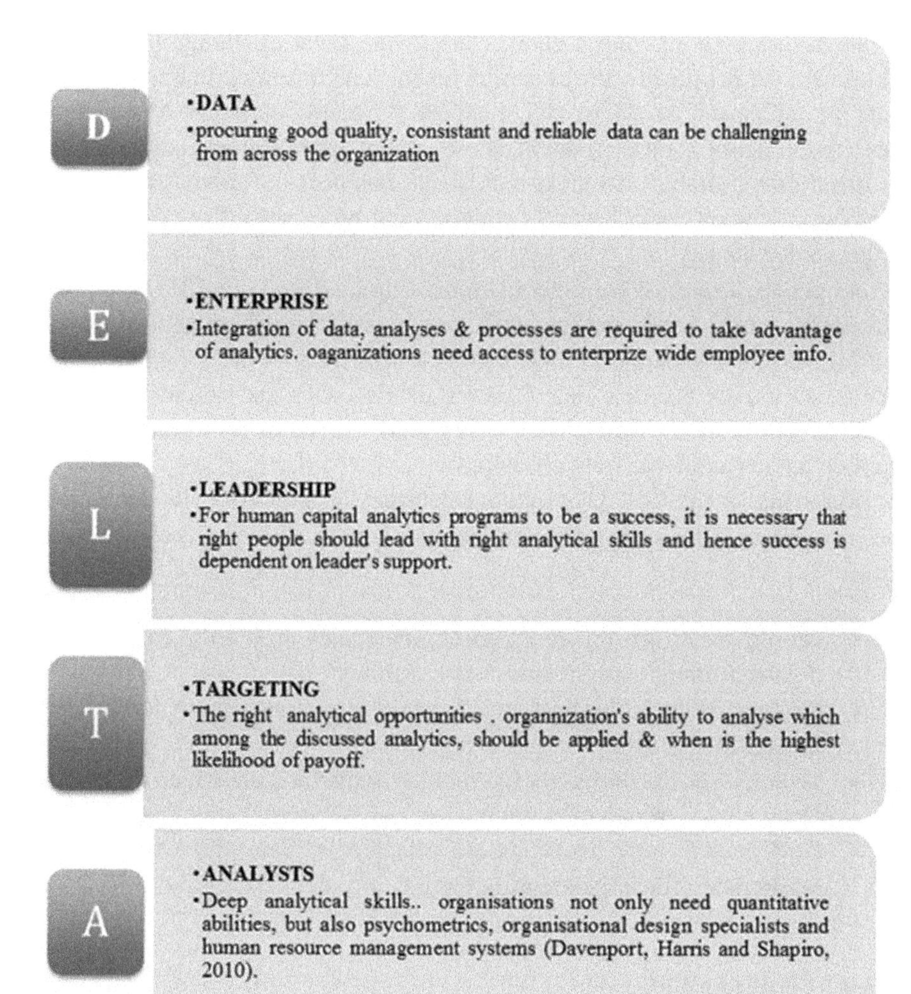

FIGURE 18.13 Essentials required for building analytical capability "DELTA" by Davenport et al. (2010).

However, an alarm is raised by Ulrich et al. (2009), for the organizations to elude from using HR analytics for reaching to a conclusion. He also advises to prevent focusing on "what is easy" and rather emphasize of measuring "what is right" and therefore, Ulrich urges that HR professionals should not only focus on activities but also on the results of the training. Finally, he extends his advice to the organizations to keep the measures as simplified as possible and to focus on the decisions.

In the opinion of Harris et al. (2010), the most challenging building blocks are the people at various and all levels that contribute in transforming data into enhanced decisions and enhanced business outcomes. As per their description about analytical talent, those people apply statistics, qualitative, or quantitative analysis to make and shape business decisions—the "math brainiacs," "quant jocks," "excel ninjas," and other activities that need to improve decisions.

As per the report of Deloitte's Human Capital Trends in 2011, the most significant step, when it comes to workforce analytics is initiating the process/ getting started. The data is already available with most of the companies and, hence, no excuse for delaying. The report also cites the statement of one of its executives that if people are being paid vide payroll mechanism then enough data is available to begin with.

Referring to the DELTA model, a comparison prevails with Deloitte's building blocks (Deloitte's Human Capital Trends, 2011) in the following style:

- People: what are the required competencies and skills for specific organization to support analytics capability?
- Process: what is the best way to make decision support tools more impactful?
- Technology: the necessary tools and systems for data-driven decisions.
- Data: how to trench maximum value out of external and internal data?
- Governance: how decisions are affected by data and who will be accountable for implementing them?

18.9 CHALLENGES

- There is a pool of suggestions made by different authors and scholars at over the time period suggesting that there are some inhibiting elements and challenges in HR analytics and the organizations need to practically synergize to eliminate and minimize the loops. Some of these challenges are briefly enumerated below:
- Barriers within an organization hinder effective use of HR analytics.
- Discomfort with the line managers and other managers with regard to discussing about HR terms, HR tests and evidences, skills required, and skills lacking to use right analysis.
- Treat human assets like widgets and making an excuse for analytics.

- Accumulating data into a centralized database with consistency and quality.
- Simple metrics used to assess employees based on grades, test score, etc., generally fails to predict success.
- Using incomplete data used by the top executives making them hesitant in the process of decisions making process.
- Managing big database—data collected and managed across the borders.
- Use of tangible measures to analyze intangibles or qualitative aspect.
- Performance cannot be easily quantified.
- Mismatch of data across sources rendering it unreliable.
- Restricting to analysis of HR efficiency only.
- Failing to focus on the impact on the performance.
- Shift from being reactive to predictive.

Data credibility or gathering data into a single, centralized database is the main challenge when it comes to data analytics.

KEYWORDS

- **human resource management**
- **performance**
- **talent**
- **organization**
- **metrics**

REFERENCES

Bersin, J. (2013). *A big data of human resources: A world of haves and have-nots.* Forbes, Oct 7, 2013. https://www.forbes.com/sites/tmobile/2020/08/06/unlocking-industry-40-understanding-iot-in-the-age-of-5g/#61b7703d7f31

Boudreau, J. W., & Ramstad, P. M. (2004). *Talent ship and human resource measurement and analysis: From ROI to strategic organizational change.*

Davenport, T. H., Harris, J., & Shapiro, J. (2010). *Essentials required for building analytical capability 'DELTA.'*

Boudreau, J. W., Ramstad, P.M. (2007). *Beyond HR: The new science of human capital,* Harvard Business Press, Boston, MA.

Fitz-Enz, Philips, Ray (2012). *The three levels of analytics.*

Gustafsson, D. (2012). *Business intelligence, analytics and human capital: Current state workforce analytics in Sweden* (Unpublished work). University of Skövde, Sweden.

Harris, J. G., Craig, E., Light, D. A. (2010). *The new generation of human capital analytics*. Accenture Institute for High Performance, Research Report.

Hoffmann, C., Lesser, E., Ringo, T. (2012). *Calculating success: How the new workplace analytics will revitalize your organization*, Harvard Business School Press.

https://hbr.org/2014/12/workforce-analytics-of-the-future-using-predictive-analytics-to-forecast-talent-needs

http://dupress.com/articles/people-and-hr-analytics-human-capital-trends-2015/?id=gx:2el:3dc:dup1136:eng:cons:hct15

http://www.inostix.com/blog/en/top-16-hr-analytics-articles-q22014/

https://www.youtube.com/watch?v=l6ISTjupi5g

Jacfitz-Enz (2010). *The new HR analytics: Predicting the economic value of your company's human capital investments*.

Kapoor, D. S. *Managing organizational intelligence: Talent. Editorial Board*, 18.

Kapoor, S. (2015). *HRM text and cases*, Taxmann Publishers.

LaValle et al. (2010). *Analytics: The new path to value*.

Levenson, A. (2005). *Harnessing the power of HR analytics: Why building HR's analytic capability can help it add bottom-line value*.

Mondore, S., Douthitt, S., & Carson, M. (2011). Maximizing the impact and effectiveness of HR analytics to drive business outcomes, *Strategic Management Decisions*, 2011 (People and Strategy Journal, Vol. 34, Issue 2).

Snell, A. (2011). Developing talent intelligence to boost business performance. *Strategic HR Review*, 10(2), 12–17.

The New Intelligent Enterprise, Analytics: The New Path to Value, Retrieved from http://sloanreview.mit.edu/feature/report-analytics-the-new-path-to-value-executive-summary).

Ulrich, D., Brockbank, W., & Younger, J. (2009). *HR transformation: Building human resources from the outside in*, McGraw-Hill Education.

Vokic (2011). HR metrics. *The relationship between the level and modality of HRM metrics*, Quality of HRM Practice and Organizational Performance.

Index

D

E

F

G

H

I

J

K

L

M

N

O

P

www.ingramcontent.com/pod-product-compliance
Ingram Content Group UK Ltd.
Pitfield, Milton Keynes, MK11 3LW, UK
UKHW021845240425
457818UK00012B/341

* 9 7 8 1 7 7 4 6 3 8 1 5 6 *